Diffusive Spreading in Nature, Technology and Society

Armin Bunde · Jürgen Caro
Jörg Kärger · Gero Vogl
Editors

Diffusive Spreading
in Nature, Technology
and Society

 Springer

Editors
Armin Bunde
Institute of Theoretical Physics
Justus Liebig University Giessen
Giessen
Germany

Jürgen Caro
Institute of Physical Chemistry and
  Electrochemistry
Leibniz University Hanover
Hanover
Germany

Jörg Kärger
Faculty of Physics and Earth Sciences
Leipzig University
Leipzig
Germany

Gero Vogl
Faculty of Physics
University of Vienna
Vienna
Austria

ISBN 978-3-319-88489-9   ISBN 978-3-319-67798-9   (eBook)
https://doi.org/10.1007/978-3-319-67798-9

Cover figure: "Exemplifying Spreading in Nature, Technology and Society"
Credit: The following images were used by Rustem Valiullin to create the cover image: "Birds" by Pixabay/Public Domain; "Monosodium Glutamate Molecule" by User:Firkin/Public Domain; "Omethoate Molecule 3D balls" by User:Amir.ahrls/Wikimedia Commons/Public Domain; Hydronium-3D-balls by User: Benjah-bmm27/Wikimedia Commons/Public Domain

Printed on acid-free paper

This Springer imprint is published by Springer Nature
The registered company is Springer International Publishing AG
The registered company address is: Gewerbestrasse 11, 6330 Cham, Switzerland

Foreword

Everybody has observed how a piece of sugar dissolves in a cup of coffee or has experienced the spreading of a rumor. Both processes propagate without external influence, even if they may be accelerated, e.g., by stirring or telecommunication. The underlying principal phenomena are called diffusion whose mechanism was theoretically explored more than 100 years ago, mainly by Fick and Einstein. Although frequently ignored in detail, diffusion is of fundamental importance for distribution of matter and information. Even the formation of structures in living systems is considered to be the result of interplay between chemical reaction and diffusion. A series of conferences entitled "Diffusion Fundamentals" was started in Leipzig in 2005. The present volume originated from the contributions presented at the sixth meeting of this series in 2015 and contains a unique overview on our knowledge ranging from the natural sciences to the humanities. It should therefore be of interest to a broad audience.

Berlin, Germany

Gerhard Ertl
Fritz-Haber-Institute, Max-Planck-Society
Nobel Prize in Chemistry 2007

Preface

Spreading phenomena are encountered almost everywhere in our world. They may concern ideas and conceptions as well as real objects and range from sub-microscopic up to galactic distances, with timescales in the blink of an eye to geological time. With the existence of atoms and molecules, nature has provided us with a miraculous playground for experiencing the fascination of spreading phenomena which, in this context, are generally referred to as diffusion. Theoretical concepts developed in the study of diffusion of atoms and molecules have proved to be of great benefit for exploring spreading phenomena with a large spectrum of objects, irrespective of the diversity of their properties and of the underlying mechanisms. Treatises and reviews of spreading phenomena therefore frequently exploit the common ground offered by the similarities in their mathematical treatment. Such concepts are well suited for developing skills in analyzing and modeling spreading data, but they often fail to provide detailed understanding at a more fundamental level.

By aiming at an insightful introduction into the fascinating diversity of spreading phenomena in nature, technology and society, the present textbook attempts to fill this gap in the existing literature.

The roots of this book may be traced back to the breakdown of the Berlin Wall and the option of a workshop series, sponsored by the Wilhelm and Else Heraeus Foundation under the auspices of the Physical Society of reunited Germany, which brought together scientists from both parts of Germany as well as from other countries. One of these workshops, organized in Leipzig in autumn 1996, was dedicated to "Diffusion in Condensed Matter." It dealt with that topic in unprecedented width and depth. Two textbooks (J. Kärger, P. Heitjans, R. Haberlandt "Diffusion in Condensed Matter," Vieweg 1998, and P. Heitjans, J. Kärger "Diffusion in Condensed Matter: Methods, Materials, Models", Springer 2005) emerged from this initiative and are still regarded as follow-ups to Wilhelm Jost's famous textbook on "Diffusion in Solids, Liquids and Gases," Academic Press 1960.

The idea of intensifying scientific exchange across the boundaries of the individual disciplines gave rise to the establishment of the "Diffusion Fundamentals" conference series, accompanied by an open-access online journal (diffusion-fundamentals.org).

Given the relevance of physical sciences for the development in the field, it was certainly not by chance that the first conference in this series was held in Leipzig in 2005 to celebrate the centenary of Albert Einstein's annus mirabilis. Subsequent meetings in L'Aquila 2007, Athens 2009, Troy 2011 and Leipzig 2013 strove to cover with increasing concern the large spectrum of diffusion and spreading phenomena until, in 2015, the Diffusion Fundamentals activities were assigned to the Saxon Academy of Sciences. As one of Germany's regional Academies of Sciences (with members of the three Central German Federal States Sachsen, Sachsen-Anhalt und Thüringen), it stands in a great tradition (with, e.g., Werner Heisenberg as one of its secretaries) and offers, with Classes of Mathematical-Natural, Philological-Historical and Engineering Sciences, ideal conditions for cross-disciplinary scientific exchange. The participants of the 6th Diffusion Fundamentals Conference in Dresden 2015 no doubt benefited from this new level of interdisciplinary contact, and we hope that, with the contributions to the present volume which largely follow the contributions to the conference, this benefit may now be passed on to the readers of the book.

With informative illustrations, we did our best to follow the saying that a single picture tells more than a thousand words. This may also be true for a single mathematical formula—provided that the actual situation does indeed allow such a description and that the underlying mathematics remains within certain limits of comfort.

In our contact with the authors of the different chapters, we became aware of the criticality of both these issues. In some types of systems, such as human societies, the available information is not sufficient to provide meaningful predictions of future developments. In such cases, the chapters have to be presented without precise mathematical formulations. Other systems, notably those accessible by investigation with the highly sophisticated techniques of measurement provided by modern physics and chemistry, offer a multitude of information so that data processing may become quite ambitious. While all our chapters start their mathematical treatment with nothing more (but, notably, with also nothing less) than school mathematics, a few of the chapters progress to more advanced topics requiring more sophisticated mathematics. We have intentionally chosen problems from the frontiers of research, i.e., beyond the "diffusion main stream." Topics such as "Phase Transitions in Biased Diffusion" and "Hot Brownian Motion" might thus appear to be somewhat challenging. We trust, however, that the interested reader will take this as an invitation to browse some of the more specialized literature.

With all chapters now in our hands, we have first to thank the authors for a most agreeable cooperation. Looking back at the genesis of the book, we have to thank all who have contributed to the development of the "Diffusion Fundamentals" activities, including Leipzig University as the venue of the first conference, Dresden Technical University as the location of the sixth conference from which this book is derived, and the many colleagues all over the world who have cooperated with us as members of the Diffusion Fundamentals Editorial Board. We appreciate generous support by the Saxon Academy of Sciences and, over the course of the whole Diffusion Fundamentals Conference series, by the German Research Foundation, the Alexander von Humboldt Foundation and the Fonds der Chemischen Industrie.

It has been a pleasure to collaborate with the staff of Springer-Verlag, notably with Dr. Claus Ascheron and Britta Rao, who handled the editing and publication with commendable efficiency. Finally, we would like to thank our wives Eva, Marion, Birge and Senta for their continued patience, tolerance and support.

Giessen, Germany   Armin Bunde
Hanover, Germany   Jürgen Caro
Leipzig, Germany   Jörg Kärger
Vienna, Austria   Gero Vogl

Contents

Part I Introduction

1 **What the Book Is Dealing With** 3
Armin Bunde, Jürgen Caro, Jörg Kärger and Gero Vogl

2 **Spreading Fundamentals** 11
Armin Bunde, Christian Chmelik, Jörg Kärger and Gero Vogl

Part II Nature

3 **Dispersal in Plants and Animals** 29
Michael Leitner and Ingolf Kühn

4 **Search for Food of Birds, Fish and Insects** 49
Rainer Klages

5 **Epicuticular Wax Formation and Regeneration—A Remarkable
Diffusion Phenomenon for Maintaining Surface Integrity and
Functionality in Plant Surfaces** 71
Wilfried Konrad, Anita Roth-Nebelsick and Christoph Neinhuis

6 **Brain Interstitial Structure Revealed Through Diffusive Spread
of Molecules** 93
Charles Nicholson

7 **Turbulent Diffusion in the Atmosphere** 115
Manfred Wendisch and Armin Raabe

8 **Hot Brownian Motion** 127
Klaus Kroy and Frank Cichos

9 **On Phase Transitions in Biased Diffusion of Interacting
Particles** 147
Philipp Maass, Marcel Dierl and Matthias Wolff

Part III   Technology

10   Diffusive Spreading of Molecules in Nanoporous Materials 171
Christian Chmelik, Jürgen Caro, Dieter Freude, Jürgen Haase,
Rustem Valiullin and Jörg Kärger

11   Nature-Inspired Optimization of Transport in Porous Media 203
Marc-Olivier Coppens and Guanghua Ye

12   NMR Versatility 233
Scott A. Willis, Tim Stait-Gardner, Allan M. Torres, Gang Zheng
and William S. Price

13   Diffusion in Materials Science and Technology 261
Boris S. Bokstein and Boris B. Straumal

**14   Spreading Innovations: Models, Designs and Research
Directions** 277
Albrecht Fritzsche

15   The Spreading of Techno-visionary Futures 295
Armin Grunwald

Part IV   Society

**16   The Neolithic Transition: Diffusion of People or Diffusion of
Culture?** 313
Joaquim Fort

**17   The Diffusion of Humans and Cultures in the Course of the
Spread of Farming** 333
Carsten Lemmen and Detlef Gronenborn

18   Modeling Language Shift 351
Anne Kandler and Roman Unger

**19   Human Mobility, Networks and Disease Dynamics on a Global
Scale** 375
Dirk Brockmann

20   Spreading of Failures in Interdependent Networks 397
Louis M. Shekhtman, Michael M. Danziger and Shlomo Havlin

Index 411

Editors and Contributors

About the Editors

Armin Bunde studied physics at the universities in Giessen and Stuttgart (Diploma in 1970 and Ph.D. in 1974). After postdoc periods in Saarbrücken, Antwerp and Konstanz (Habilitation in 1982), he was awarded a Heisenberg fellowship in 1983. In 1984, he received the Carl-Wagner prize. Between 1984 and 1987, he was Visiting Scholar in the group of Prof. H. Eugene Stanley at Boston University. In 1987, he became Professor for Theoretical Physics at Hamburg University and in 1993 Full Chair of Theoretical Solid State Physics and Statistical Physics at Giessen University. His research activities have been dedicated to many facets of interdisciplinary research, including diffusion in disordered materials, long-range correlations in literary texts, extremes in financial markets and climate variability. He has published several books, including *Fractals and Disordered Systems* (Springer 1991, with S. Havlin), *Fractals in Science* (Springer 1994, with S. Havlin), *The Science of Disasters* (Springer 2002, with J. Kropp and H. J. Schellnhuber) and *Extreme Events and Natural Hazards: The Complexity Perspective* (Wiley 2013, with A. S. Sharma, V. P. Dimri and D. N. Baker).

Jürgen Caro graduated in chemistry at Leipzig University (Ph.D. in 1977, supervised by Jörg Kärger, Habilitation in 1992). After Ph.D., he joined the Central Institute of Physical Chemistry of the Academy of Sciences of GDR in Berlin Adlershof working in porous materials as catalysts and adsorbents. Later, he was head of the Department of Functional Materials at the newly founded Institute of Applied Chemistry in Berlin Adlershof. In 2001, he became Full Chair of Physical Chemistry at Leibniz University Hannover. He is author of about 350 publications, 6 chapters and 42 patents. Caro is guest professor at 4 universities. In 2013, he received for his membrane research work together with M. Tsapatsis The Donald W. Breck Award of the International Zeolite Association and in the same year the Ostwald Medal of the Saxon Academy of Sciences. In 2016, Caro became Corresponding Member of the Saxon Academy of Sciences.

Jörg Kärger graduated in physics at Leipzig University (Diploma in 1967, Ph.D. in 1970, Habilitation in 1978), where, in 1994, he became Full Chair of Experimental Physics/Interface Physics. His research activities have been dedicated to the study of diffusion phenomena in general and the development of new experimental techniques for studying diffusion in nanoporous materials. He has published several books, including *Diffusion in Condensed Matter* (Springer 2005, with Paul Heitjans), *Leipzig, Einstein, Diffusion* (Leipziger Universitätsverlag 2007) and, more recently, *Diffusion in Nanoporous Materials* (Wiley-VCH 2012, with Douglas Ruthven and Doros Theodorou), which became standard in the field. Jointly with Paul Heitjans, he established the online journal and conference series *Diffusion Fundamentals*, with its 6th event giving rise to the present book. His work has been recognized by numerous awards, including the Donald W. Breck Award for zeolite research, the Max Planck Research Prize and election to the Saxon Academy of Sciences. Exotics among his more than 500 publications are entries in the Guinness book of records with the largest orchestra of bicycle bells and a computer game, attained during the Physics Sunday Lectures at Leipzig University.

Gero Vogl graduated as Dr. phil. (Physik) at Universität Wien, Austria, in 1965. He afterward joined the Physikdepartment of Technische Universität München, Germany, where he was habilitated in 1974 (Dr.rer.nat.habil.). In 1977, he became Professor of Experimental Physics at Freie Universität Berlin, Germany, and in 1985 Full Professor of Physics at Universität Wien, Austria. From 1999 till 2001, he served as Chairman of the Department of Structural Research at the Hahn-Meitner-Institut Berlin (Germany) before returning to Universität Wien from where he retired in 2009. For 40 years, Gero Vogl's scientific interests were centered on the dynamics of solid matter, i.e., lattice vibrations and in particular diffusion. He introduced several new methods, all based on the application of nuclear physics to the solid state (Nuclear Solid State Physics). After 2005, Gero Vogl devoted much of his capacity to interdisciplinary research in the field of spread and diffusion, publishing two booklets aiming at promoting the ideas. Presently, he attempts to make use of models from physics to understand what drives language change, a phenomenon which threats to extinguish a good part of today's languages.

Contributors

Boris S. Bokstein National University of Science and Technology "MISIS", Moscow, Russia

Dirk Brockmann Institute for Theoretical Biology, Humboldt Universität zu Berlin, Berlin, Germany; Robert Koch-Institute, Berlin, Germany

Armin Bunde Institute of Theoretical Physics, Justus Liebig University Giessen, Giessen, Germany

Jürgen Caro Institute of Physical Chemistry and Electrochemistry, Leibniz University Hanover, Hanover, Germany

Christian Chmelik Faculty of Physics and Earth Sciences, Leipzig University, Leipzig, Germany

Frank Cichos Peter Debye Institute for Soft Matter Physics, Leipzig University, Leipzig, Germany

Marc-Olivier Coppens Department of Chemical Engineering, University College London, London, UK

Michael M. Danziger Department of Physics, Bar Ilan University, Ramat Gan, Israel

Marcel Dierl Physikalisch-Technische Bundesanstalt, Berlin, Germany

Joaquim Fort Complex Systems Laboratory and Physics Department, Universitat de Girona, Girona, Catalonia, Spain; Catalan Institution for Research and Advanced Studies (ICREA), Barcelona, Catalonia, Spain

Dieter Freude Faculty of Physics and Earth Sciences, Leipzig University, Leipzig, Germany

Albrecht Fritzsche Institute of Information Management 1, Friedrich-Alexander University Erlangen-Nuremberg, Nuremberg, Germany

Detlef Gronenborn Romano-German Central Museum and Johannes Gutenberg University of Mainz, Mainz, Germany

Armin Grunwald Institute for Technology Assessment and Systems Analysis (ITAS), Karlsruhe Institute of Technology, Karlsruhe, Germany

Jürgen Haase Faculty of Physics and Earth Sciences, Leipzig University, Leipzig, Germany

Shlomo Havlin Department of Physics, Bar Ilan University, Ramat Gan, Israel

Anne Kandler Department of Human Behavior, Ecology and Culture, Max Planck Institute for Evolutionary Anthropology, Leipzig, Germany

Rainer Klages Max Planck Institute for the Physics of Complex Systems, Dresden, Germany; School of Mathematical Sciences, Queen Mary University of London, London, UK

Wilfried Konrad Institut für Botanik, TU Dresden, Dresden, Germany

Klaus Kroy Institute of Theoretical Physics, Leipzig University, Leipzig, Germany

Jörg Kärger Faculty of Physics and Earth Sciences, Leipzig University, Leipzig, Germany

Ingolf Kühn Helmholtz-Zentrum für Umweltforschung—UFZ, Halle, Germany

Michael Leitner Heinz Maier-Leibnitz Zentrum (MLZ), Technische Universität München, Garching bei München, Germany

Carsten Lemmen Institute of Coastal Research, Helmholtz-Zentrum Geesthacht, Geesthacht, Germany

Philipp Maass Fachbereich Physik, Universität Osnabrück, Osnabrück, Germany

Christoph Neinhuis Institut für Botanik, TU Dresden, Dresden, Germany

Charles Nicholson Department of Neuroscience and Physiology, New York University Langone Medical Center, New York, NY, USA

William S. Price Nanoscale Organisation and Dynamics Group, Western Sydney University, Penrith, NSW, Australia

Armin Raabe Faculty of Physics and Earth Sciences, Leipzig Institute for Meteorology, University of Leipzig, Leipzig, Germany

Anita Roth-Nebelsick State Museum of Natural History Stuttgart, Stuttgart, Germany

Louis M. Shekhtman Department of Physics, Bar Ilan University, Ramat Gan, Israel

Tim Stait-Gardner Nanoscale Organisation and Dynamics Group, Western Sydney University, Penrith, NSW, Australia

Boris B. Straumal National University of Science and Technology "MISIS", Moscow, Russia; Institute of Solid State Physics, Russian Academy of Sciences, Chernogolovka, Russia; Karlsruhe Institute of Technology (KIT), Institute of Nanotechnology, Eggenstein-Leopoldshafen, Germany

Allan M. Torres Nanoscale Organisation and Dynamics Group, Western Sydney University, Penrith, NSW, Australia

Roman Unger Technische Universität Chemnitz, Chemnitz, Germany

Rustem Valiullin Faculty of Physics and Earth Sciences, Leipzig University, Leipzig, Germany

Gero Vogl Faculty of Physics, University of Vienna, Vienna, Austria

Manfred Wendisch Faculty of Physics and Earth Sciences, Leipzig Institute for Meteorology, University of Leipzig, Leipzig, Germany

Scott A. Willis Nanoscale Organisation and Dynamics Group, Western Sydney University, Penrith, NSW, Australia

Matthias Wolff Fachbereich Physik, Universität Osnabrück, Osnabrück, Germany

Guanghua Ye State Key Laboratory of Chemical Engineering, East China University of Science and Technology, Shanghai, China

Gang Zheng Nanoscale Organisation and Dynamics Group, Western Sydney University, Penrith, NSW, Australia

Part I
Introduction

Chapter 1
What the Book Is Dealing With

Armin Bunde, Jürgen Caro, Jörg Kärger and Gero Vogl

Early in the 19th century, Robert Brown, a British botanist, found that pollen immersed in a liquid performed an unceasing motion [1]. As a careful scientist he conducted experiments under varying conditions and hence could exclude that this motion was "vitality", as argued in the beginning, and suspected that it was physics. About thirty years after Brown's observation Adolf Fick, a German physiologist, dissolved salt in water, studied the change of salt distribution in time and wrote down the equations governing the phenomenon of diffusion along a concentration gradient [2]. It was not until 1905 that Albert Einstein found an ingenious description of the "Brownian motion" and was able to connect it to Fick's equations of diffusion [3].

In this book we bring together scientists from disciplines as different as archeology, ecology, epidemics, linguistics and sociology with natural scientists from biology, chemistry, physics and technology. What is common to all these scientists is that all are interested in the motion of a certain object or phenomenon in space and time. This motion is named diffusion by physicists and chemists and sometimes

A. Bunde
Institute of Theoretical Physics, Justus Liebig University Giessen,
Giessen, Germany
e-mail: arminbunde00@googlemail.com

J. Caro
Institute of Physical Chemistry and Electrochemistry,
Leibniz University Hanover, Hanover, Germany
e-mail: caro@pci.uni-hannover.de

J. Kärger (✉)
Faculty of Physics and Earth Sciences, Leipzig University, Leipzig, Germany
e-mail: kaerger@physik.uni-leipzig.de

G. Vogl
Faculty of Physics, University of Vienna, Vienna, Austria
e-mail: gero.vogl@univie.ac.at

© Springer International Publishing AG 2018
A. Bunde et al. (eds.), *Diffusive Spreading in Nature, Technology and Society*,
https://doi.org/10.1007/978-3-319-67798-9_1

called spread by ecologists and linguists. Brown and Fick were both neither mathematician nor physicist. So why should the phenomena of endless motion and diffusion be a domain of mathematics or physics?

The moving objects can be very different: they can be particles, e.g. atoms, they can be living beings, humans, animals, plants, bacteria. But they can also be abstract terms: ideas, rumors, information, innovations or linguistic features. In the case of the spread or diffusion of abstract objects, the space wherein the objects spread may not be local, it may rather be an abstract "space", e.g. a group of people, the entity of words, the sum of preoccupations or the bulk of technological environment.

It appears daring to treat all these phenomena with related methods, to force them into the same corset, but we are definitely not the first to attempt a synopsis. In 1951, Skellam [4] who was not a physicist but rather what was called a biometrist at that time wrote: "It is apparent that many ecological problems have a physical analogue and that the solution of these problems will require treatment with which we are already very familiar." For more than thirty years there have been textbooks written mostly by mathematicians, reporting on the work on spread and diffusion not performed by themselves but rather by scientists from disciplines as different from physics as biology and ecology. The authors of these textbooks amply describe that the same analytical approaches as have been developed for the diffusion of particles in physics and chemistry can operate for treating the migration and spread of animals and plants. To the best of our knowledge the present book is, however, the first where scientists from the different very disparate disciplines report on their work on diffusion and spread in the fields of their professions. What is contained in the book you have at hand should therefore be first-hand information.

Before we enter the multifaceted world of spreading and diffusion, with numerous examples from nature, technology and society, the theoretical foundations are summarized in Chap. 2. All subsequent chapters make use of the presented formalism so that references to this chapter are "spread" throughout the book. The mathematical formalism remains within the common framework known from school and is, in addition, supported by numerous informative figures.

Already here we have to ask which framework serves best for reflecting spreading phenomena within a given system. While classical diffusion—as occurring, e.g., during dissolving a piece of sugar in a cup of coffee—is easily understood to be adequately described by using a Cartesian coordinate system, problems are immediately seen to arise if such complicated situations as the spreading of innovations over a continent (as during the transition from hunting-gathering to farming and stockbreeding in Europe) are considered.

In fact, also in such cases, complemented by "gain" and "loss" rates (associated with birth and death of the considered species), one may make use of essentially the same mathematical formalism as applicable to the spreading of sugar. The options of approaching reality along these lines are further improved by considering diffusivities and gain and birth rates as functions of space and time including, notably, their dependence on the given "concentration", i.e. the population density, rather than as by taking them as mere constants.

Alternatively, the complexity of the system may be taken into account by sub-dividing the system into various "regions" and by considering the entity of the population numbers of each individual region. Further evolution can then be predicted by considering birth and death rates within and exchange rates between the individual regions. This latter approach (with the populations of the individual regions operating as "agents") offers ideal conditions for taking account of the system's complexity by increasing the graticule subtlety. The establishment of parameter sets for reliably reflecting the internal dynamics, however, is a challenging task, with the risk of unjustified biasing.

While it is probably beyond debate that spreading phenomena occur in nature just as in technology and society, the attribution of a particular phenomenon to one of these fields is not always easy. Our attempt to sub-structure the book by attributing, in the subsequent parts, the various spreading phenomena considered to one of these fields is, therefore, not without ambiguity. In fact, for essentially any of the subsequent chapters there would be good reasons to include them in at least one of the other two parts. We assume that, on reading the different chapters, you will become aware of this issue and will take it as one of the many indications of the benefit of a comprehensive consideration of spreading phenomena!

We start our journey through spreading phenomena in **Nature** by following Michael Leitner and Ingolf Kühn in their report about the evolution of ecosystems, with special emphasis on the dispersal of newly arrived plants and animals, and an illuminating comparison of diffusive dynamics in physical sciences and ecology. It is essentially the climatic habitat which is recognized as a driving agent of advancement and invasion. Examples include the particular case of Ambrosia artemisiifolia or ragweed as an allergenic plant which, since its arrival from North America, has led to a substantial increase in medical costs for allergenic treatment. The spatio-temporal dynamics of animals is also in the focus of Rainer Klages when he describes how one is able to simulate—by analytical methods—the outcome of sophisticated experiments dealing with foraging flights of bumblebees to collect nectar and how, in the presence of predators, the dynamics are modified.

Still within the field of biology, but on a much smaller scale, Wilfried Konrad, Anita Roth-Nebelsick and Christoph Neinhuis consider transport phenomena on plant surfaces. Evolution over hundreds of millions of years gave rise to the formation of a multifunctional, self-repairing surface layer which—in parallel with ensuring proper metabolic exchange with the surroundings—protects the plant against uncontrolled evaporation. Also with Charles Nicholson we remain within microscopic dimensions—this time within the human brain. We are, in particular, invited to share the fascination of analogies in the functionality of our brain tissue with the organization of transport in a big city like New York.

The scale of observation is enhanced by more than five orders of magnitude when, in their chapter, Manfred Wendisch and Armin Raabe consider mass transfer in our atmosphere. They introduce into the complexity of the relevant phenomena, notably during turbulent motions, from which it becomes immediately understandable that weather forecast is still associated with significant uncertainties.

Complicated physical situations are likewise considered in the last two chapters of this second part of the book dedicated to spreading in nature. In fact, both phenomena considered bring us back to biology. They deal with the question how irregular thermal motion may give rise to directed transport phenomena since the occurrence of directed motions is among the prerequisites for the functionality of living organisms. Klaus Kroy and Frank Cichos consider how local heating influences the movement of particles suspended in a fluid. Such heating is immediately seen to give rise to a bias in molecular motion if it is initiated by the use of the so-called Janus particles, i.e. of particles where, by covering only one half of the surface by a gold layer, only this half of the surface gives rise to temperature enhancement.

Philipp Maas, Marcel Dierl and Matthias Wolff model the bias in motion by implying asymmetric jump probabilities, in the extreme case with jump attempts into only one direction and the (obvious) requirement that jumps are only possible if a jump attempt is directed to a site which is still unoccupied. Models of this type introduce the principles that can explain how sudden standstills on a highway may disappear suddenly, for no obvious reason, to yield regular traffic flow. The chapter provides an impression of how complicated the whole matter becomes if mutual interactions between the various particles are considered.

Part 3, scheduled for highlighting the various aspects of spreading phenomena of relevance in **Technology**, starts with an introduction by Christian Chmelik and co-authors into the relevance of diffusion for the performance of many technologies based on the application of nanoporous materials. The annual benefit worldwide by the exploitation of these technologies in petroleum refining has been estimated to at least 10 billion US dollars—and the gain in value-added products by performance can, obviously, never be faster than allowed by the rate of diffusion from the catalytically active sites in the interior of these materials to the surroundings.

Marc-Olivier Coppens and Guanghua Ye illustrate in the subsequent chapter how, on their search for optimum technologies, process engineers may gain inspiration from nature. Here, through billions of years of evolution, plants and animals have acquired highly effective transport systems, crucial to their survival. Although a chemical engineering application is different from a biological one in terms of materials and operating conditions, they share fundamental features in, notably, connecting the action at microscopic scales with those in the overall system.

There is scarcely another measuring technique that offers such a wealth of information on transport phenomena and, hence, on the key process in many technologies, as nuclear magnetic resonance (NMR). After introducing the fundamental principles of this almost universal analytical tool, William S. Price and his colleagues present an impressive array of applications, ranging from diffusion measurement in porous media to magnetic resonance imaging (MRI), currently the most powerful of the imaging techniques in medical diagnosis.

A survey over the many systems of technological relevance where mass transfer occurs under solid-state conditions is provided in their chapter by Boris S. Bokstein and Boris B. Straumal. They cover metals, amorphous alloys and polymers and include, as an attractive topic of current material research, the investigation of severe plastic deformation brought about by material straining under spatial confinement.

Spreading phenomena in technology are considered from a much more global perspective in the final two chapters of this part. Albrecht Fritzsche pursues the question how novel technologies are becoming part of the knowledge of humankind. Here technologies are understood as an embodiment of any instrumental action that occurs repetitively in our world. This rather broad definition creates the need for different operational measures for the diffusion of innovations, including the impact of both the innovating institution and public acceptance. In the subsequent chapter, Armin Grunwald takes us a step further ahead by investigating the role of visions as an established part of scientific and technological communications and by asking for the conditions under which they may indeed fulfil their potential as a major driver of scientific and technological advance. He reminds us that robots, e.g., entered society in this very way long before they came to exist in reality, where they now may be found with many of the anticipated functions and meanings. The contributions by Fritzsche and Grunwald mark the borderline with part 4 of our book, dealing with spreading phenomena in **Society**.

Here, into humanities and, notably, into archeology, analytical diffusion equations have been introduced in already the early seventies of the last century through the famous work by Ammerman and Cavalli-Sforza [5]. Standing, mathematically, on the shoulders of Fisher [6] and Skellam [4] they described the invasion of farmers, the people of the Neolithic, between 8000 and 6000 years ago, from the Near East by a one-dimensional "wave of advance" from south-east towards the north-west of Europe. But was it really people or just technology that advances —"demic" or "cultural" diffusion? Still a hot topic! At that time the authors had no more genetics at their hand than blood groups and archeological indications for adapting their diffusion equations.

The tremendous progress of genetics has recently accelerated that field. In this book the diffusion of the Neolithic is dealt with by Joaquim Fort with more and more refined diffusion equations. Already Skellam had stressed "Unlike most of the particles considered by physicists, living organisms reproduce and interact. As a result, the equations of mathematical ecology are often of a new and unusual kind." Fort has amply expanded the field of "reaction diffusion" being able to now incorporate cohabitation, cultural transmission and anisotropic expansion. Fort stresses that all models considered by him operate with a minimum of parameters. The demic model, for instance, needs only three parameters which have been estimated from ethnographic or archeological data. He mentions further on that, with such constraints, one is able largely to avoid any unjustified bias in modeling which may easily occur by the use of more parameters.

In the subsequent chapter, Carsten Lemmen and Detlef Gronenborn deal once again with the spread of Neolithic culture and the question for demic or cultural diffusion.

However, rather than applying analytical methods as done by Fort, they treat the problem with a technique, sometimes referred to as the method of cellular automaton. Here the system under study is subdivided into different regions ("cells"), with the interaction between the populations of the different cells via trade and migration taken account of by suitable mathematical expressions. Simulating the dynamics of local human populations' density and sociocultural features ("traits") needs more parameters than Fort's model, in particular a couple of habitat parameters, which—with hopefully more data from archeology—in future will have to be reconsidered

Back to modern times, Anne Kandler and Roman Unger consider the variation in the use of languages, with the focus on the retreat of the Gaelic language in North-Western Scotland. They combine their analysis with the search for reasonable measures enabling the preservation of linguistic diversity in our world [7] and with estimating the expenses and probability of success of such efforts.

Dirk Brockmann describes the spread of diseases with reaction-diffusion equations by introducing effective distances and effective times thus considering the distance for a contagion not purely geographically but rather by considering the probability for overcoming the given distances. He shows that, for a contagion on the global scale, Frankfurt may be effectively closer to New York than a village in New Jersey only 150 km away. Here we encounter the probabilistic capabilities of the use of diffusion equations and their "descendants" and, jointly with it, an impressive example of how medical sciences are already successfully applying this unconventional approach for mitigating or even preventing epidemics.

While in most contributions collected here the spreading phenomena considered refer to the movement of "objects" in space, in the concluding chapter of our book Louis M. Shekhtman, Michael M. Danziger and Shlomo Havlin talk about a totally different situation when they consider the spreading of *failures* in complex systems such as power grids, communication networks and financial networks. Here one failure is seen to easily trigger further ones, giving rise to cascades wherein the failure may spread over also other parts of the system. Failure spreading, finally, ends up in blackouts, economic collapses and other catastrophic events. Exploration for the underlying "physical" mechanisms is, once again, illustrated as an important issue in searching for suitable means for mitigating such disastrous developments.

Some of you might regard our endeavor as too daring and ask "cui bono?" if phenomena far away from physics have been treated by methods of clear provenience in physical sciences. We are aware of the risk of oversimplified approaches and the danger of reductionism is pending. But, please, have also in mind that, for science, it should be allowed if not even required as an imperative, to transgress the limits of one's own discipline. That is what we have attempted.

Throughout the book we have attempted to keep explanations simple, with as little mathematics as possible. We therefore hope that the book will appeal not only to specialists, but to anyone interested in looking for the many impressive analogies of spreading and diffusion throughout many different disciplines. We trust that, with the book in your hands, you will find plenty of motivation for enjoying the fascinating world of spreading phenomena and, notably, to track and to deeper explore them in your own disciplines.

References

1. R. Brown, Philos. Mag. **4**, 161 (1828)
2. A. Fick, Ann. Phys. Chem. **94**, 59 (1855)
3. A. Einstein, Ann. Phys. **17**, 549 (1905)
4. J.G. Skellam, Biometrika **38**, 196 (1951)
5. A.J. Ammerman, L.L. Cavalli-Sforza, Man New Ser. **6**, 674 (1971)
6. R.A. Fisher, Ann. Eugen. **7**, 355 (1937)
7. H.J. Meyer, Diffus. Fundam. Online **25**(1), 1–6 (2016)

Chapter 2
Spreading Fundamentals

Armin Bunde, Christian Chmelik, Jörg Kärger and Gero Vogl

2.1 Diffusion Step by Step

Stimulated by their thermal energy, atoms and molecules are subject to an irregular movement which, in the course of history, has become known under the term diffusion. Today, in a more generalized sense, essentially any type of stochastic movement may be referred to as diffusion.

Diffusion sensu stricto is the motion of individual objects by way of a "random walk". For simplicity we start with the one-dimensional problem: our random walker is assumed to move along only one direction (the x coordinate) and to perform steps of identical length l in either forward or backward direction. Both directions are equally probable and the direction of a given step should in no way affect the direction of a subsequent one (Fig. 2.1).

Such sequences of events are called uncorrelated. The mean time between subsequent steps is denoted by τ. Obviously, nobody can predict where exactly this random walker will have got to after n steps, this means, at time $t = n\,\tau$.

A. Bunde
Institute of Theoretical Physics, Justus Liebig University Giessen, Heinrich-Buff-Ring 16, 35392 Giessen, Germany
e-mail: arminbunde00@googlemail.com

C. Chmelik · J. Kärger (✉)
Faculty of Physics and Earth Sciences, Leipzig University, Linnéstr. 5, 04103 Leipzig, Germany
e-mail: kaerger@physik.uni-leipzig.de

C. Chmelik
e-mail: chmelik@physik.uni-leipzig.de

G. Vogl
Faculty of Physics, University of Vienna, Strudlhofgasse 4, 1090 Vienna, Austria
e-mail: gero.vogl@univie.ac.at

© Springer International Publishing AG 2018
A. Bunde et al. (eds.), *Diffusive Spreading in Nature, Technology and Society*,
https://doi.org/10.1007/978-3-319-67798-9_2

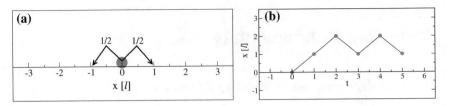

Fig. 2.1 Random walk in one dimension. **a** Step to the left and step to the right equally probable. **b** Possible displacement after five steps

The randomness of the process allows predicting probabilities only. Let us consider a large number of random walks, all beginning at the same point. The probability that at time t a random walker shall have got to position x is then simply the ratio between the number of random walks leading to this point and the total number of walks.

We are going to derive the "mean square displacement" $\langle x^2(t=n\tau)\rangle$ as a characteristic quantity of such a distribution. It denotes the mean value of the square of the net displacement after n steps, corresponding to time $t = n\tau$. Mean values are determined by summing over all values and division by the number of values considered. For our simple model we obviously have

$$
\begin{aligned}
\langle x^2(t=n\tau)\rangle &= \langle (x_1 + x_2 + x_3 + \cdots + x_n)^2 \rangle \\
&= \langle x_1^2 + x_2^2 + x_3^2 + \cdots x_n^2 + 2x_1x_2 + 2x_1x_3 + \cdots + 2x_{n-1}x_n \rangle
\end{aligned}
\tag{2.1}
$$

where x_i denotes the length of the i-th step. The magnitude of x_i can be either $+l$ (step in $(+x)$ direction, i.e. step ahead) or $-l$ (step in $(-x)$ direction), so that all of the first n terms in the second line become equal to l^2. Let us now consider the mean value of each of the subsequent "cross" terms x_ix_j, with $i \neq j$. For a given value of x_i, according to our starting assumption, the second factor x_j shall be equal to $+l$ and to $-l$ with equal probability. Hence, the resulting values x_ix_j, with $i \neq j$, will be equally often $+l^2$ and $-l^2$, leading to a mean value of zero.

Equation (2.1) is thus seen to simply become

$$
\langle x^2(t)\rangle = nl^2 = \frac{l^2}{\tau}t,
\tag{2.2}
$$

with the most important message that a diffusant departs from its origin not in proportion with time as it would be the case with directed motion. It is rather the square of the displacement which increases with time, so that the (mean) distance often called x_{rms} (rms meaning root of the mean square) increases with only the square root of time.

Quite formally, we may introduce an "abbreviation"

$$D = \frac{l^2}{2\tau} \qquad (2.3)$$

so that now Eq. (2.2) may be noted in the form

$$\langle x^2(t) \rangle = 2Dt. \qquad (2.4)$$

We shall find in the subsequent section that the thus introduced parameter D is a key quantity for quantifying the rate of the random movement which we have referred to as diffusion. We may rearrange Eq. (2.4), leading to its general definition,

$$D = \frac{\langle x^2(t) \rangle}{2t}. \qquad (2.5)$$

D is referred to as the self-diffusivity (or coefficient of self-diffusion or self-diffusion constant). The considerations may be extended to two and three dimensions, where the factor 2 on the right-hand side of Eq. (2.3) (and, correspondingly, in Eqs. (2.4) and (2.5)) has to be replaced by 4 and 6, respectively.

Abandoning the simplifying condition of equal step lengths, with essentially the same reasoning as exemplified with Eq. (2.1), Eq. (2.3) may be shown to be still valid, now with l^2 as the mean squared step length.

2.2 From Random Walk to Fluxes

Though today it is possible to follow the diffusion path ("trajectory") of an individual molecule [1], the relevance of diffusion becomes more obvious if ensembles of diffusing particles are considered. This situation is schematically presented in Fig. 2.2. In the following we shall explain that it illustrates the situation typical of the three most important ways to measure and characterize diffusion. The circles represent the diffusing particles and the lattice indicates that the process occurs within some "framework" formed by, e.g. open spaces (vacancies) in a solid state lattice, interconnected pores or territorial areas, which may serve as a reference system. Correspondingly, the scheme has to be modified (see, e.g., Chap. 13 and Fig. 13.1) when diffusion of the lattice constituents (as in solid-state diffusion) is considered.

Let us start our discourse with Fig. 2.2a, with the concentration of diffusants deliberately chosen to decay from left to right. This gradient in concentrations effects that, irrespective of the random (and, notably, undirected!) movement of each individual particle, their superposition leads to a directed flux. Macroscopically, this particle flux abolishes existing concentration gradients, following the general tendency towards equilibration in nature.

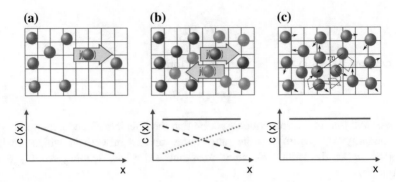

Fig. 2.2 Microscopic situation corresponding to the measurement of the transport (chemical) diffusivity (**a**) and of the self- (tracer) diffusivity (**b, c**) by either observing the flux of a labeled fraction (see e.g. [2] and Chap. 13) (**b**), or by recording the individual displacements, e.g. by methods as PFG NMR ([3] and Chap. 12), quasi-elastic neutron scattering [4, 5], Mössbauer spectroscopy [5] or X-ray photon correlation spectroscopy XPCS [6] (**c**). Examples of typical particle distributions on top and corresponding spatial dependencies of concentration below. Reproduced with permission from Ref. [7], copyright (2013) Wiley-VCH Verlag GmbH & Co. KGaA

Doubling the concentration gradient will obviously effect a doubling of the difference between the numbers of particles passing from left to right and from right to left and, hence, a doubling of the flux. This leads to the famous Fick's 1st law

$$j_x = -D_T \frac{\partial c}{\partial x}. \tag{2.6}$$

j_x denotes the flux density in x direction, where the x coordinate is chosen to indicate the direction of falling concentration and the index T indicates "transport". The flux density $j_x = \Delta N / \Delta A \cdot \Delta t$ is defined by the number ΔN of particles passing an area ΔA (perpendicular to the flux direction) during a time interval Δt, divided by ΔA and Δt. In Eq. (2.6), the concentration gradient is represented as a so-called partial derivative, which has to be introduced whenever a quantity (here the particle concentration c, i.e. the particle number per volume) is a function of various parameters, such as location (x) and time (t) in our case. This twofold dependence is expressed by the notation $c(x,t)$. Partial derivation means that one considers derivation with respect to one parameter (here x) while the other one(s) is (are) kept constant. The minus sign in Eq. (2.6) indicates that the particle flux is directed towards decreasing concentration. The factor of proportionality, D_T, is referred to as the coefficient of transport diffusion (as indicated by suffix T). Alternatively also the terms chemical or collective diffusion are used.

Let us return to Fig. 2.2, where we will now look for an option to quantify diffusion under equilibrium conditions, i.e. for uniform concentration. In this case, obviously, the irregular particle movement does not lead to any net flux. As illustrated by Fig. 2.2b, however, again a macroscopically observable effect may be generated if we are able to effect a distinction between the particles of the system without affecting their microdynamic properties. In Fig. 2.2b it is simply achieved by considering spheres in two different shades of red, with both of them assumed to behave identical and the respective concentrations given below in the figure. With this distinction, again fluxes become macroscopically observable. In complete analogy to Eq. (2.6) we may note

$$j_x^* = -D\frac{\partial c^*}{\partial x} \tag{2.7}$$

where the asterisk (*) indicates that only one sort of the differently labelled particles (i.e. either the red or the pink spheres) is considered. In experiments, such a situation may be realized by using (two) different isotopes as diffusing particles. With reference to the use of labelled molecules ("tracers"), the thus defined quantity D is referred to as the tracer diffusivity. It might come as a surprise that, at the end of this section, the thus defined tracer diffusivity will be found to coincide with the self-diffusivity introduced in the previous section.

A macroscopically existing concentration gradient (Fig. 2.2a) will generally give rise to an additional bias, as a consequence in the difference in the "surroundings" depending on whether the diffusant is moving into the direction of higher or smaller concentration. The rate of propagation of the diffusants depends on the existence of "free sites" in the range where they try to get to. While in "highly diluted" systems this should not be a problem since "free sites" can be assumed to be anywhere easily (and, hence, with equal probability) available, the situation becomes more complicated with increasing density of the diffusants. This is true, e.g., for diffusing molecules if the cavities in a porous material are occupied already by other guest molecules, for diffusion in solids where generally the concentration of free sites (vacancies) is very low, or if a new generation of farmers is forced to leave their home ground in search for new farming areas, getting into even more densely populated districts.

Such type of bias does not exist in the absence of macroscopic concentration gradients (Fig. 2.2b). Hence, reflecting two different microdynamic situations, the coefficients of tracer and transport diffusion cannot be expected to coincide quite in general. We shall return to some general rules for correlating these two types of diffusivities in Sect. 2.3. Before, however, we are going to illustrate why the coefficients of self-diffusion (as introduced by Eq. (2.5) and as resulting with a measuring procedure as illustrated by Fig. 2.2c) and of tracer diffusion (Eq. (2.7) and Fig. 2.2b) are one and the same quantity.

Figure 2.2c takes us back to Sect. 2.1 with Fig. 2.1 and Eq. (2.2) illustrating the evolution of the probability distribution of diffusing particles. Now we are going to show that this very problem may as well be treated within the frame of Fick's first law (Eq. (2.6)). For this purpose, we consider the change of the number of particles within a volume element due to diffusion. The way of reasoning is sketched in Fig. 2.3, where again we have made use of the simplifying assumption that the flux is uniformly directed into x direction (which implies uniform concentration in any

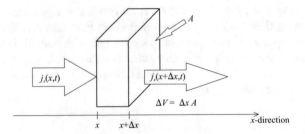

Fig. 2.3 Particle balance in a volume element $\Delta V = \Delta x \cdot A$ for diffusion in x-direction. The change in the number of particles within this volume element per time is equal to the differences of the fluxes leaving and entering. Reproduced with permission from Ref. [9], copyright (2014) Leipziger Universitätsverlag

y-z plane). For an extension to three dimensions, with the option of also orientation-dependent diffusivities, we refer to Chap. 12 and, notably, Sect. 12.5.2.

As is evident: particles entering (flux j) into a given volume must leave again or—if they do not leave again—will increase the density (c) in the volume:

$$\frac{\partial j}{\partial x} = -\frac{\partial c}{\partial t} \qquad (2.8)$$

This relation is termed the continuity equation.

Inserting Eq. (2.8) into Eq. (2.6) yields

$$\frac{\partial c}{\partial t} = D\frac{\partial^2 c}{\partial x^2} \qquad (2.9)$$

where, for simplicity, the diffusivity is assumed to be uniform anywhere in the system. Equation (2.9) represents Fick's 2nd law, stating simple proportionality between the change in concentration with time and the "gradient of the concentration gradient", i.e. the curvature of the concentration profile. We do, moreover, disregard the suffix T having in mind that our reasoning applies to both transport and tracer diffusion.

The mathematics to treat the evolution of such a system is provided by Eq. (2.9). The reader with some background in differential calculus will easily convince himself that the function

$$c(x,t) \equiv P(x,t) = \frac{1}{\sqrt{4\pi Dt}}\exp\left(-\frac{x^2}{4Dt}\right), \qquad (2.10)$$

namely a so-called Gaussian, obeys this equation (Fig. 2.4). It may be shown that, as a consequence of the central limit theorem of statistics, a Gaussian results quite generally for the distribution function of particle displacements after a sufficiently

Fig. 2.4 Evolution of the
probability distribution for the
end points of a "random
walk" starting at $t = 0$ at
$x = 0$. The curves represent
the so-called probability
density $P(x,t)$. Reproduced
with permission from Ref. [9],
copyright (2014) Leipziger
Universitätsverlag

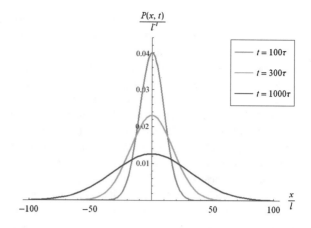

large series of uncorrelated "elementary" displacements ("steps") if they are of
identical distribution, symmetric and of finite variance, i.e. of finite mean squared
"step length" (see also Sects. 3.5.1 and 4.1 and Chap. 2 in [8]). Figure 2.1 illustrated a most simple example of such a series.

With the probability distribution given by Eq. (2.10), the mean displacement can
be noted as

$$\langle x^2(t) \rangle = \int_{x=-\infty}^{x=\infty} P(x,t)x^2 dx = \int_{x=-\infty}^{x=\infty} \frac{1}{\sqrt{4\pi Dt}} \exp\left(-\frac{x^2}{4Dt}\right)x^2 dx = 2Dt, \quad (2.11)$$

which leads to a standard integral. The analytical solution yields the expression
which has been given already by Eq. (2.4) where, via Eq. (2.3), D has been
introduced as a "short-hand expression" for $l^2/(2\tau)$ and, by Eq. (2.5), has been
defined as the self-diffusivity. This expression is now in fact seen to coincide with
the tracer diffusivity as introduced by Fick's 1st law. It was in one of his seminal
papers of 1905 [10] that Albert Einstein did find this bridge between Fick's law and
random particle movement. Thus Eq. (2.5) is often referred to as Einstein's diffusion equation. For a more profound appreciation of this achievement we refer to the
presentation of "hot" Brownian motion in Chap. 8.

Diffusive fluxes in our real world are, as a matter of course, often accompanied
by fluxes emerging from directed rather than from random motion. Such situations
do occur in also the examples considered in our book when, e.g., diffusive fluxes in
plants (Chap. 5) and turbulences in our atmosphere (Chap. 7) have to be considered
in superposition with phenomena of bulk motion, referred to as advection. The
combination of mass transfer by advection and diffusive fluxes is commonly
referred to as convection.

Throughout the book we shall be wondering about the "driving forces" giving
rise to the various types of fluxes occurring within the systems under consideration.

With Fig. 2.2. we have seen already that, under the existence of concentration gradients, diffusive fluxes emerge already as a simple consequence of random movement. In multicomponent systems of interacting particles the situation becomes more intricate. Chapter 10 gives an example that illustrates how then the gradient of the "chemical potential" may most conveniently be applied as a "driving force" of diffusion. Borrowing a conception in common use in hydrogeology, Chap. 5 deals with directed water fluxes in plants by means of Darcy's law, with the gradient in water potential as the driving force. While thus, in physical sciences and engineering, the search for the driving forces and the quantitation of fluxes is among the tasks of today, equivalent efforts on considering spreading phenomena in e.g. humanities appear to be still far before maturity.

In problems of ecology and alike and in many problems in cultural science, spreading phenomena occur in two rather than in only one dimension as considered in our introductory example. For diffusion now $\langle r(t)^2 \rangle = 4Dt$ and again the most probable place to find a "random walker" is at the origin. As Pearson [11] put it already in 1905: "The most probable place to find a drunken man who is at all capable of keeping on his feet is somewhere near his starting point." That is what can be seen from the cartoon Fig. 2.5 and has already been the message of Fig. 2.4 (which preserves its pattern in also two- and more-dimensional presentation): the maximum in the probability distribution of the location of a random walker remains in his starting point.

In two dimensions it is appropriate to use polar coordinates and Fick's 2nd law is written

$$\frac{\partial c(r,t)}{\partial t} = \frac{D}{r}\frac{\partial}{\partial r}\left(r\frac{\partial c(r,t)}{\partial r}\right) \tag{2.12}$$

with $r = \sqrt{x^2 + y^2}$ denoting the distance between the origin of the spreading process and the considered area. Just as Eq. (2.10) resulted from Eq. (2.9), the solution of Eq. (2.12) is found to be

$$c(r,t) = \frac{n}{4\pi Dt}\exp\left(-\frac{r^2}{4Dt}\right). \tag{2.13}$$

n is the number of representatives of a certain species at the origin.

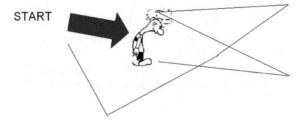

Fig. 2.5 The most probable place to find a drunken man who is at all capable of keeping on his feet is somewhere near his starting point

2.3 Interaction, Growth and Conversion

So far all our considerations were based on the simplifying assumption that the propagation probability of our diffusants is uniform all over the system under study. This implies uniformity of the medium in which the process of diffusion (spreading) occurs, as well as the absence of any interaction between the diffusants. With the lack of interaction, a distinction between equilibrium and non-equilibrium phenomena becomes meaningless [12]. The coefficients of self- and transport diffusion as considered so far do, therefore, coincide (given by Eq. (2.3) for the considered step model) and Eq. (2.9) does hold for both self- (=tracer) and transport diffusion. Due to this coincidence there was, up to this point, no real need for distinguishing between the two different types of diffusivities. On considering such interactions, however, this distinction will become necessary.

On considering molecular interactions, the diffusivity $D = D(c)$ becomes a function of the diffusant concentration c so that Fick's 2nd law is not correct anymore in the form of Eq. (2.9). Inserting Eq. (2.6) into Eq. (2.8) does now rather yield (again for the simple one-dimensional problem)

$$\frac{\partial c}{\partial t} = \frac{\partial}{\partial x}\left(D(c)\frac{\partial c}{\partial x}\right) = D(c)\frac{\partial^2 c}{\partial x^2} + \frac{\partial D(c)}{\partial c}\left(\frac{\partial c}{\partial x}\right)^2. \tag{2.14}$$

The particular dependence $D(c)$ of the diffusivity is determined by the system under study. Considering a variety of different types of random movement in nature, technology and society, the book presents a rich spectrum of possibilities for this dependence.

Starting with Eq. (2.8) we considered, so far, only the change in concentration of the diffusants in a certain range as resulting from in- and outgoing fluxes. On considering in particular biological species, however, we do have to consider a second mechanism, namely the generation of new species. In first order approximation this growth may be assumed to be proportional to the amount of species already present at a given instant of time. By correspondingly completing Eq. (2.12) we arrive at

$$\frac{\partial c(r,t)}{\partial t} = \frac{D}{r}\frac{\partial}{\partial r}\left(r\frac{\partial c(r,t)}{\partial r}\right) + \alpha c(r,t) \tag{2.15}$$

with the newly introduced parameter α referred to as the growth rate. By insertion into Eq. (2.15), the expression

$$c(r,t) = \frac{n}{4\pi Dt}\exp\left(-\frac{r^2}{4Dt} + \alpha t\right) \tag{2.16}$$

is easily seen to be its solution. We note that Eq. (2.16) differs from Eq. (2.13) in only the additional term αt in the exponential on the right hand side of Eq. (2.16).

This term gives rise to an increase in concentration with increasing time. For quantifying the speed of spreading we may now consider a distance R from the origin which we define by the requirement that there is a well-defined number of spreading species outside of a circle of this radius R, which is assumed to be negligibly small in comparison with their total amount. This radius R can now, as a second peculiarity, be shown to linearly increase with time [13, 14]. One finds $R(t) = 2\sqrt{D\alpha}t$, from which the speed of spreading following what Fisher called a "wave of advance" may be noted immediately as

$$v = 2\sqrt{D\alpha}. \tag{2.17}$$

Nature does clearly forbid unlimited growth as would occur as a consequence of Eq. (2.16) as discussed above. Most remarkably, a simple correction of the term added to Fick's 2nd law does allow a reasonable first-order description of many phenomena occurring in nature:

$$\frac{\partial c(r,t)}{\partial t} = \frac{D}{r}\frac{\partial}{\partial r}\left(r\frac{\partial c(r,t)}{\partial r}\right) + \alpha\left(1 - \frac{c(r,t)}{c_\infty(r,t)}\right)c(r,t). \tag{2.18}$$

That type of growth, eventually reaching the limiting concentration $c_\infty(r,t)$ (saturation), is termed "logistic growth". Spreading does, correspondingly, occur with concentrations eventually arriving at the limiting concentration $c_\infty(r,t)$ as schematically shown by Fig. 2.6. The propagation rate of the concentration front (speed of spreading) is still given by Eq. (2.17). A more detailed introduction into the formalism around the "logistic growth" is provided by Sect. 3.4.

If the spreading species (as e.g. molecules during a catalytic reaction) are subject to chemical conversions or reactions, these conversions as well contribute to changes in local concentration, in addition to the influence of diffusion. Equation (2.19) gives an example of the corresponding extension of Fick's 2nd law, Eq. (2.9), so-called reaction-diffusion equations, for sake of simplicity in the one-dimensional scenario:

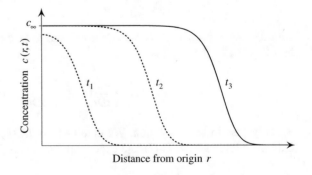

Fig. 2.6 Scheme of propagation ("wave of advance") of the concentration (number per area) of a species on spreading according to Eq. (2.17) at subsequent instants of time ($t_1 < t_2 < t_3$)

$$\frac{\partial c_1}{\partial t} = D_1 \frac{\partial^2 c_1}{\partial x^2} - k_{21}c_1 + k_{12}c_2$$

$$\frac{\partial c_2}{\partial t} = D_2 \frac{\partial^2 c_2}{\partial x^2} - k_{12}c_2 + k_{21}c_1 \qquad (2.19)$$

for a monomolecular reaction between species 1 and 2 (of local concentrations $c_i(x, t)$ with $i = 1, 2$) with the reaction rate constants k_{ij} for conversion from j to i. For simplification, the diffusivities D_i of the two species are assumed to be independent of either concentration and the diffusive fluxes on the concentration gradient of the other component. These are coupled partial differential equations which can easily be solved by computer programmes.

The idea to use coupled reaction diffusion equations and to consider interactions in addition to growth was soon applied to the spread of living beings and even to the spread of abstract objects, in particular languages (see, e.g., Chap. 18). Already more than 30 years ago Okubo [15] and a little later Murray [16] have reported on such applications. From the considerable number of more recent applications we mention the description of diffusion (demic vs. cultural) of the Neolithic transition (see e.g. [17]) and of the spread and retreat of language [18] by coupled reaction-diffusion equations.

It is obvious, however, that one reaches limits in the analytical treatment. The subsequent sections introduce into the options how these limitations may be overcome. Now spread needs not to follow the dispersal logics of the random walk, i.e. it is not necessarily of Gaussian type.

2.4 Extending the Tools

With increasing complexity of the system, in particular of the platform on which spread occurs (network or "habitat"), it becomes increasingly complicated to obtain analytical solutions as those given by Eqs. (2.10), (2.13) and (2.16), and simple reaction-diffusion models are inadequate for the description of complex, spatially incoherent spreading patterns. The global spread of epidemics, innovations etc. are processes on a complex network. In such cases it is common praxis to rely on numerical solutions of the given equations.

When a network of starting points and destinations is the basis for the spread, for the travel of individuals between nodes n and m of the network the continuity equation $\partial c/\partial t = -\partial j/\partial x$ (Eq. (2.8)) is replaced by a rate equation

$$\frac{\partial c_n}{\partial t} = \sum_{m \neq n} (p_{nm}c_m - p_{mn}c_n) \qquad (2.20)$$

where $p_{nm}c_m$ stands for the outgoing flux from node m to node n and $p_{mn}c_n$ for the flux in opposite direction. Exactly this type of analysis we shall encounter in

Chap. 19 where Brockmann applies such network logics for demonstrating the spread of diseases [19].

Another set of interesting but complex problems are the diffusional movements of animals on search for food. For randomly distributed food sources, the Lévy flight hypothesis predicts that a random search with jump lengths following a power law minimizes the search time. Such patterns end up with relations deviating from simple proportionality between the mean square displacement and the observation time. Examples of this type of motion referred to as "anomalous diffusion" may be found in Chaps. 4 ("Levy flights"), 6 (diffusion in brain "interstitials") and 10 ("single-file" diffusion).

However, Lenz et al. [20] find for bumblebees that the crucial quantity to understand changes in the bumblebee dynamics under predation risk, when the insects obviously try to avoid meeting predators, is the correlation of velocities v. These correlations correspond exactly to the sums of cross-terms in Eq. (2.1), which for the bumblebees do not cancel out. The authors reproduce these changes by a Langevin equation in one dimension adding a repulsive interaction U of bumblebee and predator:

$$\frac{\partial v(t)}{\partial t} = -\eta v(t) - \frac{dU}{dx}x(t) + \xi(t) \tag{2.21}$$

where η is a friction coefficient and $\xi(t)$ a fluctuating force (Gaussian white noise).

2.5 Agent-Based Models of Spread

An alternative possibility of modelling and eventually predicting spreading under complex conditions is Monte Carlo simulation on the basis of cells occupied by diffusants, so-called agents, which can be men, animals, plants, bacteria or even abstract concepts as e.g. innovations and ideas. The method is sometimes called cellular automaton. This has e.g. been done in ethnology for the spread of agriculturalists in the neolithicum [21], in ecology for the spread of neobiota [22] and in linguistics [23] for language competition, just to give a few examples.

The idea of Monte Carlo simulations is as follows: One reserves, in the computer, a sufficiently large number of memory cells designated i. These cells refer to the possible positions of the random walker introduced in Sect. 2.1. One considers a set of numbers $m_{i,j}$ which indicate the occupation number of cell i after time step j.

In the introductory example (Fig. 2.1), after each time step (of duration τ) the random walker was required to definitely step to one of the adjacent sites. Thus, one half of the given population of a certain cell (of number i) would have to be passed, after one step, to the next one (to cell number $i + 1$), the other to the previous one (cell number $i - 1$). In our computer simulation this would correspond to the relation

$$m_{i,\,j+1} = \frac{1}{2}m_{i-1,\,j} + \frac{1}{2}m_{i+1,\,j} \tag{2.22}$$

correlating the cell populations after subsequent steps. After 100, 300 and 1000 steps one would arrive at the occupation distributions as shown in Fig. 2.4 (where the values given in Fig. 2.4 have to be additionally multiplied by the number of agents starting at the origin).

We may come closer to the reality of the elementary steps of propagation by a modification of the simulation procedure. Rather than rigorously requiring that, after each time step τ, the agents have to definitely jump to one of the adjacent sites, one may introduce the probability $p_{i,k}$ that, during one time step, an agent gets from site k to i. This probability may include the suitability of the cell. In this case, Eq. (2.22) is replaced by a relation of the type

$$m_{i,\,j+1} = m_{i,\,j} + \sum_{k} m_{k,\,j}\,p_{i,\,k} - m_{i,\,j}\sum_{k} p_{k,\,i} \tag{2.23}$$

where the terms appearing on the right hand side, in addition to the given occupation number $m_{i,j}$, are easily recognized as population increase of cell i by agents entering from other cells k and population decrease by agent transfer from cell i to other ones.

With k equal to $i-1$ and $i+1$ and $p_{i,k} = \Delta t/2\tau$, during a time interval Δt, an agent will leave the cell with the probability $\Delta t/\tau$, with equal probabilities for both directions. This probability definition serves as a meaningful definition of a mean residence time τ.

The need for computer simulations is illustrated with the representation in Fig. 2.7, which refers to the spreading of a biological species, namely ragweed (Ambrosia artemisiifolia), a plant which has "invaded" from North America and continues to enhance its density of occurrence in Europe [22].

In the top of the figure cells populated by ragweed are shown in black. The number of cells in black will continuously increase with spread of ragweed. The simulations aim at determining the probability by which, at further instants of time, so far unpopulated areas shall become populated ("infested"). In the starting assumption that infested cells remain infested, one notably deviates from the situation considered with the introductory random walker example. In fact, by considering infestation spreading, one is already following the situation typical of growing populations as considered in Sect. 2.3.

A successful step of "spreading" (the probability of which has been just considered) is not automatically assumed to warrant infestation. In fact, environmental conditions ("habitat suitabilities") might be quite different leading to different survival probabilities. The grid on bottom left provides the numbers considered to be relevant for the given example. The products of both probabilities, representing the "total infestation probabilitys" are given on top in the middle. Whether

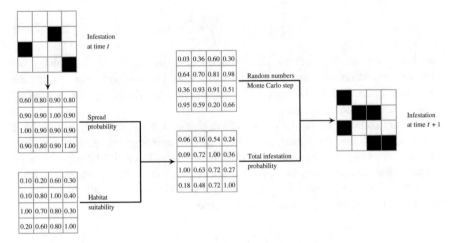

Fig. 2.7 Algorithm for determining the occurrence of a species in space at subsequent instants of time. The 4×4 squares ("grid cells") symbolize the different areas into which the space is subdivided. Redrawn from [22]

infestation will indeed occur depends on the relation between the random numbers (between 0 and 1) produced by the computer and the total infestation probabilities. Correspondingly, in the top right grid do all these cells appear in black for which the random number is exceeded by the total infestation probability.

Figure 2.8 shows as example the predicted infestation of grid cells (about 5 × 5 km) by the spread of ragweed over Austria and Bavaria.

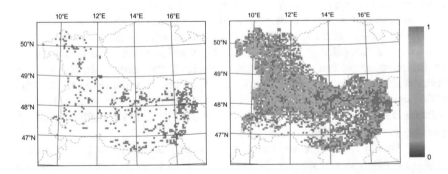

Fig. 2.8 Left: Distribution of ragweed in Austria and Bavaria in 2005. Red squares symbolize infested grid cells. Right: Predicted infestation probability indicated by colors from red (highest probability) down to blue (lowest probability) in 2050, if no action against ragweed spread is taken. Redrawn from [24]

References

1. C. Bräuchle, D.C. Lamb, J. Michaelis (eds.), *Single Particle Tracking and Single Molecule Energy Transfer* (Wiley-VCH, Weinheim, 2010)
2. H. Mehrer, *Diffusion in Solids*. Springer Series in Solid-State Science, vol. 155 (2007)
3. J. Kärger, F. Stallmach, PFG NMR studies of anomalous diffusion, in *Diffusion in Condensed Matter. Methods, Materials, Models*, ed. by P. Heitjans, J. Kärger (Springer, 2005), pp. 417–460
4. T. Springer, R. Lechner, Diffusion studies of solids by quasielastic neutron scattering, in *Diffusion in Condensed Matter. Methods, Materials, Models*, ed. by P. Heitjans, J. Kärger (Springer, 2005), pp. 93–164
5. G. Vogl, B. Sepiol, The elementary diffusion step in metals studied by the interference of gamma-rays, X-rays and neutrons, in *Diffusion in Condensed Matter. Methods, Materials, Models*, ed. by P. Heitjans, J. Kärger (Springer, 2005), pp. 65–92
6. M. Leitner, B. Sepiol, L.-M. Stadler, B. Pfau, G. Vogl, Nat. Mater. **8**, 717 (2009)
7. J. Kärger, C. Chmelik, R. Valiullin, Phys. J. **12**, 39 (2013)
8. J. Kärger, D.M. Ruthven, D.N. Theodorou, *Diffusion in Nanoporous Materials* (Wiley-VCH, Weinheim, 2012)
9. J. Kärger (ed.), *Leipzig, Einstein, Diffusion* (Leipziger Universitätsverlag, Leipzig, 2014)
10. A. Einstein, Ann. Phys. **17**, 549 (1905)
11. K. Pearson, Nature **72**, 294 and 342 (1905)
12. I. Prigogine, *The End of Certainty* (The Free Press, New York, London, Toronto, Sydney, 1997)
13. R.A. Fisher, Ann. Eugen. **7**, 335 (1937)
14. J.G. Skellam, Biometrika **38**, 196 (1951)
15. A. Okubo, *Diffusion and Ecological Problems: Mathematical Problems, Biomathematics*, vol. 10 (Springer, 1980)
16. J.D. Murray, *Mathematical Biology* (Springer, 1989; 2003)
17. J. Fort, PNAS **109**, 18669 (2012)
18. A. Kandler, R. Unger, J. Steele, Philos. Trans. R. Soc. B **365**, 3855 (2010)
19. D. Brockmann, D. Helbing, Science **342**, 1337 (2013)
20. F. Lenz, T.C. Ings, L. Chittka, A.V. Chechkin, R. Klages, Phys. Rev. Lett. **108**, 098103 (2012)
21. C. Lemmen, D. Gronenborn, K.W. Wirtz, J. Archaeol. Sci. **38**, 3459 (2011)
22. G. Vogl et al., Eur. Phys. J. Spec. Top. **161**, 167 (2008); R. Richter et al., J. Appl. Ecol. **50**, 1422 (2013)
23. D. Stauffer, X. Castello, V. M. Eguiluz, M. San Miguel, Phys. A **374**, 835 (2007); C. Schulze, D. Stauffer, S. Wichmann, Physics **3**, 271 (2008)
24. R. Richter et al., Biol. Invasions **15**, 657 (2013)

Part II
Nature

Chapter 3
Dispersal in Plants and Animals

Michael Leitner and Ingolf Kühn

3.1 Introduction

The biogeographical patterns of ecosystems and species distributions we know today are, apart from other effects such as evolution or ecological interactions, the result of a continuous progression of spatial processes since different species emerged. These spatial processes accelerated tremendously with the advent of modern humans and their effects on species and ecosystems, such as via trade, traffic, or habitat modification. Here we discuss general aspects of the resulting spatio-temporal population dynamics and review pertinent models.

Dispersal refers to the movement of individuals, either from the site of birth to the site of reproduction, or the movement between successive sites of reproduction [1]. This pertains to any group of organisms (animals, plants, fungi, bacteria) and can occur in several stages, including adult individuals as well as propagules such as spores, seeds, fruits or vegetative fragments of an individual. It is the primary mechanism of spatial gene flow, maintaining populations and fitness. These dispersal movements usually occur within the range of a species, i.e. within the geographical area where the individuals of a species are commonly encountered.

Dispersal phenomena can happen at almost any temporal and spatial scale. At evolutionary time scales, species evolve into new species and need to colonize a new range distinct from their area of origin. Further, dispersal can be tied to geological events. Consider for instance the Great American Interchange, bringing the continents of North and South America into contact, c. 10 Mio. years ago. Once both

M. Leitner (✉)
Heinz Maier-Leibnitz Zentrum (MLZ), Technische Universität München,
Lichtenbergstr. 1, 85748 Garching bei München, Germany
e-mail: michael.leitner@frm2.tum.de

I. Kühn
Helmholtz-Zentrum für Umweltforschung — UFZ,
Theodor-Lieser-Str. 4, 06120 Halle, Germany
e-mail: ingolf.kuehn@ufz.de

© Springer International Publishing AG 2018
A. Bunde et al. (eds.), *Diffusive Spreading in Nature, Technology and Society*,
https://doi.org/10.1007/978-3-319-67798-9_3

continents got connected, migration across the new continents was instantaneous compared to the time scales of continental drift. The results of these processes are the biogeographic patterns we know today with specific floristic or zoological kingdoms or realms, made up of regions with a similar composition of species, e.g. the Holarctis (temperate and cool regions surrounding the North pole), Paleotropis (Old World Tropics) or Neotropis (New World Tropics).

In our context more relevant, though, are much shorter time scales, specifically those at which demographic processes such as birth, death, colonization and extinction determine population dynamics. They happen within the life-span of an organism or within a few generations. Movement of individuals beyond their current range (that is, the region in which species successfully reproduce) and subsequent establishment will lead to range expansion. On the other hand, unfavorable conditions, for instance at range margins, can lead to range contraction. In this chapter, we will hence consider questions of how spatial patterns of populations temporally evolve on these time scales. Specifically, we will not treat evolution in the Darwinian sense. As indicated above, dispersal and the resulting range shifts are a natural aspect of population dynamics. In the face of stochastic local extinctions due to natural disasters, diseases, changing environmental conditions, increased competition or predation, it is vital for the survival of a species to be able to (re-)colonize suitable habitats as fast as possible, before the specific populations become extinct themselves.

The dominant mode of dispersal is specific to species. Many animals disperse actively, while most plant species as well as sedentary and floating aquatic animal species are dispersed passively in specific life-stages, which with regard to modeling necessitates to explicitly consider aspects such as river currents, marine currents (e.g. gulf stream/North Atlantic Drift), air currents (e.g. the jet stream) or the movement of dispersal vectors (such as animals dispersing plant seeds).

In addition, dispersal patterns depend on the properties (i.e. permeability) of the ranges (large scale) and landscapes (smaller scale) to be traversed. At large scales, climate matching [2] is crucial, i.e. the climatic conditions have to suit the individual species' preferences. Landscapes need to provide the habitat needed by a species (e.g. forest or grassland in general, or specific forests and meadows). How well these habitats are interconnected or whether there are barriers determines how fast a species can disperse and which shape the resulting range will have. All biotic and environmental conditions and resources needed by a species are jointly called "ecological niche". This niche is typically considered to be a high-dimensional region in parameter space, where each dimension is characterized by a specific condition or resource a species needs or uses [3]. Nevertheless, methods exist to account for these habitat properties in models, as well.

Already since neolithic times, when humans have started to modify land cover, but especially since the beginning of long-distance trade and travel in the Age of Discovery, range dynamics on unprecedented velocity have been happening. On the one hand, the dispersal opportunities afforded by human innovations such as intercontinental shipping traffic (on an ever-increasing volume especially in the last decades with progressing globalization [4]) result in translocations of species. This is a special case of colonization, facilitated by humans. The species in the new range are

called "alien" or "exotic", and in case the species spread rapidly and/or cause a negative impact they are called "invasive" [5, 6]. On the other hand, the historically
drastic rate of climate change, witnessed in the last decades and likely to persist [7],
entails corresponding shifts in the habitable regions for many given species [8]. This
leads to a special case in which native as well as alien species undergo range shifts
and the distinction becomes partially ambiguous. Especially this aspect of dispersing species due to global change is of special interest to ecologists: firstly to improve
the understanding of ecological processes by modeling dynamics that can currently
be observed, and secondly because such models can be the basis for management
applications in nature conservation.

The issues listed above hence motivate the large amount of attention questions of
animal and plant dispersal have been receiving within the last years. As detailed
above, this pertains primarily to dispersal on macroscopic scales which has the
potential to lead to range shifts or expansions. Our contribution will focus on these
aspects, and we will understand "dispersal" in this sense. To this end, we will follow
an order of sequential increase of model complexity, from models assuming "equilibrium conditions" to complex dynamic systems considering ecological processes
in more detail.

3.2 Comparing Diffusive Dynamics in Physics and Ecology

Random walks in their purest form belong to the realm of physics. Paradoxically,
Robert Brown, who first observed the ceaseless random motion of suspended particles in a liquid and known as Brownian motion, which we now know to be due
to stochastic collisions with the fluid's molecules, was a botanist. However, starting
with Einstein's and Smoluchowski's explanation of the phenomenon, it was primarily physicists who have studied random walks and the consequent spreading-out of
concentration gradients, which is known as diffusion. When related phenomena first
came into the focus of other fields of science, it was therefore natural to adopt the
physical concepts. However, we feel that the analogy is useful only up to a certain
point, and in this section we discuss specific aspects of diffusive dynamics in physics
and ecology and point out the differences. Note that most of the statements regarding
the ecological case hold equally also for other non-physical examples treated in this
book, be it the spreading of cultural techniques and people (Chaps. 14, 16 and 17),
languages (Chap. 18) or diseases (Chap. 19).

Mass conservation versus reactions: In physics, on the one hand, the diffusing
entities, be it atoms, molecules, or more abstract quantities such as heat, comply with
a conservation law. This means that concentration changes with time are purely due
to incoming and outgoing fluxes. Even in the case of chemistry's diffusion-reaction
phenomena, where substances that have been brought together by diffusion react,
there is a clear distinction between diffusion and reaction (for examples see Chaps.
10 and 11). Specifically in ecology of plants or sedentary animals such as corals, on

the other hand, such a distinction is not meaningful. Here spatial transfers happen by means of propagules (seeds or vegetative dispersal unit), without affecting the population density at the source. Therefore the on-site aspect of dynamics (reproduction and death) is intrinsically linked to the spatial aspect (dispersal).

Homogeneity: In physical diffusion, the governing relations in most cases do not depend explicitly on time. In contrast, the activity of living organisms is typically strongly structured by the diurnal or seasonal cycles, which naturally affects the spreading behavior. The same holds for the spatial scale, where the habitats accessible to plants or animals often display pronounced inhomogeneities, while for the physical case of diffusion in a gas, liquid, or solid, space is homogeneous. Of course there are important counter-examples, such as the case of grain-boundary diffusion in solid diffusion (Chap. 13).

Isotropy: An analogous statement can be made with regard to the directional symmetry of the governing relations: The diffusion equation of physics is invariant with respect to spatial inversion, and often it is even isotropic, which means that there are no special directions. In contrast, the non-uniformity of the wind directions leads to a bias in the transport of wind-dispersed seeds (see Chap. 7 for a discussion of the aspects specific to this problem), and the same applies for the distinction between upstream and downstream dispersal in and along rivers or upslope and downslope dispersal in steep terrain.

Linearity: In physics, the diffusion equations can often be linearized, specifically when dealing with very diluted diffusants. Only for interacting diffusants at sizable concentrations non-linear effects come into play (see Chap. 2). Ecology, on the other hand, is inconceivable without interactions. They can come in many forms: Intraspecific interaction, for instance higher reproduction success at higher densities due to increased mate-finding probabilities or increasing diffusivities at higher densities due to decreasing foraging successes corresponding to competition, or inter-species interactions as in predator-prey, host-parasite or plant-pollinator relationships. As a consequence, the governing relations in ecology are non-linear as a rule.

Continuity versus discretization: An important distinction lies in the scale of the fundamental translocations. While it is true that also in the physical case of random walks, be it the Brownian motion of suspended particles in a gas or a liquid or the jumps between neighboring sites of atoms in a crystal, there is a fundamental discretization, on length- and time-scales observable to the unaided eye the movement appears continuous. Again plant ecology provides the most striking difference, where there is apart from a few exceptions only one translocation event per generation, which can be considered to happen instantaneously. Also for animals, the granularity of the translocations leads to non-Gaussian spread kernels, as will be discussed in more detail below.

Determinism versus stochasticity: Due to the fact that in physical cases the number of considered diffusing particles is typically extremely large (a cubic meter of air holds more than 10^{25} molecules) it is neither possible nor would it be useful to track the position of every particle. Rather, the state of the system at any given time is described by particle densities. Further, the fundamental displacements' large number and microscopical scale as discussed above completely averages out their randomness. As a consequence, the particle densities follow deterministic partial differential equations. This does not hold for most ecological cases, where any detailed model has to treat the stochasticity of the fundamental processes. Specifically very rare events can have a stochastic effect on the resulting species distributions, such as long-range displacements of seeds or individuals coupled with non-linear growth leading to colonizations of hitherto unreached areas, or the introduction of diseases or predatory species, which can result in (local) extinction.

The above points can be summarized in the statement that the models that have to be used to adequately describe dispersal processes in ecology are typically more complex than those used in physics. This shall not claim that the physical problems are in any way trivial or not interesting, and in fact the physical cases treated in various chapters in this book often are counter-examples to the simplified distinctions given above. However, in this they are exceptions that capture the scientists' interest, while the complexity in the ecological cases is rather a rule.

3.3 Static Spatial Distributions

Frequently, dynamical systems are in equilibrium at the temporal or spatial scale considered. This is the case for animal species moving in order to forage or mate or for plant species where seed dispersal will not necessarily result in average range changes nor in average changes of population densities. Such stable conditions can result for species that are in equilibrium with the given environmental conditions; i.e. those that occur everywhere where environmental conditions are suitable [9]. In other cases, species are not in equilibrium with their environmental conditions but cannot realize range increases due to dispersal limitation. For instance, Svenning and Skov [10] showed that many species hardly changed their range after the last glaciation due to dispersal limitation. For the sake of modeling, though, systems are often assumed to be in equilibrium. In such cases the microscopic workings, i.e. the underlying processes, are subsumed in phenomenological dependencies, and the interest lies rather in describing the pattern. The model parameters are hence typically determined from observed population dynamics by methods of statistical inference, as will be treated below.

Nevertheless, even under equilibrium conditions, process knowledge can be rather detailed, with the resulting model covering all these details and considering all processes on a microscopic basis, so that the corresponding parameters can be determined independently. This is then called a mechanistic model. For modeling stochastic dynamics (Sect. 3.5.2), mechanistic models are often preferred.

3.3.1 Species Distribution Models (SDMs)

In principle, Species Distribution Models (also called ecological niche models, bio-climatic envelope models or habitat suitability models) describe the occurrence of species under static conditions, allowing to derive their range under specific assumptions. In their simplest form, they build on the linear predictor

$$\eta = \mathbf{X}\beta \tag{3.1}$$

understood as matrix-vector multiplication, where $\mathbf{X}$ is a matrix of environmental conditions sampled at a number of locations and β is the parameter vector of the model. These conditions can be climatic factors such as temperature or precipitation, land use or land cover such as arable land, grassland, forests, geological substrates, or anything that determines where a species can thrive, and the variation of η with a change in a given environmental condition is quantified by the respective entry in β.

Sometimes species abundance is known (such as counts of individuals, cover or biomass), but in most cases the information on species occurrences is only known in a binary way (species presence or absence). To link the linear predictor η to the expected value of the occurrences the link function g is introduced

$$g\big(E(y)\big) = \eta = \mathbf{X}\beta, \tag{3.2}$$

where y denotes the species occurrences. The expression for the link function g depends on the distribution model of the species occurrence data (e.g. binomial, Poisson, or Gamma-distributed). Technically, parameter estimation is done by any method of statistical inference (see Sect. 3.5.2). While Generalized Linear Models as introduced above are the most frequently used SDMs, there are many other approaches used.

These SDMs are frequently employed for modeling spread in cases where the environmental conditions that determine the species distribution are not static or species are not in equilibrium with environmental conditions. In the former scenario, temporally varying layers of environmental conditions (such as climate change scenario projections) are used to predict future species ranges, inherently assuming that spread (or retraction) takes place. SDMs are hence believed to predict the potential range of a species correctly and the difference of observed versus modeled range can then be used to infer spread processes. Strictly speaking, though, this is not a valid approach [9]. Non-native species have mostly not (yet) colonized all potentially possible environmental conditions [11]. Hence their ecological niche (i.e. the n-dimensional hypervolume describing all conditions they need) is frequently largely underestimated, when suitable conditions are neglected because the species has not yet colonized locations with such conditions (e.g. [12]).

As an example, the use of an SDM for predicting range shifts of a species under varying climate conditions is illustrated for the case of the common walnut (*Juglans regia*), which is native to mountain ranges of central Asia and has been introduced

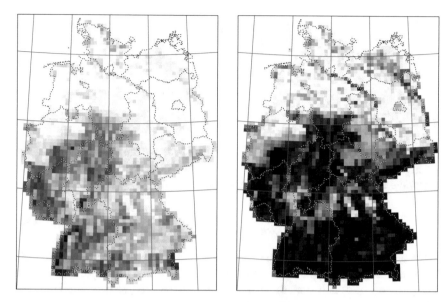

Fig. 3.1 Modeled expected value for the presence of the common walnut today (left panel) and predicted under climate change (right panel) in gray-level coding

to the Mediterranean in Antiquity. For both 38 bioclimatic parameters and 11 soil characteristics the six most important principal components were combined with four land-use parameters to produce the matrix of environmental conditions **X**, having 2995 rows (corresponding to the number of cells) and 16 columns. Given the currently realized binary distribution data, the corresponding 16-entry parameter vector β is inferred. Under the A1FI scenario [13], which assumes that future policy is concerned with stimulating economic growth as opposed to reducing carbon dioxide emissions, a temperature increase of approx. 4 °C until the end of the century is projected for Germany. The cell-wise expected value for occupation $E(Y)$ for the modified environmental condition matrix is presented in Fig. 3.1 along with the model predictions for the present conditions. It is seen that the plant will increase its naturalized range from south-western Germany to most of southern and central Germany [14].

3.4 Spread Processes: Classical Approaches

The pioneering works on the spatio-temporal dynamics of plants and animals were done before the advent of electronic computers. As a consequence, typically the aim was to capture only qualitative aspects, and modeling was done by (partial) differential equations, disregarding aspects such as spatial inhomogeneities. Here we will briefly recapitulate those concepts. As mentioned already above, in ecology reproduction and death are often integral aspects of dispersal. Therefore, we will start here with a treatment of on-site population dynamics and in a second step augment them by considering spatial dynamics.

3.4.1 On-site Dynamics

In agreement with the nomenclature of Chap. 2, we denote the local population density at time t as $c(t)$. We model the temporal evolution of the population density, that is how the number of individuals of a given species changes with time, as

$$\frac{dc(t)}{dt} = \alpha\big(c(t)\big)c(t), \tag{3.3}$$

where we allow for a density-dependent growth rate $\alpha(c)$. In this formulation, the temporal change of the population density goes to zero as the population goes to zero (provided $\alpha(c)$ is bounded), which is a sensible choice, as naturally both birth and death incidence rate will go to zero in this case.

Unbounded Growth: The most simple model of population dynamics results from choosing $\alpha(c) = \alpha$. For positive α, this has a simple exponential function as solution

$$c(t) = c(0)e^{\alpha t}, \tag{3.4}$$

corresponding to unbounded growth. After Thomas Robert Malthus, who introduced this model and pointed out its consequences [15], it is called the Malthusian Growth model. Note, however, that its main idea goes back at least to Leonardo Fibonacci's modeling of the population dynamics of rabbits [16], who, with differential calculus still five hundred years away, formulated it as a difference equation giving rise to the Fibonacci sequence.

Bounded capacities: As a rule, plants and animals produce more, sometimes much more offspring than one individuum per parent. If conditions are favorable, a large part of those survives to fertility, corresponding to potentially large population growth rates. However, at higher population densities the survival rates will decrease, due to factors such as competition for limited feeding or sunlight resources, higher incidences of diseases, or increased predation (for an explicit treatment of this effect see below). In ecological systems in equilibrium, it is indeed necessarily the case that the average number of surviving offspring is just one individuum per parent, or a growth rate of zero. The simplest model to capture this issue is due to Verhulst [17], who assumed a linearly decreasing growth rate

$$\alpha(c) = \alpha_0(1 - c/K), \tag{3.5}$$

going to zero when the population density reaches the carrying capacity K. Accordingly, the solution of the model

$$c(t) = \frac{c(0)K}{c(0) + \left(K - c(0)\right)e^{-\alpha_0 t}},\tag{3.6}$$

known as a logistic function, asymptotically approaches K (see also Fig. 16.3 and the example given in Sect. 14.2.2 for the use of the logistic function for predicting the spreading of innovations).

Inverse density effects: In addition to diminishing growth rates at high densities, this can also be the case for low densities, for instance due to difficulties of finding mating partners, if the species uses cooperative defence, hunting, or child-rearing strategies, or if it is the preferred food for a species that can sustain high numbers also when the species in focus diminishes, because it is not the only possible prey of a more generalist predator. Such an effect has first been described by Allee [18] and is therefore known as the Allee effect. The simplest corresponding expression for the growth rate is

$$\alpha(c) = \alpha_0(1 - c/K)(c - c'),\tag{3.7}$$

with a parameter c' that has the dimension of a population density. Here the sign of c' distinguishes two regimes: For $c' < 0$, the resulting qualitative features agree with the Verhulst model, only with decreased growth rates at small densities. This is called the weak Allee effect. However, a positive c' corresponds to a critical population density below which the species would die out spontaneously. This constitutes the strong Allee effect and has important implications: For instance, it suppresses the establishment of colonization foci by small founder populations, as well as severely reduces the speed of the invasion front (which will be defined below). Note that for instance also the conversion towards the locally dominating languages, leading to a decline of minority languages in linguistics (Chap. 18) can be seen as an example of the Allee effect.

Explicit interactions: Finally, the explicit interaction between the densities of different species can be considered. In the simplest case of a two-species predator-prey system, this can lead to the Lotka-Volterra model (due to Alfred James Lotka and Vito Volterra), where the population densities of a prey c_1 and a predator c_2 species are described by coupled differential equations. Specifically, the growth rate of the prey decreases linearly with the predator concentration, while the predators' growth rate grows linearly with the prey concentration [19, 20]. Under these conditions, the ecological system performs an oscillatory motion through two-dimensional phase space, where the predators multiply during periods of high prey populations, leading to their decline and a subsequent decline of the predators. Under these

low-predator conditions the prey population will recover, starting the next oscillation. Of course, in actual ecosystems species typically will interact with a number
of other species, either in a predator-prey relationship or competing for the same
resources. This can give rise to very non-linear effects, for instance when a predator
species specializes on a specific prey due to their increasing numbers. Note that for
many-species ecosystems, which indeed are the typical case, simulations show that
generic model parameters lead to chaotic dynamics, and stable situations likely arise
only due to natural selection [21].

3.4.2 Reaction-Diffusion Approaches

In the case of spatially inhomogeneous population densities, such as during the colonization of hitherto unoccupied areas (or, equivalently, range contractions), the classical approach is to augment the on-site dynamical term as discussed in the previous
section by a diffusive term, describing the fluxes due to concentration gradients via
second derivatives. This is treated in more detail in Chaps. 2 and 16, so we will state
here only the fact that the resulting equations of the form

$$\frac{\partial c(x,t)}{\partial t} = \alpha\big(c(x,t)\big)c(x,t) + D\frac{\partial^2 c(x,t)}{\partial x^2}, \tag{3.8}$$

where x is the spatial variable to account for variations in the population density, have
solutions that at long times take the form of travelling fronts. Specifically, along the
direction of propagation they can be written as

$$c(x,t) = c^*(x - vt), \tag{3.9}$$

with a constant propagation velocity v [22].

J. G. Skellam was the first to compare this model to an actual case of species
invasion: His iconic illustration of the spread of muskrats after having been imported
from Northern America and released in central Bohemia in 1905 is reproduced in
Fig. 3.2 [23]. He showed that their range grows quadratically in time, corresponding to a propagation velocity of about 11 km/year. However, a closer look reveals
already in this prototypical example deficiencies of the model. Specifically, inhomogeneities in the landscape lead to deviations from the ideally expected circular
shapes, where for instance the hill and forest chains around Bohemia retarded the
propagation, while rivers in lower Bavaria and Saxony seem to have favored the dispersal of the waterbound muskrats.

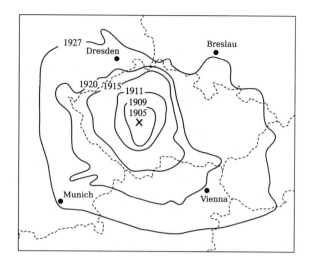

Fig. 3.2 Range populated by muskrats as function of time during their invasion of central Europe drawn after [23]. Today's political borders are dashed

3.5 Spatio-temporal Population Modeling: State of the Art

After briefly reviewing in the previous section how classically the spatio-temporal dynamics of population densities have been modeled by (partial) differential equations, here we will indicate some directions of recent improvements. Interestingly, these modern developments can be traced back to some of the differences between the physical and the ecological cases of dispersal as already discussed in Sect. 3.2. Specifically, in the following we consider finite lengths of displacement, spatial inhomogeneities, and stochasticity.

3.5.1 Spread Kernels

In Chap. 2 it was derived in some detail how the probability density of a particle performing a random walk is given by a Gaussian distribution of increasing width. Mathematically, this result is just the central limit theorem, which in its simplest form states that the probability density after a large number of uncorrelated displacements, drawn from some fixed but arbitrary distribution with finite variance, converges towards a Gaussian distribution. However, if those preconditions are not fulfilled, specifically if the macroscopic displacement is not the sum of a large number of uncorrelated microscopic displacements, there is no reason for the probability densities to have Gaussian shapes.

This is often the case in ecology (and also in the case of human mobility, as considered by way of equivalent models in Chaps. 16 and 18): When animals move, they normally do so in order to cover distance to, for instance, find genetically different mates or new feeding grounds as opposed to movement for its own sake (cp. sharks swimming to keep up the waterflow through their gills). Of course, with respect to this goal it is most efficient to move in a straight line, for which even the least developed animal lifeforms have evolved abilities. For example, nocturnal insects navigate by bright lights and some motile bacteria can move along chemical gradients (chemotaxis). In other words, their movement is correlated over time. A detailed discussion of these effects and their consequences for the statistical properties of animal motion is given in Chap. 4.

The main dispersal modes of plant seeds are either within or attached to animals, in which case the discussion above applies, or by way of wind or water, which again corresponds to correlated movement. Also dispersal mediated by human activities, be it intentionally or unintentionally, e.g. as load in shipments (for quite a number of species this is the dominating mode of long-distance transport [24, 25]) is clearly not composed of many small independent displacements.

To account for this complication on the level of the reaction-diffusion approaches, one can replace the partial differential equation (3.8) by an integro-differential equation

$$\frac{\partial c(x,t)}{\partial t} = \alpha\big(c(x,t)\big)c(x,t) + \int dy K(y)c(x-y,t) - c(x,t)\int dy K(y), \qquad (3.10)$$

where $K(y)$ is the probability density per unit time that an individual is translocated by y. Note that the last term due to the outflux of individuals can formally also be subsumed in the reaction term. Mathematically, in this equation $K(y)$ has the role of a convolution kernel. It is therefore called the spread kernel, also in frameworks that go beyond deterministic reaction-diffusion approaches as will be discussed below.

Due to the correlated movement as discussed above, in ecology spread kernels typically have positive excess kurtosis, that means, for a given standard deviation the distribution has more weight in the extreme values than a normal (or Gaussian) distribution. Such kernels are called leptokurtic or fat-tailed. Specifically the Lévy flight hypothesis would constitute a microscopic motivation for fat-tailed kernels, but its validity has recently been questioned (see Chap. 4). A different scenario resulting in non-Gaussian kernels has been proposed to explain for instance the spread of the horse chestnut leafminer moth over Germany [26, 27], with the spread kernel being composed of a small-scale contribution corresponding to dispersal of the airborne individuals by wind and a large-scale contribution due to human-mediated transportation.

3.5.2   Stochastic Dynamics

The (partial) differential equations as used in Sect. 3.4 imply a deterministic evolution of the population densities. Of course, when fitting the resulting models to observed data the agreement will not be perfect, but in a deterministic framework these deviations are understood to be due to fluctuations around some inherent "true" value, which is indeed assumed to strictly follow the governing equations. Note that this is always the implicit assumption when fitting curves to time-resolved data.

This has to be contrasted with a stochastic approach. Here one does not assume the observable data to be noisy representatives of hidden quantities that evolve deterministically, rather one admits stochasticity as an inherent property of the process. What distinguishes between the validity of either point of view are the scales of the problem: If many individuals participate and displacement happens only over small distances, a deterministic description via densities can be adequate. However, when the consequences of individual stochastic events are visible on the scales of interest, a stochastic treatment is needed. Note that even for the invasion of central Europe by the muskrat as considered above a deterministic treatment is only appropriate for the later stages: The initial release of the five individuals could equally have happened at a different location or decades earlier or later, shifting the subsequent process in space and time. Trying to model also this initial event by deterministic equations would obviously not afford much insight into the process.

3.5.2.1 Stochastic Model Formulations

To fill these concepts with life, we will now first formulate the general model and then consider a specific example: Let $\sigma(., t)$ be the state of the system at time t, specifically let $\sigma(x, t)$ be the occupancy of site x at time t. This "occupancy" shall include all information that is relevant for the considered problem, such as population densities for the considered species, optionally resolved with respect to additional state variables such as age. The model is then represented by a probability density

$$P\big(\sigma(., t+1) | \sigma(., t)\big). \tag{3.11}$$

In words, this expression corresponds to the probability for a system in state $\sigma(., t)$ to evolve to the state $\sigma(., t + 1)$ during one timestep. Note that the arguments to P are functions varying over space, not their point-wise evaluations. This makes the occupancy at some specific site x depend not only on the previous occupancy at this site, but also on all other sites. For simplicity, time is here discretized, which can be either the natural choice, for instance for annual plants, or an approximation to a temporally continuous evolution. Also the spatial variable x can be continuous or discretized, optionally with an explicit dependence of the transition probabilities on x. It can also be adequate to discretize space non-regularly, for instance in the so-called metapopulation models, where the study range decomposes naturally into patches within which movement for the species is easy, while dispersal between patches happens only rarely [28], e.g. for an aquatic species in disconnected water bodies [29] or grassland butterflies in forest-dominated regions [30]. Note finally that the random process defined by Eq. (3.11) is a Markov process, which means the transition probabilities to the next state depend only on the present state and not on earlier states. This, however, is no restriction, as any dependence on earlier states can be captured by extending the considered state variables to explicitly include those earlier states.

The degrees of freedom in modeling are on the one hand qualitative features, i.e. the mathematical form of the expressions, and on the other hand quantitative parameters.

For an exemplary realization of this general setting we consider a model of the spread of the North American weed *Ambrosia artemisiifolia* (common ragweed) across central Europe [31, 32]: Here space is discretized along lines of constant latitude or longitude into a regular grid, time is discretized into annual steps, the system state σ is the simplest conceivable, namely a given cell x at time t is either occupied ($\sigma(x, t) = 1$) or unoccupied ($\sigma(x, t) = 0$), and only cell transitions from unoccupied to occupied are considered, which furthermore are independent of each other. The transition probability density P depends on two auxiliary functions, which are the seed influx $I(x, t)$ in a given year t into cell x, and the spatially varying habitat suitability function $H(x)$, which can be seen as the probability that incoming seeds find favorable conditions and establish a stable population. The seed influx follows from the current occupancies via the spread kernel $S(y)$. All these functions depend on the model parameter vector β, which describes the width and area of the spread kernel as well as the quantitative effect of, for instance, mean precipitation, temperature or land use on the habitat suitability.

Specifically, the seed influx has the form

$$I_\beta(x, t) = \sum_y S_\beta(y)\sigma(x + y, t), \tag{3.12}$$

giving a probability for a given unoccupied cell x to stay unoccupied during one timestep of

$$p_\beta^{0 \to 0}(x, t; \sigma) = \exp\left(-H_\beta(x)I_\beta(x, t)\right) \tag{3.13}$$

and to become occupied of

$$p_\beta^{0 \to 1}(x, t; \sigma) = 1 - p_\beta^{0 \to 0}(x, t; \sigma). \tag{3.14}$$

This corresponds to transition probabilities

$$P_\beta\left(\sigma(., t + 1)|\sigma(., t)\right) = \prod_{x \in \Lambda_{t+1}\backslash\Lambda_t} p_\beta^{0 \to 1}(x, t; \sigma) \prod_{x \notin \Lambda_{t+1}} p_\beta^{0 \to 0}(x, t; \sigma), \tag{3.15}$$

where $\Lambda_t = \{x : \sigma(x, t) = 1\}$ is the set of all cells occupied at step t.

3.5.2.2 Stochastic Simulations

Given a model and an initial state, a representative sequence of future states can be generated by iteratively drawing random occupancies according to the probability density defined by Eq. (3.11). For instance, this can be used to assess how the system will evolve in the long term or to test the effect that specific variations of anthropogenic parameters have [32].

As was already briefly treated in Chap. 2, performing a stochastic simulation for a model as the one considered above is conceptually quite easy, with Fig. 2.8 as an informative example of the outcome of such investigations. This is due to the fact that the probabilities for any cell becoming occupied at a given timestep are independent of whether any other cell becomes occupied at the same timestep. Therefore, in this case the problem of finding a new high-dimensional occupancy conforming to the general probability density defined by Eq. (3.11) separates into finding a number of Bernoulli random variables. Numerically, this is done by computing a pseudo-random number r_x with uniform distribution in the range $[0, 1]$ for any unoccupied cell x and comparing it to $p_\beta^{0 \to 1}(x, t; \sigma)$. If r_x is smaller than this probability, $\sigma(x, t+1)$ is set to occupied, otherwise it is left unoccupied.

3.5.2.3 Statistical Inference

In phenomenological models, one typically has observational data about the process and some idea about the qualitative form of the governing stochastic equations, and one wants to deduce information about the parameters of the model. This problem is called statistical inference, which conceptually can be done either in the frequentist or Bayesian framework. These two schools differ in the way the notion of probability is interpreted, which with respect to statistical inference leads to two ways the model parameters are treated.

Specifically, in the frequentist approach the observation σ is considered as a random variable whose distribution is determined by an unknown but deterministic parameter vector β. The value of this unknown parameter vector is to be estimated in some way, with the most popular method being Maximum Likelihood Estimation (MLE). On the other hand, in the Bayesian approach one considers also the parameter vector β as a random variable. Here the goal is to derive a probability density for this parameter vector.

Consider a stochastic model determined by some parameter vector β. The probability for the random process to give some specific outcome σ is called the likelihood of σ with respect to β. In our setting we have

$$P(\sigma|\beta) = \prod_t P_\beta\big(\sigma(., t + 1)|\sigma(., t)\big). \tag{3.16}$$

The Maximum Likelihood estimator of β is then the value that maximizes $P(\sigma|\beta)$ for the fixed observed occupancy σ, which is an intuitive choice and numerically often quite easy. For instance, the common technique of least-squares fitting of curves to data points is justified as maximum-likelihood estimation of the curve parameters assuming Gaussian errors on the data points.

To see the potential issues with this approach, consider the problem of estimating the frequency of some rare event. If the distinct events are independent, their number within some observation interval is given by a Poissonian distribution. If it should now be the case that in a given experiment not a single event is observed, the Maximum Likelihood estimator of the Poissonian distribution's parameter would be equal to zero. However, this is not meaningful as it was known beforehand that the considered event has some finite probability.

The Bayesian framework provides a consistent way to include such prior knowledge. Here one assumes the model parameters β themselves to be realizations of a random variable, which allows to incorporate any information about the parameters via their associated probability densities. Specifically, any information on the parameters available before analyzing the experiment, such as the results of previous experiments, is conceptually encoded in the prior probability density $P_{\text{prior}}(\beta)$. This prior information is updated by the results of the considered random experiment (i.e. by the observation σ) according to Bayes' formula

$$P(\beta|\sigma) = \frac{P_{\text{prior}}(\beta)P(\sigma|\beta)}{P(\sigma)}. \tag{3.17}$$

This quantity $P(\beta|\sigma)$, that means the probability density of β given the outcome of the random experiment being the observation σ, is called the posterior distribution. Turning back to the didactical example of estimating the frequency of a rare event, we see that in the Bayesian approach an observation of zero events would only shift the posterior distribution to smaller values compared to the prior distribution (as the likelihood in this case is an exponentially decaying function), which is much more in line with common sense.

In the case of the above-mentioned study of *Ambrosia artemisiifolia*'s invasion of Europe [31] the model parameter vector β was determined by MLE. In this case MLE was preferred over Bayesian inference as the latter method's advantages are only significant when the information content of the observation is low, so that the posterior distribution is strongly influenced by the prior distribution. For high information content the posterior distribution will be very narrow and concentrated around the maximum in the likelihood, rendering it equivalent to the MLE result. On the other hand, Bayesian inference is typically computationally much more expensive: In any non-trivial case, the likelihood and in turn the posterior distribution are complicated functions, necessitating them to be sampled for further interpretation, for instance by the Metropolis-Hastings algorithm [33, 34]. In contrast, for MLE only its maximum has to be found, which for the case reported above takes a few seconds.

3.6 Conclusions

Historically, the study of processes of spread and dispersal has often been initiated from the point of view of the life sciences such as ecology and biology. However, over the course of the last century it was foremost in mathematics and physics under the subjects of random walk and diffusion, respectively, that the theoretical groundworks have been laid and experimentally verified in a quantitative manner. Only within the last decades these concepts have been transferred back to ecology for describing processes of species dispersal, which are of increasing relevance.

It is the aim of this book to bring together scientists from all disciplines where such issues are relevant, and to promote interdisciplinary transfer of ideas and concepts. The chapter at hand was written in this spirit, with the intention to present the essential aspects of dispersal in ecology, the arising problems, and the methods the community has come up with to solve them.

We have started with a point-by-point discussion of generic features of dispersal processes in ecology and compared them to the physical case. Our main message was that quite generally the ecological case is more complex, so that it typically cannot be described by simple, linear equations, let alone that explicit solutions to them could be given. We then have developed ecological models in rising degrees of sophistication, considering either only spatial variations, only temporal variations, or the full spatio-temporal problem. Further, we have treated the problem of statistical inference in some detail, as due to the scarcity of dispersal events and spatial and temporal inhomogeneities the observable data are related to the model parameters only in a stochastic sense.

We want to close with a final comment on the timely relevance of ecological spread processes. A priori, these processes have been happening naturally since the dawn of time, as we have argued in the introduction, and therefore are neither good nor bad. However, the time scales of dispersal processes have accelerated drastically during the Holocene, both due to direct human impact such as intercontinental transport as well as promoted indirectly by, e.g. land use or climate change. Formerly, dispersal events were so rare that evolutionary species differentiation was able to keep up and maintain a rich regional variation within and among ecosystems. Today, the situation is different: perhaps contrary to intuition, the introduction of alien species can lead to a loss of species diversity. Most of the introduced species will not be able to establish viable populations, but those that are able to can easily have much higher reproduction successes than established species, for instance because they have no natural enemies present, and thereby disrupt the whole native ecosystem. It is true that, if only species transport were increased under constant environmental conditions, adaptation and the introduction of additional species over time would again lead to differentiated ecosystems. However, the regional variation would still be lost [35]. In physical parlance, this corresponds to a loss of entropy, which, as the physicists know, cannot be recovered without investment of effort.

References

1. E. Matthysen, in *Dispersal Ecology and Evolution*, ed. by J. Colbert, M. Baguette, T.G. Benton, J.M. Bullock (Oxford University Press, Oxford, 2012), pp. 3–18
2. R.N. Mack, Biol. Conserv. **78**, 107 (1996)
3. G.E. Hutchinson, Cold Spring Harb. Symp. Quant. Biol. **22**, 415 (1957)
4. H. Seebens, F. Essl, W. Dawson, N. Fuentes, D. Moser, J. Pergl, P. Pyšek, M. van Kleunen, E. Weber, M. Winter, B. Blasius, Glob. Change Biol. **21**(11), 4128 (2015)
5. D.M. Richardson, P. Pyšek, M. Rejmánek, M.G. Barbour, F.D. Panetta, C.J. West, Divers. Distrib. **6**, 93 (2000)
6. M. Vilà, C. Basnou, P. Pyšek, M. Josefsson, P. Genovesi, S. Gollasch, W. Nentwig, S. Olenin, A. Roques, D. Roy, P.E. Hulme, DAISIE partners. Front. Ecol. Environ. **8**, 135 (2010)
7. T. Stocker, D. Qin, G. Plattner, M. Tignor, S. Allen, J. Boschung, A. Nauels, Y. Xia, B. Bex, B. Midgley, *Climate Change 2013: The Physical Science Basis: Working Group I Contribution to the Fifth Assessment Report of the Intergovernmental Panel on Climate Change* (Cambridge University Press, Cambridge, New York, 2013)
8. J. Settele, R. Scholes, R. Betts, S. Bunn, P. Leadley, D. Nepstad, J.T. Overpeck, M.A. Taboada, in *Climate Change 2014: Impacts, Adaptation, and Vulnerability. Part A: Global and Sectoral Aspects. Contribution of Working Group II to the Fifth Assessment Report of the Intergovernmental Panel on Climate Change*, ed. by C.B. Field, V.R. Barros, D.J. Dokken, K.J. Mach, M.D. Mastrandrea, T.E. Bilir, M. Chatterjee, K.L. Ebi, Y.O. Estrada, R.C. Genova, B. Girma, E.S. Kissel, A.N. Levy, S. Maccracken, P.R. Mastrandrea, L.L. White (Cambridge University Press, Cambridge, New York, 2014), pp. 271–359
9. R.G. Pearson, T.P. Dawson, Glob. Ecol. Biogeogr. **12**, 361 (2003)
10. J.C. Svenning, F. Skov, Ecol. Lett. **7**, 565 (2004)
11. M. Williamson, K. Dehnen-Schmutz, I. Kühn, M. Hill, S. Klotz, A. Milbau, J. Stout, P. Pyšek, Divers. Distrib. **15**, 158 (2009)
12. B.A. Bradley, Biol. Invasions **15**, 1417 (2013)
13. S. Solomon, D. Qin, M. Manning, Z. Chen, M. Marquis, K.B. Averyt, M. Tignor, H.L. Miller (eds.), *Climate Change 2007: The Physical Science Basis. Contribution of Working Group I to the Fourth Assessment Report of the Intergovernmental Panel on Climate Change* (Cambridge University Press, New York, 2007)
14. S. Pompe, J. Hanspach, F. Badeck, S. Klotz, W. Thuiller, I. Kühn, Biol. Lett. **4**, 564 (2008)
15. T.R. Malthus, *An Essay on the Principle of Population* (J. Johnson, St. Paul's Churchyard, London, 1798)
16. L. Fibonacci, *Liber Abaci* (1202/1227). Reprinted 1857 in original and 2002 in English translation
17. P.F. Verhulst, Corr. Math. Phys. **10**, 113 (1838)
18. W.C. Allee, *Animal Aggregations* (A Study in General Sociology (The University of Chicago Press, Chicago, 1931)
19. A.J. Lotka, J. Amer. Chem. Soc. **42**, 1595 (1920)
20. V. Volterra, Mem. Acad. Lincei. **2**, 31 (1926)
21. M. Kondoh, Science **299**, 1388 (2003)
22. R.A. Fisher, Ann. Eugen. **7**, 355 (1937)
23. J.G. Skellam, Biometrika **38**, 196 (1951)
24. P.E. Hulme, S. Bacher, M. Kenis, S. Klotz, I. Kühn, D. Minchin, W. Nentwig, S. Olenin, V. Panov, J. Pergl, P. Pyšek, A. Roques, D. Sol, W. Solarz, M. Vilà, J. Appl. Ecol. **45**, 403 (2008)
25. P. Lambdon, P. Pyšek, C. Basnou, M. Hejda, M. Arianoutsou, F. Essl, V. Jarošík, J. Pergl, M. Winter, P. Anastasiu, P. Andriopoulos, I. Bazos, G. Brundu, L. Celesti-Grapow, P. Chassot, P. Delipetrou, M. Josefsson, S. Kark, S. Klotz, Y. Kokkoris, I. Kühn, H. Marchante, I. Perglová, J. Pino, M. Vilà, A. Zikos, D. Roy, P.E. Hulme, Preslia **80**, 101 (2008)
26. M. Gilbert, J.C. Grégoire, J.F. Freise, W. Heitland, J. Anim. Ecol. **73**, 459 (2004)
27. G. Vogl, in *Leipzig, Einstein, Diffusion*, ed. by J. Kärger (Leipziger Universitätsverlag, 2014), pp. 99–122

28. I. Hanski, Nature (London) **396**, 41 (1998)
29. B. Facon, P. David, Am. Nat. **168**(6), 769 (2006)
30. N. Wahlberg, T. Klemetti, I. Hanski, Ecography **25**, 224 (2002)
31. M.G. Smolik, S. Dullinger, F. Essl, I. Kleinbauer, M. Leitner, J. Peterseil, L.M. Stadler, G. Vogl, J. Biogeogr. **37**, 411 (2010)
32. R. Richter, S. Dullinger, F. Essl, M. Leitner, G. Vogl, Biol. Invasions **15**, 657 (2013)
33. N. Metropolis, A.W. Rosenbluth, M.N. Rosenbluth, A.H. Teller, E. Teller, J. Chem. Phys. **21**(6), 1087 (1953)
34. W.K. Hastings, Biometrika **57**, 97 (1970)
35. M. Winter, O. Schweiger, S. Klotz, W. Nentwig, P. Andriopoulos, M. Arianoutsou, C. Basnou, P. Delipetrou, V. Didžiulis, M. Hejda, P.E. Hulme, P.W. Lambdon, J. Pergl, P. Pyšek, D.B. Roy, I. Kühn, Proc. Natl. Acad. Sci. USA **106**, 21721 (2009)

Chapter 4
Search for Food of Birds, Fish and Insects

Rainer Klages

4.1 Introduction

When you are out in a forest searching for mushrooms you wish to fill your basket with these delicacies as quickly as possible. But how do you *search efficiently* for them if you have no clue where they grow (Fig. 4.1)? The answer to this question is not only relevant for finding mushrooms [1, 2]. It also helps to understand how white blood cells kill efficiently intruding pathogens [3], how monkeys search for food in a tropical forest [4], and how to optimize the hunt for submarines [5].

In society the problem to develop efficient search strategies belongs to the realm of *operations research*, the mathematical optimization of organizational problems in order to aid human decision-making [6]. Examples are the search for landmines, castaways or victims of avalanches. Over the past two decades *search research* [5] attracted particular attention within the fields of ecology and biology. The new discipline of *movement ecology* [7, 8] studies foraging strategies of biological organisms: Prominent examples are wandering albatrosses searching for food [9–11], marine predators diving for prey [12, 13], and bees collecting nectar [14, 15]. Within this context the *Lévy Flight Hypothesis* (LFH) became especially popular: It predicts that under certain mathematical conditions on the type of food sources long *Lévy flights* [16] minimize the search time [9, 10, 17]. This implies that for a bumblebee searching for rare flowers the flight lengths should be distributed according to a power law. Remarkably, the prediction by the LFH is completely different from the paradigm put forward by Karl Pearson more than a century ago [18], who proposed to model the movements of biological organisms by simple random walks as

R. Klages (✉)
Max Planck Institute for the Physics of Complex Systems, Nöthnitzer Str. 38,
01187 Dresden, Germany
e-mail: r.klages@qmul.ac.uk

R. Klages
School of Mathematical Sciences, Queen Mary University of London,
Mile End Road, London E1 4NS, UK

© Springer International Publishing AG 2018
A. Bunde et al. (eds.), *Diffusive Spreading in Nature, Technology and Society*,
https://doi.org/10.1007/978-3-319-67798-9_4

Fig. 4.1 Illustration of a typical search problem [1, 2]: a human searcher endeavours to find mushrooms that are randomly distributed in a certain area. It would help to have an *optimal search strategy* that enables one to find as many mushrooms as possible by minimizing the search time

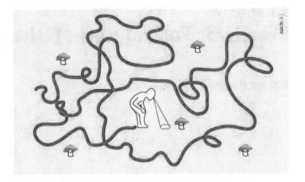

introduced in Chap. 2 of this book. His suggestion entails that the movement lengths are distributed exponentially according to a Gaussian distribution, see Eq. (2.10) in this section. Lévy and Gaussian processes represent fundamental but different classes of diffusive spreading. Both are justified by a rigorous mathematical underpinning.

More than 60 years ago Gnedenko and Kolmogorov proved mathematically that specific types of power laws, called *Lévy stable distributions* [19, 20], obey a central limit theorem. Their result generalizes the conventional central limit theorem for Gaussian distributions, which explains why Brownian motion is observed in a huge variety of physical phenomena. But exponential tails decay faster than power laws, which implies that for Lévy-distributed flight lengths there is a larger probability to yield long flights than for flight lengths obeying Gaussian statistics. Consequently, Lévy flights should be better suited to detect sparsely, randomly distributed targets than Brownian motion, which in turn should outperform Lévy motion when the targets are dense. This is the basic idea underlying the LFH. Empirical tests of it, however, are hotly debated [11, 21–24]: Not only are there problems with a sound statistical analysis of experimental data sets when checking for power laws; their biological interpretation is also often unclear: For example, for monkeys living in a tropical forest who feed on specific types of fruit it is not clear whether the observed Lévy flights of the monkeys are due to the distribution of the trees on which their preferred fruit grows, or whether the monkeys' Lévy motion represents an evolutionary adapted optimal search strategy helping them to survive [4]. Theoretically the LFH was motivated by random walk models with Lévy-distributed step lengths that were solved in computer simulations [10]. A rigorous mathematical proof of the LFH remains elusive.

This chapter introduces to the following fundamental question cross-linking the fields of ecology, biology, physics and mathematics: *Can search for food by biological organisms be understood by mathematical modeling?* [8, 17, 20, 25] It consists of three main parts: Sect. 4.2 reviews the LFH. Section 4.3 outlines the controversial discussion about its verification by including basics about the theory of Lévy motion. Section 4.4 illustrates the need to go beyond the LFH by elaborating on bumblebee flights. We summarize our discussion in Sect. 4.5.

4.2   Lévy Motion and the Lévy Flight Hypothesis

4.2.1   Lévy Flights of Wandering Albatrosses

In 1996 Gandhimohan Viswanathan and collaborators published a pioneering article in the journal *Nature* [9]. For albatrosses foraging in the South Atlantic the flight times were recorded by putting sensors at their feet. The sensors got wet when the birds were diving for food, see the inset of Fig. 4.2. The duration of a flight was thus defined by the period of time when a sensor remained dry, terminated by a dive for catching food. The main part of Fig. 4.2 shows a histogram of the flight time intervals of some albatrosses. The straight line represents a Lévy stable distribution proportional to $\sim t^{-\mu}$ with an exponent of $\mu = 2$. By assuming that the albatrosses move with an on average constant speed one can associate these flight times with a respective power law distribution of flight lengths. This suggests that the albatrosses were searching for food by performing Lévy flights.

For more than a decade albatrosses were considered to be the most prominent example of an animal performing Lévy flights. This work triggered a large number of related studies suggesting that many other animals like deer, bumblebees, spider monkeys and fishes also perform Lévy motion [4, 10, 12, 13, 17].

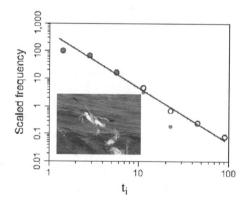

Fig. 4.2 Histogram where 'scaled frequencies' holds for the number of flight time intervals of length t_i (in hours) normalized by their respective bin widths. The data is for five albatrosses during 19 foraging bouts (double-logarithmic scale). Blue open circles show the data from Ref. [9]. The straight line indicates a power law $\sim t^{-\mu}$ with exponent $\mu = 2$. The red filled circles are adjusted flight durations using the same data set by eliminating times that the birds spent on an island [11]. The histogram is reprinted by permission from *Macmillan Publishers Ltd: Nature Ref.* [11], *copyright 2007*. The inset shows an albatross catching food; reprinted by permission from *Macmillan Publishers Ltd: Nature Ref.* [5], *copyright 2006*

4.2.2   The Lévy Flight Hypothesis

In 1999 the group around Gandhimohan Viswanathan published another important article in *Nature* [10]. Here the approach was more theoretical by posing, and
addressing, the following general question:

"What is the *best statistical strategy* to adapt in order to search *efficiently* for
randomly located objects?"

To answer this question they introduced a special type of what is called a *Lévy
walk* [20] in two dimensions and studied it both by computer simulations and by
analytical approximations. Their model consists of point targets randomly distributed in a plane and a (point) forager moving with constant speed. If the forager spots
a target within a pre-defined finite vision distance, it moves to the target directly.
Otherwise the forager chooses a direction at random with a jump length ℓ randomly
drawn from a Lévy stable distribution $\sim \ell^{-\mu}$, $1 \leq \mu \leq 3$. While the forager is moving it constantly looks out for targets within the given vision distance. If no target is
detected, the forager stops after the given distance and repeats the process.

Although these rules look simple enough, there are some subtleties that exemplify
the problem of mathematically modeling a biological foraging problem:

1. Here we have chosen what is called a *cruise forager*, i.e., a forager that senses
   targets whenever it is moving. In contrast, a *saltatory forager* would not sense
   a target while moving. It needs to land close to a target within a given radius of
   perception in order to find it [26].
2. For a cruise forager a jump is terminated when it hits a target, hence this model
   defines a *truncated* Lévy walk [13].
3. One has to decide whether a forager eliminates targets when it finds them or not,
   i.e., whether it performs *destructive* or *non-destructive* search [10]. As we will
   see below, whether a monkey eats a fruit thus effectively eliminating it, at least
   for a long time, or whether a bee collects nectar from a flower that replenishes
   quickly defines mathematically different foraging problems.
4. We have not yet said anything about the *density of the targets*.
5. We have deliberately assumed that the targets are *immobile*, which may not
   always be realistic for a biological foraging problem (e.g., marine predators [12,
   13]).
6. If we ask about the *best* strategy to search *efficiently*, how do we define *optimality*?

These few points illustrate the difficulty to relate abstract mathematical random
walk models to biological foraging reality. Interestingly, the motion generated by
these models often sensitively depends on right such details: In Ref. [10] foraging efficiency was defined as the ratio of the number of targets found divided by
the total distance traveled by a forager. Different definitions are possible, depend on
the type of forager and may yield different results [26]. The foraging efficiency was
then computed in Ref. [10] under variation of the exponent μ of the above Lévy

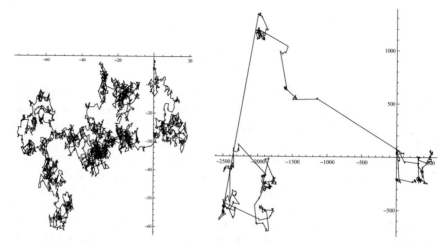

Fig. 4.3 Brownian motion (left) versus Lévy motion (right) in the plane, illustrated by typical trajectories

distribution generating the jump length. The results led to what was coined the **Lévy Flight Hypothesis** (LFH), which we formulate as follows:

Lévy motion provides an *optimal search strategy* for *sparse, randomly distributed, immobile, revisitable targets in unbounded domains.*

Intuitively this result can be understood as follows: Fig. 4.3 (left) displays a typical trajectory of a Brownian walker. One can see that this dynamics is 'more localized' while Lévy motion shown in Fig. 4.3 (right) depicts clusters interrupted by long jumps. It thus makes sense that Brownian motion is better suited to find targets that are densely distributed while Lévy motion outperforms Brownian motion when targets are sparse, since it avoids oversampling due to long jumps. The reason why the targets need to be revisitable is that the exponent μ of the Lévy distribution depends on whether the search is destructive or not, cf. the third point on the list of foraging conditions above: For non-destructive foraging $\mu = 2$ was found to be optimal while for destructive foraging $\mu = 1$ maximized the foraging efficiency, which corresponds to the special case of ballistic flights [20]. The reason for these different exponents is that destructive foraging changes the distribution and the density of the targets thus selecting a different foraging strategy to be optimal.

4.3 Lévy or Not Lévy?

4.3.1 Revisiting Lévy Flights of Wandering Albatrosses

Several years passed before the results by Viswanathan et al. were revisited in another *Nature* article led by Andrew Edwards [11]: When analyzing new, larger and more precise data for foraging albatrosses the old results of Ref. [9] could not be recovered, see Fig. 1 in Ref. [11]. This led the researchers to reconsider the old albatross data. A correction of these data sets yielded the result shown in Fig. 4.2 as the red filled circles: One can see that the Lévy stable law with an exponent of $\mu = 2$ for the flight times is gone. Instead the data now seems to be fit best with a gamma distribution.

What happened is explained in Ref. [21]: For all measurements the sensors were put onto the feet of the albatrosses when the birds were sitting on an island, and at this point the measurement process was started. However, to this time the sensors were dry; and in Ref. [9] these times were interpreted as Lévy flights. The same applied to the end of a foraging bout when the birds were back on the island. Subtracting these erroneous time intervals from the data sets eliminated the Lévy flights.

However, in Ref. [27] yet new albatross data was analyzed, and the old data from Refs. [9, 11] was again reanalyzed: This time truncated power laws were used for the analysis, and furthermore data sets for individual birds were tested instead of pooling together the data for all birds. In this reference it was concluded that some individual albatross indeed do perform Lévy flights while others do not.

4.3.2 The Lévy Flight Paradigm

The debate about the LFH created a surge of publications testing it both theoretically and experimentally; see Refs. [8, 17, 20, 25] for reviews. But experimentally it is difficult to verify the mathematical conditions on which the LFH formulated in Sect. 4.2.2 is based. Often the LFH was thus interpreted in a much looser sense by ignoring any mathematical assumptions in terms of what one may call the **Lévy Flight Paradigm** (LFP):

Look for *power laws* in the probability distributions of step lengths of foraging animals.

We illustrate virtues and pitfalls related to the LFP by data from Ref. [13] on the diving depths of free-ranging marine predators. Impressively, in this work over 12 million movement displacements were recorded and analyzed for 14 different species. As an example, Fig. 4.4 shows results for a blue shark: Plotted at the bottom are probability distributions of its diving depths, called move step length frequency distribution, where a step length is defined as the distance moved by the shark per unit time. Included are fits to a *truncated* power law and to an exponential distribution. Since here Lévy distributions were used whose longest step lengths were cut off, the fits do not consist of straight lines but are bent off, in contrast to Fig. 4.2.

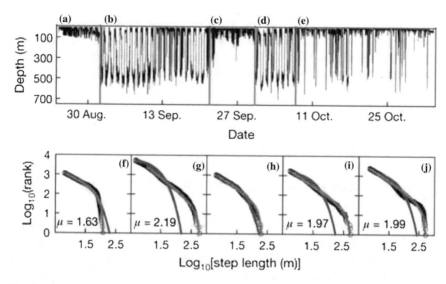

Fig. 4.4 Top: time series of the diving depth of a blue shark. The red lines split the data into different sections (**a–e**), where the shark dives deep or the diving depth is more constrained. These sections match to the shark being off-shelf or on the shelf, respectively. Bottom: double-logarithmic plots of the move step length frequency distribution ('rank') as a function of the step length, which is the vertical distance moved by the shark per unit time, with the notation (**f–j**) corresponding to the primary data shown in sections (**a–e**). Black circles correspond to data, red lines to fits with truncated power laws of exponent μ, blue lines to exponential fits. This figure is reprinted by permission from *Macmillan Publishers Ltd: Nature Ref.* [13], *copyright 2010*

The top of this figure depicts the corresponding time series from which the data was extracted, split into five different sections. Each section is characterized by profoundly different average diving depths. These different sections correspond to the shark being in different regions of the ocean, i.e., either on-shelf or off-shelf. It was argued that on-shelf, where the diving depth of the shark is very limited, the data can be better fitted with an exponential distribution (sections f and h) while off-shelf the data displays power-law behavior with an exponent close to two (sections g, i and j). Figure 4.4 thus suggests a strong dependence of the foraging dynamics on the environment in which it takes place, where the latter defines the food distribution. Related switching behavior between power law-like Lévy and exponential Brownian motion search strategies was reported for microzooplankton, jellyfish and mussels.

The power law matching to the data in the off-shelf regions was interpreted in support of the LFH. However, note the periodic oscillations displayed by the time series at the top of Fig. 4.4. Upon closer inspection they reveal a 24 h day-night cycle: During the night the shark hovers close to the surface of the sea while over the day it dives for food. For the move step length distributions shown in Fig. 4.4 the data was averaged over all these periodic oscillations. But the distributions in sections g, i and j all show a 'wiggle' on a finer scale. This suggests to better fit the data by a superposition of two different distributions [14] taking into account that day and night define two very different phases of motion, instead of using only one function by averaging over all times. Apart from this, one may argue that this analysis does not test for the original LFH put forward in Ref. [9]. But this requires a bit more knowledge about the theory of Lévy motion; we will come back to this point in Sect. 4.3.5.

4.3.3 Two Different Lévy Flight Hypotheses

Our discussion in the previous sections suggests to distinguish between *two different* LFHs:

1. The first is the 'conventional' one that we formulated in Sect. 4.2.2, originally put forward in Ref. [9]: It may now be further specified as the **Lévy Search Hypothesis** (LSH), because it suggests that under certain conditions Lévy flights represent an *optimal search strategy*. Here optimality needs to be defined rigorously mathematically. This can be done in different ways given the specific biological situation at hand that one wishes to model [26]. Typically optimality within this context aims at minimizing the search time for finding targets. The interesting biological interpretation of the LSH is that it has been evolved in biological organisms as an *evolutionary adaptive* strategy that maximizes the success for survival. The LSH version of the LFH became most popular.
2. In parallel there is a second type of LFH, which may be called the **Lévy Environmental Hypothesis** (LEH): It suggests that Lévy flights *emerge* from the interaction between a forager and a food source distribution. The latter may be scale-free thus directly inducing the Lévy flights. This is in sharp contrast to the LSH, which suggests that under certain conditions a forager performs Lévy flights irrespective of the actual food source distribution. Emergence of novel patterns and dynamics due to the interaction of the single parts of a complex system with each other, on the other hand, is at the heart of the theory of complex systems. The LEH is the hypothesis that to some extent was formulated in Ref. [9], but it became more popular rather later on [4, 12, 13].

Both the LSH and the LEH are bound together by what we called the Lévy Flight Paradigm (LFP) in Sect. 4.3.2. The LFP extracts the formal essence from both these different hypotheses by proposing to look for power laws in the probability distributions of foraging dynamics by ignoring any conditions of validity of these two hypotheses. Consequently, in contrast to the LSH and LEH the mathematical, physical and biological origin and meaning of power laws obtained by following the LFP is typically not clear. On the other hand, the LFP motivated to take a fresh look at foraging data sets by not only testing for exponential distributions. It widened the scope by emphasizing that one should also check for power laws in animal movement data.

4.3.4   Intermittent Search Strategies as an Alternative to Lévy Motion

Simple random walks as introduced in Sect. 2.1 represent examples of *unimodal* types of motion if the random step lengths are sampled from only one specific distribution. For example, choosing a Gaussian distribution we obtain Brownian motion while a Lévy-stable distribution produces Lévy flights. Combining two different types of motion like Brownian and Lévy yields *bimodal motion*. A simple example is shown in Fig. 4.5: Imagine you have lost your keys at home, but you have a vague idea where to find them. Hence, you are running straightforwardly to the location where you expect them to be. This may be modeled as a *ballistic flight* during which you quickly relocate, say, from the kitchen to the study room. However, when you arrive in your study room you should switch to a different type of motion, which is suitably adapted to locally search the environment. For this mode you may choose, e.g., Brownian motion. The resulting dynamics is called *intermittent* [25]: It consists of two different phases of motion mixed randomly, which in our example are ballistic relocation events and local Brownian motion.

This type of motion can be exploited to search efficiently in the following way: You may not bother to look for your keys while you are walking from the kitchen to the study room. You are more interested to get from point A to point B as quickly as possible, and while doing so your search mode is switched off. This is called a *non reactive* phase in Fig. 4.5. But as you expect the keys to be in your study room, while switching to Brownian motion therein you simultaneously switch on your scanning abilities. This defines your local search mode called *reactive* in Fig. 4.5. Correspondingly, for animals one may imagine that during a fast relocation event, or flight, they are unable to detect any targets while their sensory mechanisms become active during slow local search. This is close to what was called a saltatory forager in Sect. 4.2.2, but this forager did not feature any local search dynamics.

Intermittent search dynamics can be modeled by writing down a set of two coupled equations, one that generates ballistic flights and another one that yields Brownian motion. The coupling captures the switching between both modes.

Fig. 4.5 Illustration of an intermittent search strategy: A human searcher looks for a target (key) by alternating between two different modes of motion. During fast, ballistic relocation phases the searcher is not able to detect any target (non reactive). These phases are interrupted by slow phases of Brownian motion during which a searcher is able to detect a target (reactive) [25]

One furthermore needs to model that search is only performed during the Brownian motion mode. By analyzing a respective ballistic-Brownian system of equations it was found that this dynamics yields a minimum of a suitably defined search time under parameter variation if a target is *non-revisitable*, i.e., it is destroyed once it is found. Note that for targets that are non-replenishing the Lévy walks of Ref. [10] did not yield any non-trivial optimization of the search time. Instead, they converged to pure ballistic flights as being optimal. The LSH, in turn, only applies to revisitable, i.e., replenishing targets. Hence intermittent motion poses no contradiction. A popular account of this result was given by Michael Shlesinger in his Nature article 'How to hunt a submarine?' [5].

4.3.5 Theory of Lévy Flights in a Nutshell

We now briefly elaborate on the theory of Lévy motion. This section may be skipped by a reader who is not so interested in theoretical foundations. We recommend Ref. [16] for an outline of this topic from a physics point of view and Chap. 5 in Ref. [19] for a more mathematical introduction. We start from the simple random walk on the line introduced in Chap. 2 of this book,

$$x_{n+1} = x_n + \ell_n \quad , \tag{4.1}$$

where x_n is the position of a random walker at discrete time $n \in \mathbb{N}$ moving in one dimension, and $\ell_n = x_{n+1} - x_n$ defines the jump of length $|\ell_n|$ between two positions. In Chap. 2 the special case of constant jump length $|\ell_n| = \ell$ was considered, where the sign of the jump was randomly determined by tossing a coin with, say, plus for heads and minus for tails. The coin was furthermore supposed to be *fair* in the sense of yielding equal probabilities for heads and tails. This simple random walk can be generalized by considering a bigger variety of jumps. Mathematically this is modeled by drawing the random variable ℓ_n from some more general probability distribution than featuring only probability one half for each of two outcomes. For example, instead we could draw ℓ_n at each time step n randomly from a uniform distribution, where each jump between $-L$ and L is equally possible given by the probability density $\rho(\ell_n) = 1/(2L)$, $-L \le \ell_n \le L$ and zero otherwise. Alternatively, we could allow arbitrarily large jumps by drawing ℓ_n from an unbounded Gaussian distribution, see Eq. (2.10) in Chap. 2 (by replacing x therein with ℓ_n and setting t constant). For both generalized random walks Eq. (4.1) would still reproduce in the long time limit the fundamental diffusive properties Eq. (4) discussed in Chap. 2, i.e., the linear growth in time of the mean square displacement, and Eq. (2.10) in Chap. 2, the Gaussian probability distribution for the position x_n of a walker at time step n. This follows mathematically from the conventional central limit theorem.

We now further generalize the random walk Eq. (4.1) in a more non-trivial way by randomly drawing ℓ_n from a *Lévy α-stable distribution* [19],

$$\rho(\ell_n) \sim |\ell_n|^{-1-\alpha} \, (|\ell_n| \gg 1) \, , \, 0 < \alpha < 2 \, , \tag{4.2}$$

characterized by power law tails in the limit of large $|\ell_n|$. This functional form is in sharp contrast to the exponential tails of Gaussian distributions and has important consequences, as it violates one of the assumptions on which the conventional central limit theorem rests. However, for the range of exponents α stated above it can be shown that these distributions obey a *generalized central limit theorem*: The proof employs the fact that these distributions are *stable*, in the sense that a linear combination of two random variables sampled independently from the same distribution reproduces the very same distribution, up to some scale factors [16]. This in turn implies that Lévy stable distributions are *scale invariant* and thus *self-similar*. Physically one speaks of ℓ_n sampled independently and identically distributed from Eq. (4.2) as *white Lévy noise*. As by definition there are no correlations between the random variables ℓ_n the stochastic process generated by Eq. (4.1) is *memoryless*, meaning at time step $(n + 1)$ the particle has no memory where it came from at any previous time step n. In mathematics this is called a *Markov processes*, and Lévy flights belong to this important class of stochastic processes.

What we presented here is only a very rough, mathematically rather imprecise outline of how to define an α-stable Lévy process generating Lévy flights. Especially, the function in Eq. (4.2) is not defined for small ℓ_n, as the given power law diverges for $\ell_n \to 0$. A rigorous definition of Lévy stable distributions is obtained by using the characteristic function of this process, i.e., the Fourier transform of its probability distribution, which is well-defined analytically. The full probability distribution can then be generated from it [16, 19]. For $\alpha = 2$ this approach reproduces Gaussian distributions, hence Lévy dynamics suitably generalizes Brownian motion [16, 19].

Another important property of Lévy stable distributions is that the mean squared flight length of a Lévy walker does not exist,

$$\langle \ell_n^2 \rangle = \int_{-\infty}^{\infty} d\ell_n \, \rho(\ell_n) \ell_n^2 = \infty \, . \tag{4.3}$$

The above equation defines what is called the second moment of the probability distribution $\rho(\ell_n)$. Higher moments are defined analogously by $\langle \ell^k \rangle$, $k \in \mathbb{N}$, and for Lévy distributions they are also infinite. This means that in contrast to simple random walks generating Brownian motion, see again Chap. 2, Lévy motion does not have any characteristic length scale. However, since moments are rather easily obtained from experimental data this poses a problem to Lévy flights as a viable physical model to be validated by experiments.

This problem can be solved by using the very related concept of *Lévy walks* [20]: These are random walks where again jumps are drawn randomly from the Lévy stable distribution Eq. (4.2). But as a penalty for long jumps the walker spends a time t_n proportional to the length of the jump to complete it, $t_n = v\ell_n$, where the proportionality factor v, typically chosen as $|v| = const.$, defines the velocity of the Lévy walker. This implies that both jump lengths ℓ_n and flight times t_n are distributed according to the same power law. In contrast, for the Lévy flights introduced above a walker makes a jump of length $|\ell_n|$ during an integer time step of duration $\Delta n = 1$, which implies that contrary to a Lévy walker a Lévy flyer can jump instantaneously over arbitrarily long distances with arbitrarily large velocities.

Lévy walks belong to the broad and important class of *continuous time random walks* [19, 28, 29], which further generalize ordinary random walks by allowing a walker to move by non-integer time steps. We do not discuss all the similarities and differences between Lévy walks and Lévy flights, see Ref. [20] for details, but instead highlight only one important fact: While for Lévy flights the mean square displacement $\langle x^2 \rangle$, see Eq. (1) in Chap. 2, does not exist, which follows from our discussion above, for Lévy walks it does. This is due to the finite velocities, which truncate the power law tails in the probability distributions for the positions of a Lévy walker. However, in contrast to Brownian motion where it grows linearly in time as shown in Chap. 2, see Eq. (2), for Lévy walks it grows faster than linear,

$$\langle x^2 \rangle \sim t^\gamma \ (t \to \infty), \tag{4.4}$$

with $\gamma > 1$. If $\gamma \neq 1$ one speaks of *anomalous diffusion* [19, 28]. The case $\gamma > 1$ is called *superdiffusion*, since a particle diffuses faster than Brownian motion, correspondingly $\gamma < 1$ refers to *subdiffusion*. There is a wealth of different stochastic models exhibiting anomalous diffusion, and while superdiffusion appears to be more common among foraging biological organisms than subdiffusion the whole spectrum of anomalous diffusion is found in a variety of different processes in the natural sciences, and even in the human world [19, 28, 30].

Often the difference between Lévy walks and flights is not quite appreciated in the experimental literature, see, e.g., Fig. 4.4, where move step length frequency distributions were plotted. By definition a move step length x per unit time corresponds to what we defined as a jump length ℓ_n by Eq. (4.1) above, $x = \ell_n$. Hence, a truncated power law fit $\sim x^{-\mu}$ to the distributions plotted in Fig. 4.4 corresponds to a fit with a truncated form of the jump length distribution Eq. (4.2) with exponent $\mu = 1 + \alpha$ testing for truncated Lévy flights [20]. The truncation cures the problem of infinite moments exhibited by random walks based on ordinary Lévy flights mentioned above. However, this analysis does not test the LFH put forward in Ref. [10], which was derived from Lévy walks. But checking for Lévy walks requires an entirely different data analysis [3, 20].

4.4 Beyond the Lévy Flight Hypothesis: Foraging Bumblebees

The LFH and its variants illustrated the problem to which extent biologically relevant search strategies may be identified by mathematical modeling. What we then formulated as the LFP in Sect. 4.3.2 motivated to generally look for power laws in the probability distributions of step lengths of foraging animals. Inspired by the long debate about the different functional forms of move step lengths probability distributions, and by further diluting the LFP, an even weaker guiding principle would be to assume that the foraging dynamics of biological organisms can be understood by analyzing such probability distributions alone. In the following we discuss an experiment, and its theoretical analysis, which illustrate that one may miss crucial information by studying only probability distributions. In that respect, this last section provides a look beyond the LFH that focuses on such distributions.

4.4.1 Bumblebees Foraging Under Predation Risk

In Ref. [31] Thomas Ings and Lars Chittka reported a laboratory experiment in which environmental foraging conditions were varied in a fully controlled manner. The question they addressed with this experiment was whether changes of environmental conditions, in this case exposing bumblebees to predation threat or not, led to changes in their foraging dynamics. This question was answered by a statistical analysis of the bumblebee flights recorded in this experiment on both spatial and temporal scales [14].

The experiment is sketched in Fig. 4.6: Bumblebees (*Bombus terrestris*) were flying in a cubic arena of ≈75 cm side length by foraging on a 4 × 4 vertical grid of artificial yellow flowers on one wall. The 3D flight trajectories of 30 bumblebees, tested sequentially and individually, were tracked by two high frame rate cameras. On the landing platform of each flower nectar was given to the bumblebees and replenished

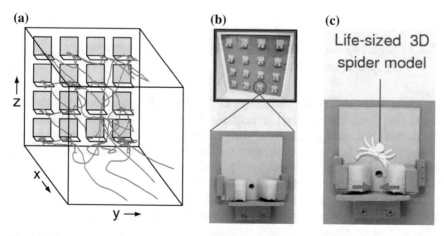

Fig. 4.6 Illustration of a laboratory experiment investigating the dynamics of bumblebees foraging under predation risk: **a** Sketch of the cubic foraging arena together with part of the flight trajectory of a single bumblebee. The bumblebees forage on a grid of artificial flowers on one side of the box. While being on the landing platforms, they have access to nectar. All flowers can be equipped with spider models and trapping mechanisms simulating predation attempts as shown in (**b**), (**c**) [14, 31]

after consumption. To analyze differences in the foraging behavior of the bumble-bees under threat of predation, artificial spiders were introduced. The experiment was staged into several different phases of which, however, only the following three are relevant to our analysis:

1. spider-free foraging
2. foraging under predation risk
3. a memory test one day later

Before and directly after stage 2 the bumblebees were trained to forage in the presence of artificial spiders, which were randomly placed on 25% of the flowers. A spider was emulated by a spider model on the flower and a trapping mechanism, which briefly held the bumblebee to simulate a predation attempt. In stages 2 and 3 the spider models were present but the traps were inactive in order to analyze the influence of previous experience with predation risk on the bumblebees' flight dynamics; see Ref. [31] for full details of the experimental setup and staging.

It is important to observe that neither the LSH nor the LEH can be tested by this experiment, as the flight arena is too small: The bumblebees always sense the walls and may adjust their flight behavior accordingly. However, there is a cross-link to the LEH in that this experiment studies the interaction of a forager with the environment, and its consequences for the dynamics of the forager, in a very controlled way. The weaker guiding principle derived from the LFP that we discussed above furthermore suggests that the main information to understand the foraging dynamics may be contained in the probability distributions of flight step lengths only. On this basis one may naively expect to see different step lengths probability distributions emerging by changing the environmental conditions, which here is the predation risk.

4.4.2 Velocity Distributions Versus Velocity Correlations: Experimental Results

Figure 4.7 shows a typical probability distribution of the horizontal velocities parallel to the flower wall (cf. the y-direction in Fig. 4.6a) for a single bumblebee. This distribution is in analogy to the move step length frequency distributions of the shark shown in Fig. 4.4, which also represent velocity distributions if the depicted step lengths are divided by the corresponding constant time intervals of their measurements as discussed in Sect. 4.3.5. The distribution of bumblebee flights per unit time is characterized by a peak at low velocities. Only a power law and a Gaussian distribution can immediately be ruled out by visual inspection as matching functional forms. However, a mixture of two Gaussian distributions and an exponential function appear to be equally possible. Maximum likelihood fits supplemented by respective information criteria yielded the former as the most likely functional form matching the data. This result can be understood biologically as representing two different flight modes near a flower versus far away from it, which is confirmed by spatially separated data analysis [14]. That the bumblebee switches to a specific distribution

Fig. 4.7 Semi-logarithmic plot of the distribution of velocities v_y parallel to the y-axis in Fig. 4.6a (black crosses) for a single bumblebee in the spider-free stage 1. The different lines represent maximum likelihood fits with a Gaussian mixture (red line), exponential (blue dotted), power law (green dashed), and single Gaussian distribution (violet dotted) [14]

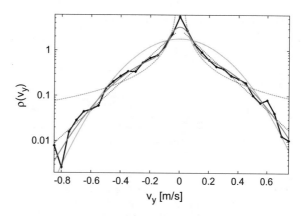

of lower velocities when approaching a flower reflects a spatially adapted flight mode to accessing the food sources. As a result, here we encounter another version of intermittent motion: In contrast to the temporal switching between different flight modes discussed in Sect. 4.3.4 this one is due to switching in different regions of space.

Surprisingly, when extracting the velocity distributions of single bumblebees at the three different stages of the experiment and comparing their best fits with each other, qualitatively and quantitatively *no differences* could be found in these distributions between the spider-free stage and the stages where artificial spider models were present [14]. This means that the bumblebees fly with the very same statistical distribution of velocities irrespective of whether predators are present or not. The answer about possible changes in the bumblebee flights due to changes in the environmental conditions is thus not given by analyzing the probability distributions of move step lengths, as one may infer from our diluted LFP guiding principle. We will now see that it is provided by examining the correlations of horizontal velocities $v_y(t)$ parallel to the wall for all bumblebee flights. They can be measured by the *velocity autocorrelation function*

$$v_y^{ac}(\tau) = \frac{\langle (v_y(t) - \mu)(v_y(t + \tau) - \mu) \rangle}{\sigma^2} . \qquad (4.5)$$

Here μ and σ^2 denote the mean and the variance of the corresponding velocity distribution of v_y, respectively, and the angular brackets define an average over all bumblebees and over time. This quantity is a special case of what is called a *covariance* in statistics. Note that velocity correlations are intimately related to the mean square displacement introduced in Chap. 2 of this book: While the above equation defines velocity correlations that are normalized by subtracting the mean and dividing by the variance, unnormalized velocity correlations emerge straightforwardly from the right hand side of Eq. (2.1) in Chap. 2 by rewriting it as products of velocities. This yields the *(Taylor-)Green-Kubo formula* expressing the mean square displacement exactly in terms of velocity correlations [32]. Note that the velocity autocorrelation

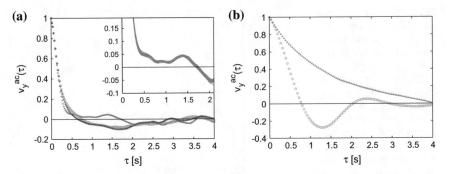

Fig. 4.8 Velocity autocorrelation function Eq. (4.5) for bumblebee velocities v_y parallel to the wall at three different stages of the experiment shown in Fig. 4.6: **a** Experimental results for stage 1 without spiders (red), 2 under predation threat (green), and 3 under threat a day after the last encounter with the spiders (blue). The data show the effect of the presence of spiders on the bumblebee flights. The inset presents the resampled autocorrelation for the spider-free stage in the region where the correlation differs from the stages with spider models, which confirms that the positive autocorrelations are not a numerical artifact. **b** Theoretical results for the same quantity obtained from numerically solving the Langevin equation (4.6) by switching off (red triangles, upper line)/on (green circles, lower line) a repulsive force modeling the interaction of a bumblebee with a spider. These results qualitatively reproduce the experimental findings in (**a**)

function is defined by an average over the product between the initial velocity at time $\tau = 0$ and the velocity at time lag τ along a trajectory: By definition it is maximal and normalized to one at $\tau = 0$, because the initial velocity is maximally correlated with itself. It will decay to zero if on average all velocities at time τ are randomly distributed with respect to the initial velocities. Physically this quantity thus measures the *correlation decay* in the dynamics over time τ by giving an indication to which extent a dynamics loses memory. For example, for a simple random walk as defined in Chap. 2 and by Eq. (4.1) in our section the velocity correlations would immediately jump to zero from $\tau = 0$ to $\tau \neq 0$, which reflects that these random walks are completely memory-free. This property was used in Chap. 2 to derive Eq. (2.2) from Eq. (2.1) by canceling all cross-correlation terms.

Figure 4.8a shows the bumblebee velocity autocorrelations defined by Eq. (4.5) for all three stages of the experiment. While for the spider-free stage the correlations remain positive for rather long times, in the presence of spiders they quickly become negative. This means that the velocities are on average anti-parallel to each other, or anti-correlated. In terms of flights, when predators are not present the bumblebees thus fly on average more often in the same direction for short times while in the presence of predators on average they often reverse their flight directions for shorter times. This result can be biologically understood as reflecting a more careful search under predation threat: When no predators are present, the bumblebees forage with more or less direct flights from flower to flower. However, under threat the bumblebees change their direction more often in their search for food sources, rejecting flowers with spiders. Mathematically this means that the *distributions* of velocities remain the same, irrespective of whether predators are present or not, while the *topology*, i.e., the shape of the bumblebee trajectories changes profoundly being on average more 'curved'.

In order to theoretically reproduce these changes we model the dynamics of v_y by a *Langevin equation* [33]. It may be called Newton's Law of stochastic physics, as it is based on Newton's Second Law: $F = m \cdot a$, where m is the mass of a tracer particle in a fluid moving with acceleration $a = d^2x/dt^2$ at position $x(t)$ (for sake of simplicity we restrict ourselves to one dimension). To model the interaction of the tracer particle with the surrounding fluid, the force F on the left hand side is written as a sum of two different forces, $F = F_S + F_b$: a friction term $F_S = -\eta v = -\eta\, dx/dt$ with Stokes friction coefficient η, which models the damping by the surrounding fluid; and another term F_b that mimicks the microscopic collisions of the tracer particle with the surrounding fluid particles, which are supposed to be much smaller than the tracer particle. The latter interaction is modeled by a stochastic force $\xi(t)$ of the same type as we have described in Sect. 4.3.5 for which here one takes Gaussian white noise. Interestingly, the stochastic Langevin equation can be derived from first principles starting from Newton's microscopic equations of motion for the full deterministic dynamical system of a tracer particle interacting with a fluid consisting of many particles [32].

At first view it may look strange to apply such an equation for modeling the motion of a biological organism. However, for a bumblebee the force terms may simply be reinterpreted: While the friction term still models the loss of velocity due to the surrounding air during a flight, the stochastic force term now mimicks both the force actively exerted by the bumblebee to perform a flight and the randomness of these flights due to the surrounding air, and to sudden changes of direction by the bumblebee itself. In addition, for our experiment we need to model the interaction with predators by a third force term. This leads to Eq. (2.21) stated in Chap. 2, which for bumblebee velocities v_y we rewrite as

$$\frac{dv_y(t)}{dt} = -\eta v_y(t) - \frac{dU(y(t))}{dy} + \xi(t) \,. \tag{4.6}$$

Here we have combined the mass m with the other terms on the right hand side. The term $F_i = -dU(y(t))/dy$ with potential U mimicks an interaction between bumblebee and spider, which can be switched on or off depending on whether a spider is present or not. Data analysis shows that this force is strongly repulsive [14]. Computing the velocity autocorrelation function Eq. (4.5) by solving the above equation numerically for a suitable choice of a repulsive force qualitatively reproduces a change from positive to negative correlations when switching on the repulsive force, see Fig. 4.8b.

These results demonstrate that velocity correlations can contain crucial information for understanding foraging dynamics, here in the form of highly non-trivial correlation decay *emerging* from the interaction of a forager with predators. This experiment could not test the LSH, as the mathematical assumptions on its validity were not fulfilled. However, conceptually these results are in line with the idea underlying the LEH: Theoretically the interaction between forager and environment was modeled by a repulsive force, to be switched on in the presence of predators, which qualitatively reproduced the experimental results. Together with the spatially intermittent dynamics when approaching the food sources as discussed before, these findings illustrate a complex spatio-temporal adjustment of the bumblebees both to the presence of food sources and predators. This is in sharp contrast to the scale-free dynamics singled out by the LFH.

Of course, modeling bumblebee flights by a Langevin equation like Eq. (4.6) ignores many fine details. A more sophisticated model that reproduces bumblebee flights far away from the flowers more appropriately has been constructed in Ref. [15] based on the same data as discussed above.

4.5   Lévy Flights Embedded in Movement Ecology

The main theme of our chapter was the question posed to the end of the introduction: *Can search for food by biological organisms be understood by mathematical modeling?* While about a century ago this question was answered by Karl Pearson in terms of simple random walks yielding Brownian motion, about two decades ago the LFH gave a different answer by proposing Lévy motion to be optimal for foraging success, under certain conditions. Discussing experimental results testing it, we arrived at a finer distinction between two different types of LFHs: The LSH captured the essence of the original LFH by stating that under certain conditions Lévy flights represent an optimal search strategy for finding targets. In contrast the LEH stipulates that Lévy flights may emerge from the interaction between a forager and possibly scale-free food source distributions. A weaker version of these different hypotheses we coined the LFP, which suggests to look for power laws in the probability distributions of move step lengths of foraging organisms. An even weaker guiding principle derived from it is to assume that the foraging dynamics of biological organisms can generally be understood by analyzing step length probability distributions alone. We thus have a hierarchy of different LFHs that have all been tested in the literature, in one way or the other.

By elaborating on experimental results, exemplified by selected publications, we outlined a number of problems when testing the different LFHs: miscommunication between theorists and experimentalists leading to incorrect data analysis; the difficulties to mathematically model a specific foraging situation by giving proper credit to all relevant biological details; and problems with an adequate statistical data analysis that really tests for the theory by which it was motivated. We highlighted that there are alternative stochastic processes, such as intermittent search strategies, that may outperform Lévy strategies under certain conditions, or at least lead to similar results, such that it may be hard to clearly distinguish them from Lévy motion. We also discussed an experiment on foraging bumblebees, which showed that relevant information to understand a biological foraging process may not always be contained

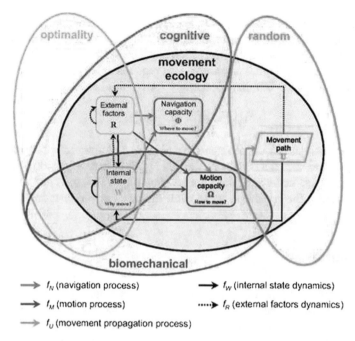

> $\longrightarrow$ f_N (navigation process) $\longrightarrow$ f_W (internal state dynamics)
>
> $\longrightarrow$ f_M (motion process) $\cdots\!\blacktriangleright$ f_R (external factors dynamics)
>
> $\longrightarrow$ f_U (movement propagation process)

Fig. 4.9 Sketch of the *Movement Ecology Paradigm*: It cross-links four other existing paradigms representing different scientific disciplines, which describe specific aspects of the movements of biological organisms. The aim is to mathematically model the dynamics emerging from the interplay between these different fields by an equation like Eq. (4.7); from [7], *copyright (2008) National Academy of Sciences, U.S.A.*

in the probability distributions that are at the heart of all versions of the LFH. These experimental results suggested that biological organisms may rather perform a complex spatio-temporal adjustment to optimize their search for food sources, which results in different dynamics on different spatio-temporal scales. This is at variance to Lévy motion, which by definition is scale-free.

However, these results are well in line with another, more general approach to understand the movements of biological organisms, called the *Movement Ecology Paradigm* [7]: This theory aims at more properly embedding the movements of biological organisms into their biological context as shown in Fig. 4.9. In this figure, the theory centered around the LFH is rather represented by the region labeled 'random', which focuses on analyzing movement paths only. However, movement paths of organisms cannot properly be understood without embedding them into their biological context: They are to quite some extent determined by the cognitive abilities of the organisms and their biomechanical abilities, see the respective two further regions in this diagram. Indeed, only on this basis the question about optimality may be asked, cf. the fourth region in this diagram, which here is rather understood in a biological sense than as purely mathematical efficiency. Physicists and mathematicians are used to think of diffusive spreading, which underlies foraging, primarily in terms of moving point particles; however, living biological organisms are not point particles but interact with the surrounding world in a very different manner. The aim of this approach is to model the interaction between the four core fields sketched in this diagram by a *state space approach*. This requires to identify relevant variables, cf. the diagram, by establishing functional relationships between them in form of an equation

$$\mathbf{u}_{t+1} = F(\boldsymbol{\Omega}, \boldsymbol{\Phi}, \mathbf{r}_t, \mathbf{w}_t, \mathbf{u}_t),\tag{4.7}$$

where $\mathbf{u}_t$ is the location of an organism at time t. A simple, boiled-down example of such an equation is the Langevin equation Eq. (4.6) that we proposed to describe bumblebee flights under predation threat. Here $du_{t+1}/dt = v_y(t)$ and the potential term is related to the variable r_t above while all the other variables are ignored.

4.6   Conclusions

The discussion about the LFH is still very much ongoing. As an example we refer to research on movements of mussels, where experimental measurements seemed to suggest that Lévy movement accelerates pattern formation [22]; however, see the discussion that emerged about these findings as comments and replies to the above paper, which mirrors our discussion in the previous sections. A second example is the debate about a recent review by Andy Reynolds [24], in which yet another new version of a LFH was suggested; again, see all the respective comments and the authors' reply to them. While these two articles are in support of the LFH, we refer to a recent review by Graham Pyke [23] as an example of a more critical appreciation of it.

We conclude that one needs to be rather careful with following power law hypotheses, or paradigms, for data analysis, here applied to the problem of understanding the search for food by biological organisms. These laws are very attractive because of their simplicity, and because in certain physical situations they represent underlying universalities. While they clearly have their justification in specific settings, these are rather simplistic concepts that ignore many details of the biological situation at hand. This can cause problems when biological processes are more complex. What we have outlined represents not an entirely new scientific lesson; see, e.g., the discussion about power laws in self-organized criticality. On the other hand, the LFH did pioneer a new way of thinking that goes beyond applying simple traditional random walk schemes to understand biological foraging.

Financial support of this research by the MPIPKS Dresden and the Office of Naval Research Global is gratefully acknowledged.

References

1. M. Chupeau, O. Bénichou, R. Voituriez, Nat. Phys. **11**, 844 (2015)
2. R. Klages, Phys. J. **14**, 22 (2015)
3. T. Harris, E. Banigan, D. Christian, C. Konradt, E.T. Wojno, K. Norose, E. Wilson, B. John, W. Weninger, A. Luster, Nature **486**, 545 (2012)
4. G. Ramos-Fernández, J.L. Mateos, O. Miramontes, G. Cocho, H. Larralde, B. Ayala-Orozco, Behav. Ecol. Sociobiol. **55**, 223 (2003)
5. M. Shlesinger, Nature **443**, 281 (2006)
6. L. Stone, *Theory of Optimal Search*, 2nd edn. (Informs, Hanover, MD, 2007)
7. R. Nathan, W.M. Getz, E. Revilla, M. Holyoak, R. Kadmon, D. Saltz, P.E. Smouse, Proc. Natl. Acad. Sci. **105**, 19052 (2008)
8. V. Méndez, D. Campos, F. Bartumeus, *Stochastic Foundations in Movement Ecology* (Springer Series in Synergetics (Springer, Berlin, 2014)
9. G. Viswanathan, V. Afanasyev, S. Buldyrev, E. Murphy, P. Prince, H. Stanley, Nature **381**, 413 (1996)
10. G. Viswanathan, S. Buldyrev, S. Havlin, M. da Luz, E. Raposo, H. Stanley, Nature **401**, 911 (1999)
11. A. Edwards, R. Phillips, N. Watkins, M. Freeman, E. Murphy, V. Afanasyev, S. Buldyrev, M. da Luz, E. Raposo, H. Stanley, G. Viswanathan, Nature **449**, 1044 (2007)
12. D. Sims, E. Southall, N. Humphries, G.C. Hays, C.J.A. Bradshaw, J.W. Pitchford, A. James, M.Z. Ahmed, A.S. Brierley, M.A. Hindell, D. Morritt, M.K. Musyl, D. Righton, E.L.C. Shepard, V.J. Wearmouth, R.P. Wilson, M.J. Witt, J.D. Metcalfe, Nature **451**, 1098 (2008)
13. N. Humphries, N. Queiroz, J. Dyer, N. Pade, M. Musy, K. Schaefer, D. Fuller, J. Brunnschweiler, T. Doyle, J. Houghton, G. Hays, C. Jones, L. Noble, V. Wearmouth, E. Southall, D. Sims, Nature **465**, 1066 (2010)
14. F. Lenz, T.C. Ings, L. Chittka, A.V. Chechkin, R. Klages, Phys. Rev. Lett. **108**, 098103/1 (2012)
15. F. Lenz, A.V. Chechkin, R. Klages, PLoS ONE **8**, e59036 (2013)
16. M. Shlesinger, G. Zaslavsky, J. Klafter, Nature **363**, 31 (1993)
17. G. Viswanathan, M. da Luz, E. Raposo, H. Stanley, *The Physics of Foraging* (Cambridge University Press, Cambridge, 2011)
18. K. Pearson, Biometric ser. **3**, 54 (1906)
19. R. Klages, G. Radons, I. Sokolov (eds.), *Anomalous Transport* (Wiley-VCH, Berlin, 2008)
20. V. Zaburdaev, S. Denisov, J. Klafter, Rev. Mod. Phys. **87**, 483 (2015)
21. M. Buchanan, Nature **453**, 714 (2008)
22. M. de Jager, F.J. Weissing, P.M.J. Herman, B.A. Nolet, J. van de Koppel, Science **332**, 1551 (2011)
23. G. Pyke, Meth. Ecol. Evol. **6**, 1 (2015)
24. A. Reynolds, Phys. Life Rev. **14**, 59 (2015)
25. O. Bénichou, C. Loverdo, M. Moreau, R. Voituriez, Rev. Mod. Phys. **83**, 81 (2011)
26. A. James, J.W. Pitchford, M.J. Plank, Bull. Math. Biol. **72**, 896 (2009)
27. N. Humphries, H. Weimerskirch, N. Queiroz, E. Southall, D. Sims, PNAS **109**, 7169 (2012)
28. R. Metzler, J. Klafter, Phys. Rep. **339**, 1 (2000)
29. J. Klafter, I. Sokolov, *First Steps in Random Walks: From Tools to Applications* (Oxford University Press, Oxford, 2011)
30. R. Metzler, J. Klafter, J. Phys. A: Math. Gen. **37**, R161 (2004)
31. T.C. Ings, L. Chittka, Curr. Biol. **18**, 1520 (2008)
32. R. Klages, *Microscopic Chaos, Fractals and Transport in Nonequilibrium Statistical Mechanics*, vol. 24 (Singapore, Advanced Series in Nonlinear Dynamics (World Scientific, 2007)
33. F. Reif, *Fundamentals of Statistical and Thermal Physics* (McGraw-Hill, Auckland, 1965)

Chapter 5
Epicuticular Wax Formation and Regeneration—A Remarkable Diffusion Phenomenon for Maintaining Surface Integrity and Functionality in Plant Surfaces

Wilfried Konrad, Anita Roth-Nebelsick and Christoph Neinhuis

5.1 Introduction

Diffusion processes are ubiquitous in organisms, varying from being essential short-distance transport phenomena to posing threats, such as uncontrolled leakages of substances. A membrane, consisting of a phospholipid double layer with integrated proteins and other additional functional molecules envelops all cells, the smallest units of life. This membrane represents the device to reconcile the need of protecting the cell interior from its environment while maintaining intracellular conditions with the necessary exchange of substances with the surroundings (see Fig. 5.1). This exchange occurs—apart from processes such as endo-/exocytosis where whole membrane patches are used as transport vehicles—often via "controlled diffusion", involving pores or channels formed by proteins. In this manner, the membrane is semipermeable allowing diffusion of water and uncharged small molecules whereas other substances are hindered from passing through.

Controlled diffusion processes are thus central for managing cell metabolism and—in the end—the metabolism of multicellular plants and animals. Originally, water was the only immediate surroundings of single- and multi-cellular organisms, since there is general agreement that life evolved within the oceans and therefore within an aquatic environment. During evolution, however, life moved on land and was confronted with the problem of a strong humidity gradient, namely the difference between water-saturated cells and tissues and the much drier air [2]. The water

W. Konrad (✉) · C. Neinhuis
Institut für Botanik, TU Dresden, Dresden, Germany
e-mail: wilfried.konrad@tu-dresden.de

C. Neinhuis
e-mail: christoph.neinhuis@tu-dresden.de

A. Roth-Nebelsick
State Museum of Natural History Stuttgart, Stuttgart, Germany
e-mail: anita.rothnebelsick@smns-bw.de

© Springer International Publishing AG 2018
A. Bunde et al. (eds.), *Diffusive Spreading in Nature, Technology and Society*,
https://doi.org/10.1007/978-3-319-67798-9_5

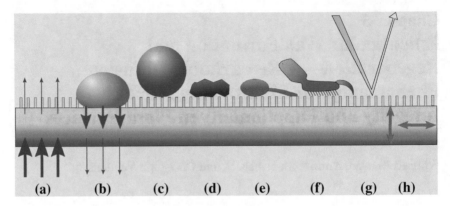

Fig. 5.1 Schematic summary of the most prominent functions of the cuticle as represented by a hydrophobic microstructured plant surface. **a** Transport barrier: limitation of uncontrolled water loss or leaching from the interior and **b** against foliar uptake. **c** Water repellency: control of surface water status. Anti-adhesive, self-cleaning properties: **d** reduction of contamination, **e** pathogen attack and **f** control of attachment and locomotion of insects. **g** Spectral properties: protection against harmful radiation. **h** Mechanical properties: resistance against mechanical stress and maintenance of physiological integrity (modified after [1])

vapor deficit even at high relative humidity is huge and results in desiccation within a short time, lest there are any means to prevent this from happening. In fact, plants and animals are enveloped by desiccation barriers, which hinder water from rapid and uncontrolled loss into the atmosphere. There are, however, some exceptions, notably desiccation-tolerant organisms, such as mosses, some ferns and a few seed plants, which can dry out and recover upon wetting without damage. All other organisms are necessarily equipped with a kind of "skin" preventing rapid desiccation.

To conserve water by suppressing water vapor diffusion into the atmosphere, the envelope has to be hydrophobic. Terrestrial plants developed a hydrophobic layer covering the outermost cells called epidermis [3]. This hydrophobic layer consists of two main components, the polymer cutin and soluble waxes, described further below, and is termed "cuticle" [4, 5] (see Figs. 5.1 and 5.2).

Due to its key importance for maintaining the hydrated state, the cuticle evolved during early stages of land plant evolution [7]. Cuticles are already present in 400 million year old plant fossils from the Lower Devonian, a time during which the vegetation consisted of quite small and leafless axes (Fig. 5.3). In fact, cuticle-like remains can be found in much older fossil material, dating back to the earliest times of land plant evolution from which only microfossil fragments are preserved [8].

Also terrestrial animals need a protective cover against uncontrolled evaporation. In this respect, arthropods are interesting since they show a hydrophobic cover similar in many aspects to the plant cuticle. This is particularly the case for insects, since both plants and insects exchange gases with the atmosphere via their body surface. Land plants cannot be completely isolated from the atmosphere since they have to absorb CO_2 for photosynthesis and to evaporate water to maintain internal

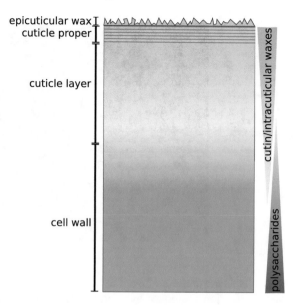

Fig. 5.2 Simplified scheme of the structural features of the plant cuticle and their major components (modified after [6])

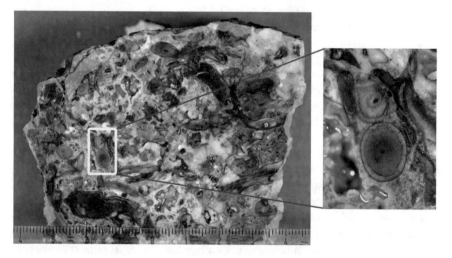

Fig. 5.3 A section through a piece of Rhynie Chert sediment (Scotland, near Aberdeen), containing axes of early land plants from the Lower Devonian, approximately 400 million years old. On the right, two plant axes in cross-section are shown in detail. They belong to so-called "rhyniophytic" plants thriving during that time on land. These plants were up to 20 cm high and consisted of leafless axes already covered by a cuticle

Fig. 5.4 A scanning
electron microscope image
of the lower leaf surface of
Helleborus niger, showing
stomata (bar: 100 μm).
These micropores can be
closed to control the
exchange of CO_2 and water
vapor between atmosphere
and plant interior

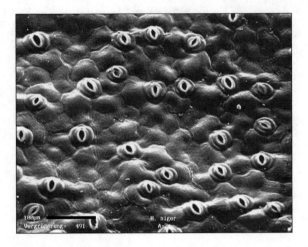

transport processes. Plants are therefore forced to allow for a limited and controlled gas exchange and transpiration, facilitated by micropores, called stomata, which can be closed, according to the environmental conditions (Fig. 5.4). How micropore-based technologies in chemical engineering have, quite generally, been inspired by nature is referred to in great detail in Chap. 11.

The cuticle thus reconciles two conflicting tasks, namely suppression of outward diffusion of water vapor and uptake of CO_2. The solution is to pierce the isolating cover, the cuticle, with pores whose aperture can be regulated.

Terrestrial insects are very much under the same constraint, and consequently also developed a hydrophobic cuticle, which simultaneously serves as gateway for respiration by being equipped with openings, the spiracles, leading to an internal tubing system, the tracheae. In fact, plants and terrestrial arthropods share many similarities with respect to evolutionary solutions against desiccation [9].

Conspicuous for both groups is the occurrence of wax blooms deposited upon the cuticles [1, 10, 11] (see Fig. 5.5). For both, essential functional roles are indicated.

For different plant species, cuticles can show quite different thicknesses, with plants from arid environments often showing considerably thicker cuticles than plants from humid habitats. The reason for that is not fully understood since the suppression of water vapor loss appears to be not dependent on cuticle thickness but on its chemical composition [12]. A thick cuticle can also contribute to mechanical stabilization [13] whereas wax crystals on plant cuticles are often associated with the famous Lotus effect, forming structured hydrophobic surfaces resulting in vigorous water-repellency (contact angle $\gtrsim 150°$) [14].

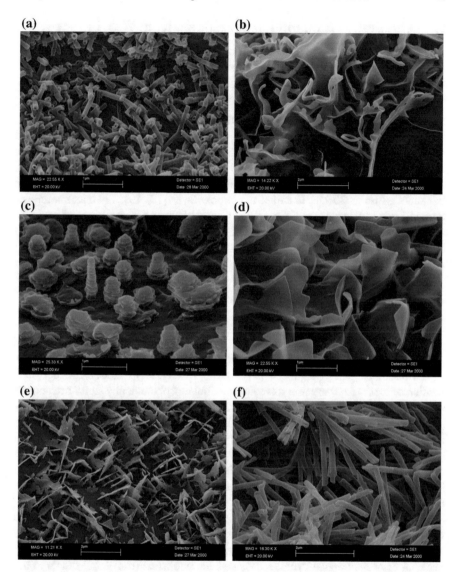

Fig. 5.5 Scanning electron microscopic images of epicuticular waxes from plants. **a** Nonacosanol-based tubules (bar: 1 μm). **b** Irregularly shaped wax crystals (bar: 2 μm). **c** Transversely rigid rodlets based on palmitone (bar: 1 μm). **d** Membraneous platelets (bar: 1 μm). **e** Irregularly shaped platelets (bar: 2 μm). **f** Tubules based on β-diketones (bar: 2 μm). Photographs: Institut für Botanik, TU Dresden

5.2   Qualitative Explanation of Self-repairing Wax Layers

5.2.1   Chemical Properties of the Cuticle

The cuticle may be regarded as a natural composite comprising two major hydropho-bic components: an insoluble polymer fraction composed of cutin and, in some species, cutan as well as soluble lipids of diverse chemistry, collectively called waxes. In addition, a certain amount of polysaccharides is present (overview in [15, 16]). The outer very thin region (usually less than 100 nm), called cuticle proper, contributes for 99% of the barrier efficiency [17], while the region determining the thickness of up to 20 μm, is called the cuticle layer [18, 19]. Chemical composi-tion and internal structure of the cuticle seems to show a high degree of variabil-ity during ontogeny and among different plant species and organs. Whereas intra-cuticular waxes may be either amorphous or crystalline, epicuticular waxes (Fig. 5.5) are assumed to be of crystalline nature [20–22].

5.2.2   Wax Transport and Cuticle Self-repair

The crystal nature of epicuticular waxes implies self-assembly as the driving force for the formation of such structures. This has been proven after extraction and recrystal-lization of waxes from organic solvents revealing morphologically similar structures as compared to the plant surface [20, 21, 23–29]. To allow selfassembly of com-plex three-dimensional structures, the individual molecules must be mobile within a suitable matrix or solvent in which they are free to find an energetically favorable position, which also includes phase separation of different components or component classes found in wax mixtures. Recrystallisation of extracted waxes from a solution is considerably influenced by temperature, chemical nature of the solvent and the underlying substrate resulting in a large structural variability [20, 22].

The most intriguing problem, however, was the process of wax deposition onto the surface, as the molecules have to move from inside the cell through a hydrophilic cell wall and the hydrophobic cuticle and finally onto the ridges and edges of the growing crystals. Several hypotheses have been published from ectodesmata to the involvement of transport proteins [30–32]. One obvious hypothesis is the existence of some kind of channels or pathways, but no evidence of trans-cuticular structures that could serve as pathways for wax molecules have yet been found in the plant cuticle by Scanning Electron Microscopy (SEM), Transmission Electron Microscopy (TEM) or Atomic Force Microscopy (AFM) investigations [28, 33].

While studying the various phenomena related to water-repellency and self-cleaning of natural and artificial surfaces, one particular interest was the ability of plants to reestablish these properties after damaging of the surface. Wax crystals are weak structures very susceptible to mechanical influences and therefore easily altered or completely removed. Since plants are able to maintain the functionality of

Fig. 5.6 Exposed cuticular surface after applying and peeling off the glue prior to regeneration of wax layers

the surface over quite a considerable period of time, the question turned up whether at all and if, how quick and to which extent wax layers and structures are re-established. To address this question, we performed a number of experiments with different plant species, from bud break to senescence, i.e. when the leaves are shed. The experimental setup was rather simple: during one vegetation period we applied glue, not containing organic solvents to the leaves, let it dry and peeled it off (Fig. 5.6). Subsequently samples were taken on a regular basis to check for regeneration of wax layers and crystals.

Generally, a wide range of species was able to regenerate waxes after removal within a few days or up to 2 weeks. Many species could achieve that only in young stages of leaf development, while others maintained this capability during the whole lifespan. Only very few species showed wax regeneration confined to later stages of development. Interestingly the regeneration was confined to the area from which the wax layer has been removed, independently of cell borders, meaning that the reestablishment of a wax cover happened within the area of one cell (Fig. 5.7).

During these experiments we faced the problem that very young and delicate leaves were destroyed during the attempt to peel of the glue. So we waited for the leaves to expand expecting that the glue would fall of by itself. However, since the material was highly elastic, the polymer film expanded together with the expanding leaf without being dropped. So we waited even longer expecting that the glue would be separated by the emerging wax cover that should act like a separating agent.

However, to our biggest surprise, this did not happen as well but now the wax cover emerged on the surface of the glue (Fig. 5.8).

This accidental and unexpected result of our experiment allowed only one conclusion: the transport of wax molecules must be independent of the living cell, since no transporters, channels or other cellular components can be involved in the movement through the polymer (i.e. the glue).

As a consequence we isolated the cuticle enzymatically to remove every component of the living protoplast, covered it with a pure polyurethane film and span the resulting specimen over a diffusion chamber filled with water (Fig. 5.9).

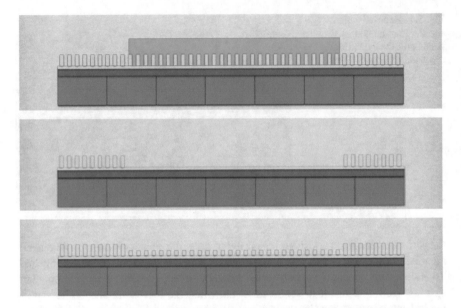

Fig. 5.7 Upper: Glue applied to the leaf surface. Centre: Leaf surface after removing wax by peeling off the glue. Lower: Wax regeneration occurring in the areas only where wax has been removed, independently of cell borders

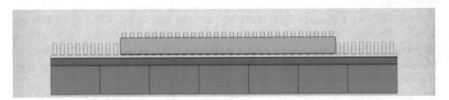

Fig. 5.8 Wax regeneration—the observation on young leaves. Waxes move through the glue and crystallize similar in size and distribution as on the leaf surface

The result was basically the same as were observed on leaves. The wax moved through the polymer membrane following the transpiration gradient built up by evaporating water. The structures formed on the polymer surface again were virtually identical with those found on leaves in situ. In a final approach we also removed the cuticle and replaced the latter by a polymer membrane alone that was applied to a filter paper, replacing the cell wall. The latter was loaded with wax and the sandwich again placed on top of a diffusion chamber (Fig. 5.10).

The experiment again revealed the same result. Waxes moved through the polymer and crystallized on the surface. These results were independent of the type of polymer (e.g. PU, PP, PE, PC) or the used wax (e.g. plant waxes, montan waxes, artificial waxes). In case of plant waxes the size, individual morphology and distribution was virtually the same as on the leaf surface (Fig. 5.11).

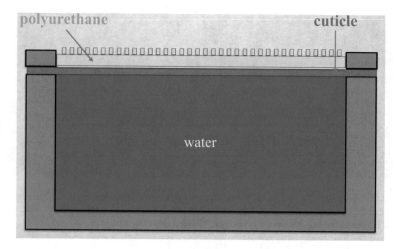

Fig. 5.9 In vitro experiments with isolated cuticles in a diffusion chamber

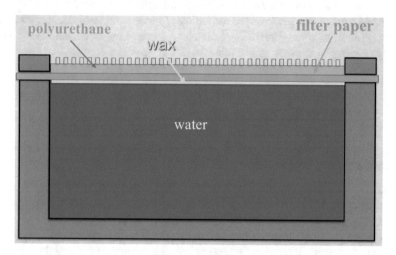

Fig. 5.10 Completely artificial approach in which a filter paper was loaded with wax of different origin that moved through a polyurethane membrane together with water

Neinhuis et al. [34] consequently proposed a co-transport of wax components with water that constantly is lost via the cuticle, although in very small amounts. Assuming such a process is appealing since no pathways, carrier molecules or sensors are needed. Since cuticular waxes are the main permeability barriers, the transport to the outside slows down while more wax is deposited on the surface, so it is self-regulating. In addition it easily explains the intriguing phenomenon of wax regeneration. Since removal of epicuticular wax also partly removes the water barrier, more wax is able to move through the cuticle in this particular spot and builds up a new layer without affecting neighbouring area. AFM in situ demonstrated the rather quick

Fig. 5.11 Waxes tubules, based on β-dicetones recrystallized after movement through artificial polyurethane membranes. Size and distribution are not distinguishable from plant surfaces (left), while a higher density is achieved by a longer diffusion time (right)

reassembly of new wax layers after their removal under environmental conditions in vivo. AFM time-series pictured the formation of mono- and bi-molecular wax films and the growth of three-dimensional platelets, either directly on the cuticle or on already existing wax layers within minutes [28, 35].

5.2.3   Summary of Sect. 5.2

The qualitative explanation of cuticle self-repair and wax transport to the plant surface can be summarized as follows:

- Intact cuticles are very efficient barriers against evaporation of water from the plant interior. Hence, if the wax layer is degraded, evaporation from this zone increases, generating a current of liquid water from the plant interior.
- This water current transports the wax molecules from the epidermal cells (where they are presumably produced) towards the outer fringe of the cuticle. There the water evaporates. Being much heavier than the water molecules, the wax molecules do not evaporate, they rather form wax crystals rebuilding hereby the damaged cuticle layer by layer.
- As this repair process proceeds, both evaporation and the evaporation driven water current decrease and smaller amounts of wax molecules are transported to the damaged cuticle. Finally, the cuticle attains its original thickness and the repair process comes to a halt.

The advantages of this self-regulating model over other hypotheses are:

- Neither distinct pathways (such as micro-channels or ectodesmata) nor the existence of lipid transfer proteins have to be postulated. The waxes move through the cuticle due to the presence of the water flow, hence neither organic solvents nor special receptors are necessary.

- It explains almost all phenomena we have observed, including some which were hitherto hard to explain, such as the wax regeneration and the appearance and the distribution patterns of epicuticular wax in distinct leaf areas.

5.3 Quantitive Explanation of Self-repairing Wax Layers

In this section we present a condensed version of a quantitive model of cuticle repair, i.e. notably of the movement of the wax molecules deployed for this purpose. It emerged from the qualitative scenario outlined in the previous section. A detailed account can be found in [36].

5.3.1 Equation of Mass Transfer Through the Cuticle

We make use of a few assumptions which keep the mathematics manageable, hereby providing insight into the model structure: We employ the porous medium approximation, allowing to restrict the mathematics to one dimension (the z-direction in Fig. 5.12), thus largely following the introductory remarks on diffusive movement in Chap. 2, Eqs. (2.6)–(2.11); we assume that the properties of the biological structures along the z-axis are (approximately) constant within each of the four different layers depicted in Fig. 5.12; and we assume stationary conditions, that is, none of the transport processes involved depends explicitly on time.

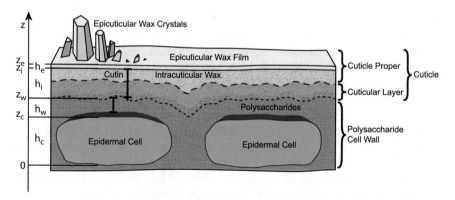

Fig. 5.12 Plant cuticle structure. Schematic diagram highlighting the major structural features of the cuticle and underlying epidermal cell layer. h_e, h_i, h_w and h_c denote the thicknesses of the various layers, z_e, z_i, z_w and z_c the z-coordinates of their outer fringes. (Typical) numerical values of these quantities are given in Table 5.1. (Not drawn to scale, modified after [3]). For photographs of epicuticular waxes see Fig. 5.5

Table 5.1 List of variables and numerical values. Subscripts c, w, i, e refer to the different structural layers depicted in Fig. 5.12. Numerical data for diffusion coefficients and thicknesses of cutin layer and wax film layer are partly based on Tables 2 and 3 in [37] for cultivar "Elstar" and partly derived by educated guessing. Similarly, the value of K_c is based on [38]. The diffusion coefficient of the polysaccharide layer has been set arbitrarily to one tenth of the diffusion coefficient of the cutin layer

Quantity	Value	Quantity	Value
h_c	16 μm	R	8.314 J/mol/K
h_w	0.5 μm	T	20 °C
h_i	11.93 μm	g	9.81 m/s^2
h_e	4.14 μm	V_w	18.07 × 10^{-6} m^3/mol
S_c	4.33 × 10^{-12} m^2/s	V_{wax}	404 × 10^{-6} m^3/mol
S_w	7.16 × 10^{-11} m^2/s	ρ_w	18.07 × 10^{-6} m^3/mol
S_i	7.16 × 10^{-10} m^2/s	w_{rel}	0.6
S_e	3.03 × 10^{-10} m^2/s	ψ_{leaf}	−204 m
K_c	1 × 10^{-14} m/s	ξ	4 /s
K_w	1.69 × 10^{-15} m/s	c_s	10 mol/m^3
K_i	1.69 × 10^{-14} m/s	c_t	5.53 mol/m^3
K_e	7.18 × 10^{-15} m/s		

In the framework described in Sect. 5.2, the wax molecules are transported from the places where they are formed (presumably the epidermal cells depicted in Fig. 5.12) to the epicuticular wax layer where they are deployed by "swimming" passively in the midst of (liquid) water molecules. These vaporize at the plant surface into the atmosphere, causing hereby the flow of the liquid water molecules which is ultimately fed by soil water ascending through the plant's vascular water system. In addition to "swimming" with the flux of water, wax molecules are also subject to the transport mechanism emerging from their Brownian motion, i.e. to a diffusive flux in the direction of decreasing wax concentrations (see Fig. 2.2a). With Fig. 5.12, wax concentration is easily understood to assume its maximum value, namely its saturation concentration c_s, at $z = z_e$, given the immediate vicinity of the epicuticular wax crystals. Wax concentration is thus expected to decrease from the leaf surface into its interior, giving rise to a diffusive flux just opposite to the flux of the water molecules. As it turned out, both transport mechanisms are equally important and indispensable in order to formulate a coherent mathematical model.

Considering both transport mechanisms, the flux $j(z)$ of wax molecules of concentration $c(z)$ is given by the expression (see e.g. [39, 40])

$$j = -S\frac{dc}{dz} + cJ. \tag{5.1}$$

The first term on the right hand side describes diffusion. $S = Dn/\tau$ denotes the (effective) diffusion coefficient in a porous medium, n and τ are porosity and tortuosity

obtained by multiplying pressure head (units m) by $\rho_w g \approx 9.81 \times 10^3 \, \text{kg/m}^2/\text{s}^2$, as can be seen from (5.6)).

The water flux equation is derived from the continuity equation for liquid water which reduces due to our assumptions to $0 = dJ/dz$. Insertion of (5.5)—while keeping in mind the assumption that $K(z)$ is constant within each layer—yields the differential equation

$$0 = \frac{d^2\psi}{dz^2}. \tag{5.7}$$

Similarly as in the case of (5.4), this equation has to be solved separately for each layer. Each of the four solutions of equation (5.7) contains two arbitrary constants. These are determined from the condition of continuity for the water potential $\psi(z)$ and the water flux $J(z)$ at the layer margins at z_c, z_w and z_i and from two boundary conditions for $\psi(z)$: We require $\psi(0) = \psi_{leaf}$ and $\psi(z_e) = \psi_{wv}$ with ψ_{wv} as given in (5.6).

Application of this procedure is straightforward. It results, however, in lengthy expressions for $\psi(z)$; since we do not need them in what follows we give here only what results if we insert $\psi(z)$ into expression (5.5) for $J(z)$:

$$J = \frac{\psi_{leaf} - \psi_{wv}}{\dfrac{h_c}{K_c} + \dfrac{h_w}{K_w} + \dfrac{h_i}{K_i} + \dfrac{h}{K_e}}. \tag{5.8}$$

h_e, h_i, h_w and h_c denote the thicknesses of the various layers, as indicated in Fig. 5.12, and K_e, K_i, K_w and K_c are the respective water conductivities. $J > 0$ indicates a water flux towards positive z-values, i.e. towards the plant surface. In what follows, h_e denotes the thickness of the intact epicuticular wax film while h denotes its actual thickness during any stage of the repair process (thus, $0 \leq h \leq h_e$).

Several features of expression (5.8) are noteworthy:

- It shows a close analogy to Ohm's law in electrodynamics: if water flux J is identified with electric current and water potential difference $\psi_{leaf} - \psi_{wv}$ (the driving force of water flux) with voltage, then the four terms in the denominator of the right hand side of (5.8) represent four resistances connected in series.
- J is independent of z, simplifying the solution of the differential equation (5.4) for the wax flux j considerably. (This property was to be expected from the physics of the situation: no water sources or sinks are present).
- The water flux J depends roughly reciprocally on the thickness h of the epicuticular wax film. This corroborates the qualitative conception developed in Sect. 5.2: the water flux decreases while the repair process proceeds (i.e. $h \rightarrow h_e$) and the wax layer regains its original thickness h_e.

5.3.3 Solution of the Transport Equation

By inserting the expression (5.5) determined for the water flux J into Eq. (5.4), we are now able to determine the concentration $c(z)$ of the wax molecules and, subsequently, by inserting into Eq. (5.1), their flux j. Solving a differential equation of second order in four adjacent layers gives rise to eight integration constants which may be determined by taking account of the respective boundary conditions, ending up in a number of quite lengthy expressions. All these relations and how they have been determined may be found in [36]. Here we confine ourselves to the graphical representation of the solution in three characteristic situations as resulting with the parameters summarized in Table 5.1

Figure 5.13 displays the wax concentration $c(z)$ and the wax flux $j(z)$ along the pathway of wax molecules between epidermal cell and epicuticular wax film (cf. Fig. 5.12) and the wax production rate $q = \xi \left[c_t - c(z) \right]$ within the epidermal cells as resulting as a solution of equations (5.4) and (5.1).

Subfigure (c) shows the (net) wax flux $j(z)$. It is the sum of the diffusive component (represented by the first term in expression (5.1)) and of the advective component (the second term in (5.1)). These two are displayed in subfigure (d); the upper three curves represent advective components, cJ, the lower three curves depict the diffusional parts, $-S\, dc/dz$. Positive fluxes are directed towards the cuticle, negative fluxes point to the leaf interior. Blue curves are related to a damaged cuticle (the outer fringe is located at $z = z_i$), green curves represent an intact cuticle (the outer fringe is at $z = z_e$), and red curves represent the fictitious case of an epicuticular wax film which is twice as thick as it ought to be (the outer fringe is at $z = z_e + h_e$).

Comparison between the blue and green curves allows to visualise the repair scenario:

- As long as the cuticle is undamaged, the green curves in subfigures (a), (c) and (d) terminate at $z = z_e$, and the green curves representing advection and diffusion (subfigure (d)) have for all points with $z > z_w$ the same distance to the z-axis, thus adding up to a vanishing net wax flux (green curve in subfigure (c)).
- When the cuticle is damaged the repair process begins. This is illustrated by the blue curves which terminate at $z = z_i$: the absolute values of both advection and diffusion flux have increased, compared to the intact cuticle (see subfigure (d)), but now results a net flux towards the cuticle (see subfigure (c)).
- During cuticle regrowth all blue curves "migrate" towards the green curves, that is, the absolute values of advection and diffusion flux decrease and converge slowly until they have merged with the green curves; then the net flux ceases and the repair process is completed.

Notice, that the model predicts also what happens to (fictitious) protrusions of height $h > h_e$, extending from the epicuticular wax film: This case is represented by the red curves. The one representing the net flux (subfigure (c)) runs for $z > z_w$ below the z-axis, indicating a negative net flux directed towards the plant interior; this means that the protrusions are dissolved and transported to the leaf interior. This process

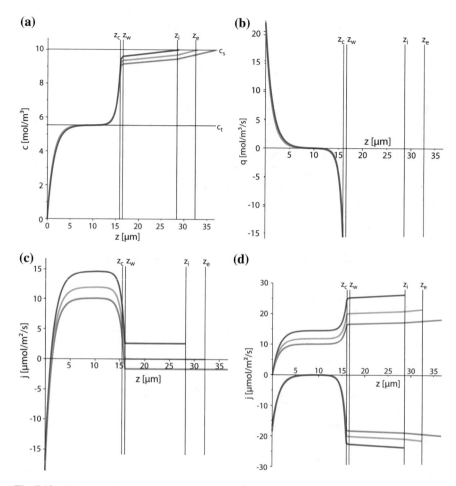

Fig. 5.13 Wax concentration (**a**) and wax fluxes (**c, d**) along the pathway of wax molecules between epidermal cell and epicuticular wax film (cf. Fig. 5.12). The (net) wax flux $j(z)$ in subfigure (**c**) is the sum of the diffusive component (lower three curves in subfigure (**d**)) and of the advective component (upper three curves in subfigure (**d**)). Positive fluxes are directed towards the cuticle, negative fluxes point to the leaf interior. (For detailed explanation see text.) Vertical lines delineate the tissue layers defined in Fig. 5.12; the horizontal lines in subfigure (**a**) denoted c_s and c_t mark the saturation and the threshold wax concentrations. Subfigure (**b**) depicts the wax insertion (or removal) rate $q = \xi \left[c_t - c(z) \right]$ within the epidermal cells. Positive values indicate insertion, negative values indicate removal of wax molecules. Notice that the graph depicts three nearly identical curves. Blue curves are related to a damaged cuticle (the outer fringe is located at $z = z_i$), green curves represent an intact cuticle (the outer fringe is at $z = z_e$), and red curves represent the fictitious case of an epicuticular wax film which is twice as thick as it ought to be. Numerical values are as in Table 5.1

stops when the cuticle has been eroded to thickness h_e and the red curve has migrated to and merged with the green curve.

Comparison of subfigures (b) and (c) of Fig. 5.13 illustrates the continuity equation (5.3) which states that the gradient of the net wax flux equals the injection (or removal) of wax molecules: The region $0 < z \lesssim 8 \, \mu m$ acts as a wax source (indicated by $q > 0$). Wax molecules that are generated in the region $z \lesssim 2 \, \mu m$ flow towards the plant interior (indicated by $j < 0$, subfigure (c)), those produced in the interval $2 \, \mu m \lesssim z \lesssim 8 \, \mu m$ flow a short distance towards the cuticle (indicated by $j > 0$). In the case of an intact cuticle (green curves), all of them are removed from the cell liquid in the region $8 \, \mu m \lesssim z < z_c$ which acts as wax sink ($q < 0$). If the cuticle is damaged (blue curves), however, a certain fraction of the injected wax molecules reaches and repairs the cuticle.

5.3.4 Restoration of the Wax Layer as a Function of Time

Provided the restoration proceeds slowly, compared to the travel time τ of a wax molecule between epidermal cell and epicuticular wax layer, the solution of the wax transport equation (5.1) can be exploited to derive the temporal development of the wax layer repair, although it has been derived under the assumption of stationarity. The values given in Table 5.1 imply for the velocity of the water current $J \approx 2.17 \, \mu m/s$ and thus $\tau = z_e/J \approx 15 \, s$ for the travel time of a wax particle between epidermal cell and cuticle. Hence, if the repair process lasts perhaps 1 h, this approach is certainly justified.

In order to calculate the temporal development of the epicuticular wax layer restoration we assume that it has been eroded completely before the restoration process begins. That is, at the starting point $t = 0$ of the restoration process the outer fringe of the cuticle is located at $z = z_i$, equivalent to $h = 0$ (h denotes the actual thickness of the wax layer, h_e its thickness when it is intact, cf. Fig. 5.12).

The water brought there by the water flux J evaporates from the eroded area, leaving behind the much heavier wax molecules that came by the wax flux j. The wax molecules organize themselves as crystals, thus restoring the wax layer until it reaches its original thickness h_e whereupon the wax flux j ceases.

If V_{wax} denotes the molar volume of the wax molecules, the thickness h of the wax layer regrows with the velocity $dh/dt = V_{wax} j(h)$. In view of the structure of the expressions for $j(z)$ and $J(z)$ (cf. (5.8)), this is an ordinary but non-linear differential equation for $h(t)$.

Its non-linearity precludes a straightforward solution but an approximation approach (for details see [36]) allows to calculate the thickness of the wax layer h as a function of time, resulting in

$$h(t) = h_e \left[1 - e^{\left(j_1 V_{wax} t \right)} \right], \tag{5.9}$$

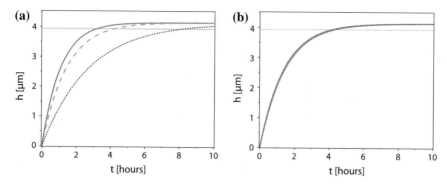

Fig. 5.14 Growth of the wax layer with time according to expression (5.9). The intersections with the grey, horizontal line indicate the time it takes to rebuild the wax layer to 95% of its original thickness of $h_e = 4.14\,\mu m$. **a** Temperature is kept constant at $T = 293\,K = 20\,°C$ while the relative atmospheric humidity w_{rel} and the threshold concentration c_t of wax molecules in epidermal cell assume the values $(w_{rel}, c_t) = (0.8, 7.78\,mol/m^3)$ (blue, dotted line), $(w_{rel}, c_t) = (0.6, 5.53\,mol/m^3)$ (green, broken line) and $(w_{rel}, c_t) = (0.4, 3.48\,mol/m^3)$ (red, continuous line). The related time spans are $t_{95} = 8.13\,h$ (blue line), $t_{95} = 4.20\,h$ (green line) and $t_{95} = 3.19\,h$ (red line). **b** $w_{rel} = 0.6$ is kept constant, T and c_t assume the values $(T, c_t) = (30\,°C, 5.65\,mol/m^3)$ (blue line), $(T, c_t) = (20\,°C, 5.53\,mol/m^3)$ (green line) and $(T, c_t) = (10\,C, 5.42\,mol/m^3)$ (red line). The three curves are nearly indistinguishable; their common t_{95} time amounts to $t_{95} = 4.20\,h$. Other numerical values are as in Table 5.1. t_{95} is defined in the text

with $j_1 := (\partial j / \partial h)|_{h=h_e}$. Notice the implication $h(0) = 0$, that is, the cuticle layer started to (re-)grow at time $t = 0$. Its original thickness h_e approaches the wax crystal layer asymptotically, as $t \to \infty$. Thus, the repair process lasts—in principle— infinitely long; the time which is necessary to rebuild for instance 95% of the layer is, however, finite and amounts to the value $t_{95} := \ln(20)/(-j_1 V_{wax})$.

Figure 5.14 illustrates the result (5.9) for two different cases:

- In subfigure (a), temperature is kept constant and the relative atmospheric humidity w_{rel} adopts three different values. The time spans t_{95} increase with increasing w_{rel}: this is to be expected because the water potential difference $\left| \psi_{leaf} - \psi_{wv} \right|$ which is the driving force of evaporation decreases if w_{rel} is increased, according to (5.6). Accordingly, the wax supply for restoration decelerates.
- In subfigure (b), relative atmospheric humidity is kept constant and temperature is varied ($T = 10, 20$ and $30\,°C$). The related curves are nearly indistinguishable.

5.4 Conclusions

The model presented here corroborates, extends and quantifies the conjecture of Neinhuis et al. [34] which explains almost all phenomena observed in connection with cuticle repair.

They proposed the co-transport of wax components with water which relies on comparatively simple physics instead of postulating sophisticated living structures such as carrier molecules or specialised pathways for wax molecules; they were also able to confirm their hypothesis qualitatively by carrying out experiments with isolated cuticles and artificial membranes.

Adding diffusion as a transport mechanism which counteracts water transport of the wax components leads to a further clarification of the observations: the presence of two antagonistic transport mechanisms allows for the scenario that the two driving forces are balanced in the case of intact cuticles and that a damaged cuticle causes an imbalance resulting in net wax transport and cuticle self-repair which lasts until balance is readjusted.

The model explains these findings in detail: its mathematical structure allows, for instance, to conclude that the thickness of the epicuticular wax layer and the typical restoration time after degradation (which are the result of two physical processes that are independent of living structures) are nonetheless controlled by living structures, namely the epidermal cells which generate the wax molecules. In the framework of the model the cells have two degrees of freedom at their disposal to regulate the wax production: they can predefine both the thickness h_e and the restoration time t_{95} of the epicuticular wax layer by fine-tuning the parameters c_t and ξ of expression (5.2).

References

1. H. Bargel, K. Koch, Z. Cerman, C. Neinhuis, Funct. Plant Biol. **33**, 893 (2006)
2. J.A. Raven, D. Edwards, *The Evolution of Plant Physiology*, chap. Physiological Evolution of Lower Embryophytes: Adaptations to the Terrestrial Environment (Academic Press, 2004), pp. 17–41
3. T.H. Yeats, J.K. Rose, Plant Physiol. **163**, 5 (2013)
4. J. Schönherr, F. Kerler, M. Riederer, Dev. Plant Biol. **9**, 491 (1984)
5. M. Riederer, L. Schreiber, J. Exp. Bot. **52**, 2023 (2001)
6. P. Holloway, *Pesticide Science* (1993)
7. J.A. Raven, Bot. J. Linn. Soc. **88**, 105 (1984)
8. C.H. Wellman, J. Gray, Philos. Trans. R. Soc. Lond. B: Biol. Sci. **355**, 717 (2000)
9. J. Raven, Philos. Trans. R. Soc. Lond. B: Biol. Sci. **309**, 273 (1985)
10. E. McClain, M.K. Seely, N.F. Hadley, V. Gray, Ecology **66**, 112 (1985)
11. S. Naidu, J. Insect Physiol. **47**, 1429 (2001)
12. L. Schreiber, T. Kirsch, M. Riederer, *Plant Cuticles—An Integrated Functional Approach*, chap. Diffusion Through Cuticles: Principles and Models (Bios: Oxford, England, 1996), pp. 109–118
13. H. Bargel, C. Neinhuis, J. Exp. Bot. **56**, 1049 (2005)
14. W. Barthlott, C. Neinhuis, Planta **202**, 1 (1997)
15. C.E. Jeffree, *The Cuticle, Epicuticular Waxes and Trichomes of Plants, with Reference to their Structure, Functions and Evolution* (Edward Arnold, London, 1986), pp. 23–63
16. P. Kolattukudy, *Polyesters in Higher Plants* (Springer, Berlin, 2001), pp. 4–49
17. M. Riederer, L. Schreiber, *Waxes-The Transport Barriers of Plant Cuticles*, vol. 6 (The Oily Press, Dundee, Scotland, 1995), pp. 131–156
18. P. Holloway, *Plant Cuticles: Physicochemical Characteristics and Biosynthesis, NATO ASI Series* (Springer, Berlin, 1994), pp. 1–13

19. C.E. Jeffree, *The Fine Structure of the Plant Cuticle*, vol. 23 of *Annual Plant Reviews* (Blackwell, 2006)
20. K. Koch et al., Planta **223**, 258 (2005)
21. H.J. Ensikat, B. Boese, W. Mader, W. Barthlott, K. Koch, Chem. Phys. Lipids **144**, 45 (2006)
22. K. Koch, A. Domisse, W. Barthlott, Cryst. Growth Des. **6**, 2571 (2006)
23. C. Jeffree, E. Baker, P. Holloway, New Phytol. **75**, 539 (1975)
24. R. Jetter, M. Riederer, Planta **195**, 257 (1994)
25. R. Jetter, M. Riederer, Bot. Acta **108**, 111 (1995)
26. I. Meusel, C. Neinhuis, C. Markstadter, W. Barthlott, Can. J. Bot. Revue Can. De Bot. **77**, 706 (1999)
27. I. Meusel, W. Barthlott, H. Kutzke, B. Barbier, Powder Diffr. **15**, 123 (2000)
28. K. Koch, C. Neinhuis, H.J. Ensikat, W. Barthlott, J. Exp. Bot. **55**, 711 (2004)
29. K. Koch, A. Dommisse, A. Niemietz, W. Barthlott, K. Wandelt, Surf. Sci. **603**, 1961 (2009)
30. W. Franke, Pestic. Sci. **1**, 164 (1970)
31. L. Schreiber, Ann. Bot. **95**, 1069 (2005)
32. L. Samuels, R. Jetter, L. Kunst, Plant Biosyst. **139**, 65 (2005)
33. D. Canet, R. Rohr, A. Chamel, F. Guillain, New Phytol. **134**, 571 (1996)
34. C. Neinhuis, K. Koch, W. Barthlott, Planta **213**, 427 (2001)
35. K. Koch, A. Dommisse, C. Neinhuis, W. Barthlott, *Self-assembly of Epicuticular Waxes on Living Plant Surfaces by Atomic Force Microscopy* (American Institute of Physics, Melville (NY, USA), 2003), pp. 457–460
36. W. Konrad, C. Neinhuis, A. Roth-Nebelsick, *diffusion-fundamentals.org* **25**, 1 (2016)
37. E.A. Veraverbeke, P. Verboven, N. Scheerlinck, M.L. Hoang, B.M. Nicolaï, J. Food Eng. **58**, 285 (2003)
38. L. Schreiber, M. Riederer, Plant. Cell Environ. **19**, 1075 (1996)
39. P.A. Domenico, F.W. Schwartz, *Physical and Chemical Hydrogeology* (Wiley, New York, 1998)
40. R.A. Freeze, J.A. Cherry, *Groundwater* (Prentice-Hall, 1977)
41. P.S. Nobel, *Physicochemical and Environmental Plant Physiology*, 3rd edn. (Elsevier Academic Press, Amsterdam, 2005)

Chapter 6
Brain Interstitial Structure Revealed Through Diffusive Spread of Molecules

Charles Nicholson

6.1 Introduction

Freshly removed from the skull, the human brain looks like a cauliflower with the consistency of a crème caramel flan (Fig. 6.1a). This mundane object conceals the most complex structure ever discovered. When the interior of the brain is examined with a resolution of a few micrometers, using appropriate staining techniques and light microscopy, it is seen to be composed of a vast number of cells (Fig. 6.1b), although the apparent space between the cells in this illustration is misleading. The full complexity of the ensemble of cells is only finally revealed at the submicron level using electron microscopy (Fig. 6.1c). This chapter will show how the application of methods and models based on diffusion can lead to an understanding of how brain cells pack together and some of the remarkable properties of the narrow spaces that separate them.

Brain cells comprise two types: neurons and glia; some examples of neurons are shown in Fig. 6.1b. The neurons form vast networks that convey and process electrical signals. The connections in the network mostly take place at junctions called synapses that provide a separation of about 20 nm. At these junctions the electrical signals are converted into packets of chemicals that diffuse across the narrow space. The diffusion properties of the synapse could be the basis for the whole chapter but will not be discussed further, except to note that this physical separation of neurons ensures the structural isolation of each cell.

Glia cells lack electrical signaling capability and are something of an enigma. They appear to support neurons in several ways and new properties are being revealed at a rapid pace. Neurons and glia are intermingled in the brain; they are present in roughly equal numbers and both cell types have many different forms in

C. Nicholson (✉)
Department of Neuroscience and Physiology, New York University
Langone Medical Center, 550 First Avenue, New York, NY 10016, USA
e-mail: charles.nicholson@nyu.edu

© Springer International Publishing AG 2018
A. Bunde et al. (eds.), *Diffusive Spreading in Nature, Technology and Society*,
https://doi.org/10.1007/978-3-319-67798-9_6

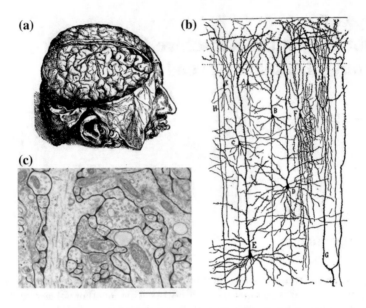

Fig. 6.1 a The human cortex exposed in a drawing by Versalius published in his book De Fabrica in 1543. **b** Nerve cells (neurons) of the cortex taken from layers 1–3 of the precentral gyrus of a 1-month-old human infant. The cells and their extensions have dimensions that range from about 50–300 μm. Cells stained with the Golgi method and drawn by S. Ramón y Cajal; published in his two-volume work Histologie du Systéme Nerveux in 1909. **c** Electron micrograph of a small region of the cerebral cortex of a rat with a prominent synapse. The black areas between cells indicate the interstitial space (IS), which may have been reduced in size as a consequence of the histological processing. The scale bar under the figure represents a distance of about 1 μm. Electron micrograph kindly supplied by Dr. C. B. Jaeger. Figures reproduced from [1]

different brain regions. All this complexity will be ignored here where the focus will be on the spaces between the cells. This space is called the interstitial space (IS) while the entire space that lies outside the cell and includes the IS as well as blood vessels and ventricles is the extracellular space. Often the distinction between the terms is ignored and 'extracellular space' is used even when 'interstitial space' would be more accurate.

The IS is filled with a salt solution that closely resembles the cerebrospinal fluid (CSF) that is found in the cavities in the center of the brain and the narrow canal that runs down the middle of the spinal cord. In human CSF, the predominant ion is Na^+ (~150 mM); other critical ions include K^+ (~3 mM) and Ca^{2+} and Mg^{2+} (~1.2 mM each). Theses cations are largely balanced by the anions Cl^- (~120 mM) and bicarbonate (~20 mM). There are many other compounds, often in very small amounts but which nonetheless serve important functions as neuro-modulators and signals. In addition to the salt solution there is an extensive ex-tracellular matrix of long-chain glycosaminoglycan and proteoglycan molecules. Many are anchored in cell membranes and among the predominant molecular species are chondroitin sulfate and heparan sulfate, both of which carry numerous negative charges. A third major component of the matrix is hyaluronan, which is often linked to other components of the matrix. The distribution and the function of the matrix are still uncertain but it resembles a hydrogel insofar as it does not seem to greatly impede the movement of molecules, although components of the matrix may interact with specific substances (see Sect. 6.7).

There are several reasons why there is a significant IS in the brain. Neurons require a reservoir of Na^+, K^+ and Ca^{2+} on the outside of their membranes to maintain a resting potential across the membrane and enable action potentials and synaptic transmission. The IS also functions as a communication channel where the signals are mediated by neuroactive substances; this communication mode is often known as volume transmission. More controversially, it has been suggested that the IS is a conduit for the removal of waste products from the brain [2, 3].

6.2 Biophysical Properties of Interstitial Space

Imagine looking down on a large city in the late afternoon when people are leaving their offices and workplaces (Fig. 6.2). The streets become crowded because the volume available for movement is limited to the spaces between the buildings. These spaces have been engineered to be sufficient to allow a reasonable density of people and other forms of traffic, such as cars and buses but the travelers still may become quite concentrated. Suppose the buildings were replaced by an open plaza and the same people were introduced, then the concentration would be less—it would be reduced in proportion to the *volume fraction* defined as the ratio of the area occupied by the streets to the area of an open plaza with the same perimeter.

Now think about the destination of the people leaving work. Many head to rail stations, bus terminals, car parks or some may be close enough to walk home. They cannot walk directly the way they might in the open plaza but must follow the

Fig. 6.2 North-facing view of New York City from the observation deck of the Empire State Building. Photo: C. Nicholson

streets and so they walk further and take more time to reach their destination than if they moved in a straight line. Sometimes they may encounter an obstruction and have to retrace their steps. Thus the motion of the people is hindered by the structure of the city. This hindrance can be measured by a parameter called *tortuosity* that will be defined below.

Other fates may befall our travelers. After a hard day they may stray into a bar and relax with a drink or two before continuing to their destination. This introduces an additional delay in their passage. To an external observer they appear to be held for a short time in the bar before being released. So long as the time in the bar is brief compared to the overall journey this process may be likened to fast equilibrium binding. If the sojourn in the bar accounts for most of the traveler's journey time then the journey will be dominated by the binding kinetics and the rest of the trajectory will be less important.

In all the cases described so far, the number of travelers in play remains constant. This may not always hold true, however. Returning to the bar scenario, if the person enters but does not leave or if some other fate befalls them that removes them from the moving population, then there will be irreversible loss and the traveler will never arrive home.

These metaphors may be translated into concrete terms for brain tissue. To begin we define a Representative Elementary Volume (REV) that contains a sufficient number of cellular elements and IS to allow the average properties to be reproducible (Fig. 6.3a).

We start with volume fraction, represented by α in the discipline of brain biophysics (more often called porosity and represented by ϕ in porous media research):

$$\alpha = V_{IS}/V_{Total} \tag{6.1}$$

where V_{Total} is the volume of the whole REV and V_{IS} is the volume of the IS contained in the REV. It follows that $0 \leq \alpha \leq 1$. Typically α is about 0.2; in other words 20% of the brain volume is found in the gaps between cells. How this is measured will be explained later.

Tortuosity is a more complex parameter than volume fraction. The operational definition is simple; take a small 'probe' molecule with a hydrodynamic diameter that is much less than the typical width of the IS and measure the effective diffusion coefficient, D^*, of the molecule in brain tissue. Measure the free diffusion coefficient, D, of the same molecule in water or a very dilute gel. Then the tortuosity, denoted by λ, is given by:

$$\lambda = \sqrt{D/D^*}. \tag{6.2}$$

Note that in porous media theory, tortuosity may be represented by a different symbol and may simply be the ratio of the two diffusion coefficients, not the square root. The volume fraction may also enter the definition, depending on how concentration is measured (see, e.g., Chap. 11 and Eq. (11.9)). Sometimes use of the diffusion permeability, $\theta = D^*/D$, is preferable [4].

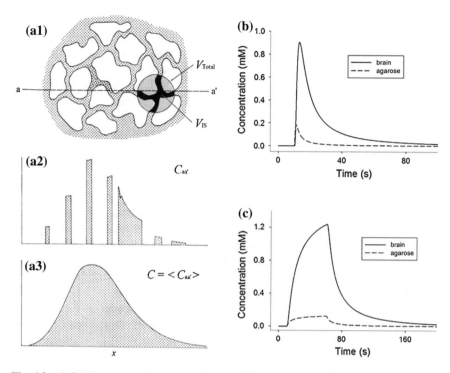

Fig. 6.3 a1–3 Volume averaging in brain tissue. **a1** Depicts the intracellular phase (no shading) and the IS phase (dots). A line, aa′, drawn in the medium alternately intersects the two phases. **a2** For a substance that only distributes in the IS phase (dots), the profile is discontinuous. To remove the problem, a REV (V_{Total}, Panel **a1**) is selected and the IS concentration averaged over the IS volume (V_{IS}, Panel **a1**). **a3** As the averaging volume moves along the line, the average concentration $\langle C_{aa'} \rangle$ now varies continuously. The volume averaging process also yields the macroscopic parameters α and λ. Reproduced from [1]. **b** Plot of concentration of TMA$^+$ (tetramethylammonium ions) in brain (solid line) and agarose (a polysaccharide polymer material in common use as a low concentration gel to provide an anti-convection medium, dotted line) computed for a location $r = 100$ µm from a micropipette that initiates an instantaneous source at 10 s and computed using Eqs. (6.5), (6.6). Here $U = 25$ pL and $C_f = 100$ mM. **c** Similar plots of TMA$^+$ concentration using iontophoresis and Eqs. (6.7), (6.8). Here $I = 50$ nA, $n_t = 0.5$ and duration of current is 50 s, commencing at 10 s. Common parameters for plots in (**b**) and (**c**) are as follows. $D = 1.31 \times 10^{-5}$ cm^2 s^{-1} for TMA$^+$ at 37 °C. For agarose $\alpha = 1$, $\lambda = 1$, $k' = 0$; for brain $\alpha = 0.2$, $\lambda = 1.6$, $k' = 0.005$ s^{-1}. Panels **b** and **c**, unpublished data from C. Nicholson

The subtlety in the tortuosity arises from the fact that more than just geometry may contribute to this parameter. As an example, fast equilibrium binding will be indistinguishable from an increase in tortuosity (Sect. 6.7) provided that the binding and unbinding kinetics (typically a bimolecular reaction) are much faster than the local diffusion process. Obviously this requires a more rigorous definition but the concept is clear.

Irreversible loss is a separate process, often represented by a first order kinetic process, with constant k' where the loss is proportional to the concentration of a molecule in the IS. A well-known example is loss of molecules into the numerous blood vessels that run through the brain [5]. This loss occurs across the so-called blood-brain barrier that surrounds each blood vessel in the brain and only allows certain classes of molecules to cross. Other types of loss may occur: irreversible binding to cell surfaces or extracellular matrix, enzymatic degradation or active transport into cells. These processes may be subject to complex kinetics, sometimes involving a Michaelis-Menten formulation [6].

Finally there is another transport process that might affect the behavior of probe molecules in brain tissue: bulk flow. Returning to the metaphor of the city, if the city lies close to water and the water rises excessively because of heavy rain, a hurricane or a tsunami, flooding will occur. As the water rushes down the streets it may carry the people with it. It has been postulated for many years that bulk flow may occur in the IS [7]. Recently there has been renewed interest in the topic leading to the concept that waste product may be removed from the brain by bulk flow based on the so-called glymphatic pathway [2, 3]. Here, the rate of flow versus the rate of diffusion will be important; this may be assessed to a first approximation using the appropriate Péclet number [8] (see also Sect. 11.4). Over periods of tens of minutes diffusion is likely to dominate. If the molecule is very large, so that diffusion is slow, or the period of observation involves a timescale of hours, then flow may have a significant role [9]. The glymphatic hypothesis has been subject to critical discussion and modeling [8, 10, 11].

6.3   Diffusion Analysis Reveals Properties of Interstitial Space

These considerations lead us to the central question: how can a study of the diffusion of small probe molecules in the brain reveal the properties described in Sect. 6.2? To answer this requires an appropriate diffusion equation. The grounds for this equation will not be discussed here. Suffice to say that the use of volume averaging justifies its use (Fig. 6.3a1–a3; [1, 12]). One important definition is that of C, which is the volume-averaged concentration of the probe molecule referenced to the IS. In other contexts, the concentration may be referenced to the whole tissue. These two definitions differ by a factor α. The reason for using the concentration in the IS is that this is the concentration actually experienced by a molecular receptor or transporter in the cell membrane or at the blood-brain barrier.

The modified diffusion equation is:

$$\frac{\partial C}{\partial t} = \frac{D}{\lambda^2}\nabla^2 C + \frac{Q}{\alpha} - k'C - \frac{f(C)}{\alpha} + \mathbf{v}\cdot\nabla C \tag{6.3}$$

The symbol ∇^2 represents the three-dimensional second spatial derivative in whatever coordinate system is being used. The new variables appearing in Eq. (6.3) are a source term Q, representing the release of molecules into the IS, a function $f(C)$ that accounts for reversible binding or other kinetics and which, under some conditions, may be absorbed into λ. The bulk flow velocity vector is $\mathbf{v}$ and forms a scalar product with the concentration gradient, if bulk flow is deemed significant.

If several terms are set to zero, namely Q (which often may be accounted for via suitable boundary conditions), $f(C)$, k' and $\mathbf{v}$, Eq. (6.3) becomes

$$\frac{\partial C}{\partial t} = D^* \nabla^2 C \tag{6.4}$$

where $D^* = D/\lambda^2$, the effective diffusion coefficient in the tissue. This is the diffusion equation (Chap. 2 and Eq. (2.9)) otherwise known as Fick's Second Law.

The solutions to Eq. (6.3) are well-known for a variety of situations [1, 12, 13]. Here only two will be illustrated; the first is for a source consisting of molecules released instantaneously at a point in the tissue. It is assumed that $f(C) = 0$ and $\mathbf{v} = 0$ but there is loss ($k' > 0$). Then the molecules diffuse in a spherically symmetric cloud so the solution may be written in terms of the radial distance from the source, r, and time, t:

$$C(r, t) = \frac{Q}{\alpha} \frac{\lambda^3}{(4Dt\pi)^{3/2}} \exp\left(-\frac{\lambda^2 r^2}{4Dt} - k't\right). \tag{6.5}$$

The source term Q may be written as:

$$Q = UC_f, \tag{6.6}$$

where U is the volume of molecules injected and C_f is the concentration. This implies that a finite volume of molecules is released from an infinitesimal point. In practice, for small spherical release volumes of radius r_f there is little error in the concentration predicted by Eq. (6.5) so long as $r \geq 2r_f$. For precise work or closer distances, an extension of Eq. (6.5) is available that takes into account the finite radius of the injected volume [1, 14].

The plot of this equation for realistic parameters is shown in Fig. 6.3b. Equation (6.5) is essentially a Gaussian curve (see Eq. (2.10) for the one-dimensional version) reminding us that the diffusion equation considered here may be thought of as being generated by molecules leaving the source and executing random walks in three dimensions in the IS subject to occasional destruction represented by k'. Note how much smaller the curve in a free medium (agarose, a polysaccharide polymer material in common use as a low concentration gel to provide an anti-convection medium) is compared to that in the brain. This mainly reflects the difference in volume fraction although the shape of the curve is also altered by the difference in tortuosity in the two media.

If the point source is switched on at time zero and emits molecules at a constant rate thereafter, then the solution to Eq. (6.3) is arrived at by integrating Eq. (6.5) from time zero up to t, yielding:

$$C(r,t) = \frac{Q}{\alpha} \frac{\lambda^2}{8\pi Dr} \left[\text{erfc}\left(\frac{r\lambda}{2\sqrt{Dt}} + \sqrt{k't} \right) \exp\left(r\lambda\sqrt{\frac{k'}{D}} \right) + \text{erfc}\left(\frac{r\lambda}{2\sqrt{Dt}} - \sqrt{k't} \right) \exp\left(-r\lambda\sqrt{\frac{k'}{D}} \right) \right].$$
(6.7)

In Eq. (6.7) 'erfc' refers to the complementary error function, which relates to the area under a Gaussian curve. Equation (6.7) represents the results of a source that is activated at time zero and continues indefinitely. In reality, the source is terminated after some time (typically 50 s) and the falling phase of the diffusion curve is arrived at by subtracting a delayed version Eq. (6.7) after 50 s (see [1, 13]) because the principle of linear superposition applies to solutions to the diffusion equation. This may be thought of as activating a delayed 'virtual sink' to terminate the real source. A plot of this equation, with both rising and falling phases, is shown in Fig. 6.3c. Again there is a striking difference between agarose and brain.

The source term Q in Eq. (6.7) is usually generated by iontophoretic release from a micropipette. A current I is applied to the ionic solution in the micropipette but only a fraction n_t of the current expels the ion of interest (the rest of the current moves the counter-ion in the opposite direction); n_t is called the transport number. Then:

$$Q = In_t/zF$$
(6.8)

Table 6.1 Selected values of α and λ obtained with radiolabel, RTI, RTP and IOI methods

Molecule	M_r	d_H (nm)	$D \times 10^6$ (cm^2 s^{-1})	°C	Species	Region	α	λ	Method	Refs.
Sucrose	342	1	7.0	37	Rabbit	Caudate	0.21	1.60	Radio	[15]
TMA$^+$	74	<1	11.1	37	Rat	Cortex	0.18–0.22	1.54–1.65	RTI	[13]
TMA$^+$	74	<1	11.1	37	Mouse	Cortex	0.23	1.67	RTI	[18]
TMA$^+$	74	<1	9.82	22	Turtle	cb. ml	0.31	1.44, 1.95, 1.58	RTI	[19]
α-NS$^-$	174	<1	7.60	37	Rat	cb. ml	0.18	1.54	RTI	[12]
Ca^{2+}	40	<1	9.4	34	Rat	Cortex	n/a	2.05*	RTP	[20]
AF488	547	<1	5.19	34	Rat	Cortex	n/a	1.54	IOI	[20]
Dex3	3000	3	2.2	37	Rat	Cortex	n/a	2.04	IOI	[21]
Dex70	70,000	14	0.47	37	Rat	Cortex	n/a	2.69	IOI	[21]
EGF	6600	3.7	1.7	34	Rat	cortex	n/a	1.79	IOI	[22]
BSA	66,000	7.4	0.83	34	Rat	Cortex	n/a	2.26	IOI	[23]
Lactoferrin	80,000	9.3	0.71	37	Rat	Cortex	n/a	3.50*	IOI	[24]
IgG	150,000	10.2	0.65	37	Rat	Cortex	n/a	3.11	IOI	[25]

M_r, relative molecular weight; d_H hydrodynamic diameter; D, free diffusion coefficient. Molecules: TMA$^+$, tetramethylammonium; α-NS,$^-$ α-naphthalenesulfonate; AF488, Alexa Fluor 488 hydrazide (Molecular Probes, Invitrogen, Carlsbad, California, USA); Dex3, dextran 3000 M_r; Dex70, dextran 70,000 M_r; EGF, epidermal growth factor; BSA, bovine serum albumin; IgG, immunoglobulin G. Brain regions: caudate, caudate nucleus; cortex, various areas of neocortex; cb. ml. cerebellum, molecular layer. Methods: Radio, radiolabel; RTI, real-time iontophoresis; RTP, real-time pressure; IOI, integrative optical imaging. n/a, not available with this method. *Ca^{2+} and lactoferrin interact with the extracellular matrix and this increases their effective tortuosity

where z is the valency of the ion, and F is Faraday's Electrochemical Equivalent that relates mass to charge. Equations (6.3), (6.5) and (6.7) are readily generalized to accommodate anisotropic diffusion [1, 13].

Equations (6.5) and (6.7) both represent the consequences of the spread of molecules from a point source. This 'point-source paradigm' has led to the development of experimental methods that will be described in Sects. 6.4 and 6.6.

In an earlier phase of research into diffusion in the brain, use was made of radiolabeled compounds that were perfused into the ventriculocisternal spaces of the brain [1, 5, 15]. Animals were sacrificed after various periods and the penetration of the tracer determined. Both the volume fraction and tortuosity were obtained (the latter from the one-dimensional solution to the diffusion equation). Sucrose gave the best results (Table 6.1) and the data are in good agreement with the more recent studies. The primary advantage of the radiotracer method is that it may be used to study the diffusion of a wide variety of substances, so long as they can be radiolabeled. The main shortcomings are that there is only one time point per animal and the method has low spatial resolution.

6.4 Measuring Volume Fraction and Tortuosity with Real Time Iontophoresis

The solutions to Eq. (6.3) may be put to good use to provide a means to measure α, λ and k'. Equations (6.7), (6.8) form the basis of the Real-Time Iontophoresis (RTI) method. This method was introduced in detail in 1981 by Nicholson and Phillips [12] and has evolved over the years (Fig. 6.4), however the concept has remained the same. Two micropipettes are inserted into brain tissue. The first is filled with a solution containing a suitable small ion, typically tetramethylammonium (TMA), a cation with about twice the molecular weight of K^+. The TMA^+ is released by passing a current through the micropipette according to Eq. (6.8) and the counter-ion is usually Cl^-. The second micropipette is placed about 100 µm from the first and contains an ion-exchanger that makes it an ion-selective microelectrode (ISM). With an appropriate choice of exchanger this micropipette can be made highly selective to TMA^+ versus other major ions in the IS of the brain (Na^+, K^+, Ca^{2+}, Mg^{2+} and Cl^-). The TMA^+ ions emitted from the first micropipette diffuse throughout the IS and a few arrive at the second micropipette where they are sensed and, providing the ISM has been properly calibrated, the local IS concentration of TMA, C, is measured as a function of time. Although TMA is the most commonly used probe ion with the RTI method, other cations and anions may be employed [12].

Using non-linear curve fitting, Eq. (6.7) may be fitted to the concentration versus time curves and the three parameters α, λ, and k' extracted. The first two parameters are the ones of interest and some representative values are shown in Table 6.1. The table lists a variety of molecules along with their relative molecular weight (M_r) and hydrodynamic diameter (calculated from D, the free diffusion coefficient, using the Stokes-Einstein-Sutherland Equation—see Eq. (6) in Ref. [13] and Chap. 12,

Eq. (12.3)). Table 6.1 goes on to show values of α and λ measured in different brain regions and in different species. Earlier radiotracer measurements favored large brain regions next to the ventricular cavities, such as the caudate nucleus of the rabbit. Most measurements, however, have been made in the cerebral cortex of rats using a point-source paradigm (e.g. RTI, RTP or IOI methods; see later for further descriptions). Measurements demonstrating diffusion anisotropy were made in the molecular layer of the turtle cerebellum.

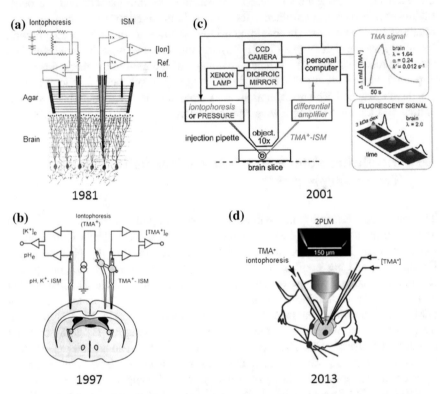

Fig. 6.4 Evolution of the RTI-TMA Method. **a** Original design showing iontophoresis micropipette on left and TMA ion-selective microelectrode (ISM) on the right lowered into cerebellar cortex of the rat. In practice the two micropipettes were glued together with a known spacing between tips. A trough of dilute agarose sits above the brain with known $\alpha = 1, \lambda = 1$ to enable D and n_t to be measured. Reproduced from [12]. **b** Implementation by Syková and co-workers in rat cerebral cortex showing RTI micropipettes on the right and an ISM measuring K^+ or pH on the left. Reproduced from [16]. Reprinted by permission of SAGE Publications, Ltd. **c** Adaptation to brain slices. Micropipettes introduced independently with robotic manipulators under a microscope and water-immersion objective enabling precise measurement of spacing between micropipette tips. Use of the microscope also permitted the IOI Method to be simultaneously employed (see Sect. 6.6). Reproduced from [17]. Reprinted with permission from Elsevier. **d** RTI-TMA method employed in awake mouse cortex. Micropipettes inserted independently and tip spacing measured with a two-photon light microscope (2PLM). Reproduced from [3]. Reprinted with permission from AAAS

Table 6.1 shows that small molecules, such as sucrose (uncharged), TMA^+ (monovalent cation) and α-NS^- (monovalent anion) all reveal tortuosities of 1.54 – 1.67 in isotropic brain regions. Small molecules in strongly anisotropic brain regions deviate from this range (TMA^+ in turtle cerebellar molecular layer). Molecules with much larger hydrodynamic diameters (e.g. Dex70, BSA and IgG) show larger tortuosities, most likely because of significant interaction with the cell walls that form the boundaries of the IS. The divalent cation Ca^{2+} and the protein lactoferrin both interact with the extracellular matrix and this increases their effective tortuosity (see later). The small molecules also reveal that the volume fraction of the IS ranges between 0.18–0.23; the value in the anisotropic brain region is higher, although we do not have an explanation for this. The parameter k' is only measured with RTI to account for the small loss of TMA^+ from the IS during the measurement; this increases the accuracy of the method. Typically for TMA^+ in the rodent brain $k' = 5 \times 10^{-3}$ s^{-1} (Table 5 in [1]).

The IS probably varies in width but in many parts it may be very narrow ($\sim$40 nm—see Sect. 6.6) and the tip of an iontophoresis micropipette or ISM is 1–5 μm in diameter. Hence, it may be asked why the measured value of C represents the concentration in the IS. The answer is that the tip of each micropipette creates a small cavity in the tissue, of the same order of magnitude as the tip, that very rapidly equilibrates with the local IS [12].

Although the basic RTI method has remained unchanged, the supporting software has evolved considerably as the power of personal computers has increased. The present software consists of two custom programs written in MATLAB [26, 27]. The first is called Wanda and is responsible for controlling the experiment and acquiring and storing the data. The second program is called Walter and performs the curve-fitting and parameter extraction. The RTI-TMA method is visualized and described in detail elsewhere [28].

The RTI-TMA method has been used extensively to interrogate the IS structure in various brain regions and species under normal and pathophysiological conditions (see [13] for a comprehensive review). In the anesthetized animal and in brain slices the typical parameter values are $\alpha \sim 0.2$ (20% volume fraction) and $\lambda \sim 1.6$ (Table 6.1). This means that 20% of brain tissue resides in the IS and that the diffusion coefficient of a small molecule will be reduced to about 40% of its value in water. A recent study suggests that the volume fraction in the cortex of the awake mouse is about 14% and expands to around 23% when the animal sleeps, possibly facilitating the clearance of waste products from the brain [3]. It is thought that brain pathways involving adrenergic inputs are responsible for the changes in α between sleep and wakefulness [3] and support for this conjecture has been obtained in brain slices [29]. The finding that there is an appreciable IS in all brain regions and species so far studied (including invertebrates with sufficiently large brain mass—see [30]) implies that the IS is essential for brain function.

There are at least two examples of systematic deviation from the parameters listed in the last paragraph. Just after birth, the rat cortex has a volume fraction of 40%, which declines to the adult value of 20% by postnatal day 21 [31]. Interestingly, during this developmental period the tortuosity has already reached the adult value ($\sim$1.6). The second example is when the brain undergoes ischemia (loss of blood flow) or severe hypoxia (loss of oxygen supply). Under these circumstances volume fraction falls to about 5% and tortuosity measured with TMA increases to about 2.0 [13].

Equations (6.5), (6.6) may also be employed with TMA$^+$ or another ion, such as Ca^{2+} (see Sect. 6.7) using the Real-Time Pressure (RTP) method. In this technique a brief pulse of nitrogen or other inert gas is used to expel the ion from the micropipette. Because the volume released is difficult to quantify only the tortuosity is obtained by the curve-fitting method. The RTP method is useful, however, when iontophoresis is not reliable or when there is reason to expel more than one substance from the micropipette (e.g. [32]).

6.5   Modeling with MCell to Test Hypotheses About Structure

The finding of widespread similarities in the values of volume fraction and tortuosity has led to attempts to construct models of brain cell aggregates that would yield these values. Unfortunately the results have not been always consistent and some models have been overly simplistic (see [13] for a brief survey of models). One approach to resolving these issues is to use Monte Carlo simulation methods to estimate the value of λ for a given α based on ensembles of cells with shapes that are capable of packing three-dimensional space while maintaining a uniform IS width. This packing property ensures that α can be varied in the range $0 \leq \alpha \leq 1$. The simplest cell type that meets these requirements is the cubic cell (Fig. 6.5) and Tao and Nicholson [33] used the MCell program (www.mcell.org) to perform this type of simulation. This software was originally developed by Stiles and Bartol [34] for modeling diffusion of transmitter molecules at the neuromuscular junction where nerve fibers connect to muscles. Surprisingly, modeling brain tissue it was found that the tortuosity never exceeded $\lambda = 1.225$ even when $\alpha \rightarrow 0$. The simulations were repeated with cells in the shape of truncated octahedra and with mixtures of rhombicuboctahedra, cubes and tetrahedra [33]. Both of these choices pack three-dimensional space, however the results were always the same and could be well-represented by the equation:

$$\lambda = \sqrt{(3-\alpha)/2}. \tag{6.9}$$

This is related to the result obtained by James Clerk Maxwell in 1873 for the resistivity of a dilute suspension of non-conducting spheres in a conducting medium (see [33]).

In light of these findings there had to be another explanation of the higher tortuosity routinely seen in the brain. On the basis of experiments involving ischemic tissue [35] it was postulated that brain tissue harbored dead-space

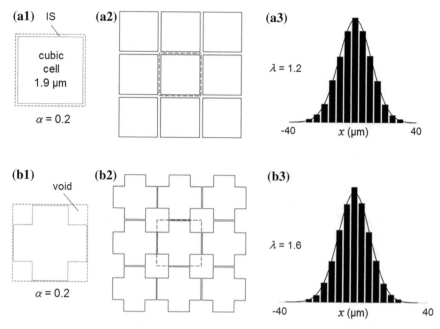

Fig. 6.5 Monte Carlo simulations using MCell. **a1** Basic cell represented by a cube of side 1.9 μm with an 'atmosphere' of IS (red dashed line). **a2** These pack together with uniform separation of 147 nm to give a volume fraction $\alpha = 0.2$. TMA molecules are released from a point location in the IS and diffuse randomly between cells. **a3** After $t = 0.1$ s the distribution of molecules is measured and the mean square distance of all the molecules calculated. Using Eq. (6.11) the effective diffusion coefficient, D^*, is calculated and the appropriate distribution curve generated. This is seen to accurately fit the histogram of particle position confirming that it is a Gaussian, however λ, calculated from Eq. (6.9) is smaller than the value measured experimentally. **b1** By introducing dead spaces by creating cavities at each corner of every cube, while reducing the separation between cubes to 50 nm, it is possible to keep $\alpha = 0.2$. **b2** The simulation is run with an ensemble of modified cubes. **b3** Again, a Gaussian fit is obtained but now $\lambda = 1.6$, in conformity with experimental data. Unpublished data from C. Nicholson

microdomains that transiently trapped diffusing molecules as they diffused through the brain so delaying them [4, 36] and leading to a lower effective diffusion coefficient. This was duly tested using MCell with models that incorporated cavities or invaginations in the cell wall, meaning that cells were no longer convex. Other models featured local enlargements of the IS, meaning that the IS no longer had a uniform width [4, 36]. It was shown that such dead-spaces could indeed generate the tortuosity seen in brain tissue (Fig. 6.5). These studies resulted in a semi-empirical extension of Eq. (6.9) that estimated λ in the presence of dead-spaces [36]:

$$\lambda = \left(\frac{3-\alpha}{2}\right)^{1/2}\left(\frac{\alpha}{\alpha-\alpha_c}\right)^{1/\beta}.$$ (6.10)

where α_c is the volume fraction of the cavity or void space and β is an empirical 'exit factor' that informally captures the probability that molecules leave the cavity. Usually, $2 < \beta < 3$ (see Table 1 in [36]). To calculate α_c, α_0 is defined as the volume fraction in the absence of dead-spaces, then $\alpha = \alpha_0 + \alpha_c$.

In the initial studies of λ using Monte Carlo simulations and the MCell software [33, 36], this parameter was estimated by counting all the molecules in a series of concentric boxes that included many cubic cells, at a sequence of times. A simple integration of Eq. (6.5) with $k' = 0$ provided the required estimate of D^* [27, 33, 36] and hence λ. This approach was also used by Kinney et al. [37] in their study of a reconstruction of a block of brain tissue visualized with electron microscopy. Later work [27] utilized the mean square position, $\langle r^2 \rangle$, of all diffusing molecules at different times and used this to estimate D^* from the well-known equation for an ensemble of molecules undergoing a random walk (Chap. 2, Eq. (2.5)):

$$\langle r^2 \rangle = 2nD^*t$$ (6.11)

where n is the dimension of the space (typically $n = 3$). This approach easily lends itself to computing the tensor form of D^* in an anisotropic medium and also appears more accurate than the integral method.

Further work from Hrabětová and co-workers has suggested that glial cells in the part of the brain called the cerebellum may wrap around neurons also producing an IS geometry that delays the movement of molecules [38]. This study combined experimental measurements using the IOI technique that will be detailed in Sect. 6.6 together with MCell modeling and showed that over short times and distances the diffusion of molecules was anomalous. Anomalous diffusion is described by the equation:

$$\langle r^2 \rangle = 2nDt^{\frac{2}{d_w}}$$ (6.12)

where d_w is the anomalous diffusion exponent. When $d_w > 2$, the phenomenon is classified as anomalous subdiffusion; when $d_w = 2$ the process is classical diffusion as described in Eq. (6.11). Xiao et al. [37] found that d_w was as high as 4.8 in the granule cell layer of the rat cerebellum. This was likely because of the unusual glomerular anatomy of this brain region [38, 39].

6.6   Measuring Tortuosity with Macromolecules Using Integrative Optical Imaging

The RTI method may be used with a few small ions other than TMA, including some anions (Table 6.1; [12]) but is restricted to compounds for which an ISM may be fabricated. Many biologically important molecules are much larger than TMA and yet still move through the IS. Consequently it was important to devise a method of measuring the diffusion properties of macromolecules. This was achieved when Nicholson and Tao introduced the Integrative Optical Imaging (IOI) method [39]. The concept is to take a macromolecule to which has been attached a fluorescent dye, release it from a micropipette by using a brief pressure pulse (because many macromolecules are not charged) and then image the diffusing cloud of molecules as they spread through the IS, using an epifluorescent microscope (Fig. 6.6). The three-dimensional image is projected onto the two-dimensional sensor of a suitable digital camera and it was shown that fitting a solution of the diffusion equation based on Eq. (6.5) to this projected image would accurately extract the effective diffusion coefficient and hence λ [40, 41]. The IOI data are acquired and analyzed with custom MATLAB programs [26, 27]. As with the RTP method, the volume fraction α cannot be obtained because pressure injection is used.

The IOI method has been employed to measure λ for dextran molecules with M_r ranging from 3000–525,000 [13, 42] and for a variety of albumins, including bovine serum albumin (BSA) with $M_r = 66,000$ [23]. Recently Thorne and co-workers [25] measured the diffusion of Immunoglobulin G (IgG) antibody ($M_r = \sim 150,000$) in rat cortex. For molecules with an approximately spherical shape the value of λ increases with size (Table 6.1) suggesting that interaction with the 'walls' of the IS may become a factor. This conjecture was tested in a study of the anesthetized rat cortex that included quantum dots with a hydrodynamic diameter of 35 nm. Using the theory of restricted diffusion in pores [43] it was shown that the typical width of the IS was between 38 and 64 nm, depending on whether the IS was modeled as a set of intersecting planes or intersecting tubes [21]. As the MCell simulations (Sect. 6.5) and many electron micrographs suggest, the spaces are probably not uniform. Some support for the non-uniformity of the IS has come from simulating molecular diffusion in a three-dimensional reconstruction of electron micrographs [37]. Previously, electron micrographs had been interpreted to imply that the IS was about 20 nm wide but it is now recognized that the IS is almost always greatly reduced in the preparation of such material [21, 44, 45]. Based on the quantum dot data mentioned above and restricted diffusion theory together with data suggesting reptation of dextrans [42], a consensus is beginning to emerge that the typical width of the IS is about 40 nm [46], although the precise meaning of this value is not clear because some regions of the IS are likely to be expanded into dead-spaces.

Attempts have been made to apply two-photon microscopy for IOI imaging [47] but the small volume sampled by this technique seems to lead to a poor signal-to-noise ratio so the results have been disappointing to date.

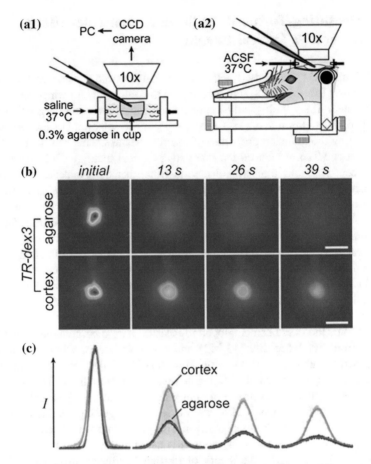

Fig. 6.6 a1,2 Experimental setup for IOI diffusion measurements. Images of fluorescent probe diffusion were captured by a cooled charge-coupled device (CCD) camera through a microscope with a ×10 water-immersion objective after pressure ejection from a micropipette into either dilute agarose (**a1**) or somatosensory cortex (**a2**). **b** Dextran diffusion in rat cortex. Representative pseudo-color images (red indicates high concentration, blue low) after 3 kDa dextran labelled with Texas Red (TR-dex3) ejection into agarose or cortex. Scale bars 200 μm. **c** Fluorescence intensity profiles and theoretical fits for the images yielding $D = 2.3 \times 10^{-6}$ cm^2 s^{-1} and $D^* = 4.5 \times 10^{-7}$ cm^2 s^{-1}. Reproduced from [21]

6.7 Probing the Extracellular Matrix

The interstitial space contains not only a salt solution but also an extracellular matrix of glycosaminoglycans and proteoglycans. Typical constituents are chondroitin sulfate, heparan sulfate and hyaluronan along with numerous proteins that may link elements of the matrix together or anchor them to cell membranes. An important feature of the extracellular matrix is that it has many negative charges

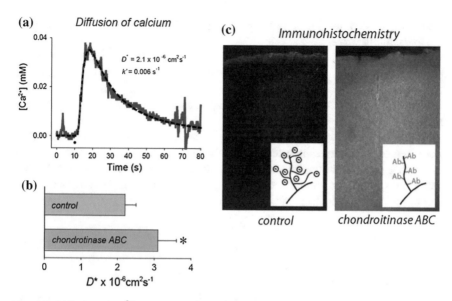

Fig. 6.7 Diffusion of Ca^{2+} before and after chondroitin sulfate had been cleaved with enzyme. **a** Control record of Ca^{2+} diffusion curves in brain slice from neocortex using RTP method (blue line data, black line fit with Eq. (6.5)) under normal conditions. **b** Comparison of measurements of D^* in control conditions and after the enzyme chondroitinase ABC had been applied to the brain slice. Diffusion in treated slice is significantly faster. **c** Immunohistochemical staining of oligosaccharide stubs confirms the cleavage of chondroitin sulfate component of extracellular matrix with chondroitinase ABC. Reproduced from [20]

associated with the sulfate groups and an obvious question is whether these affect the diffusion of ions or charged molecules.

In a study that used the Real-Time Pressure (RTP) method, divalent Ca^{2+} ions were pressure ejected from a micropipette and their diffusion measured with Ca^{2+}-selective ISMS (Fig. 6.7; [20]). The analysis was based on Eqs. (6.5), (6.6) (Fig. 6.7a). It was found that D^* was unusually low (i.e. tortuosity was unusually high) but was increased when the charged sites on the chondroitin sulfate molecules were removed with a suitable enzyme (Fig. 6.7b, c). In contrast, application of the enzyme did not affect the diffusion of the monovalent TMA^+. This suggested that the negative sites on the matrix normally may be screened by the high concentration of Na^+ in the IS (~150 mM), but the higher charge density of divalent Ca^{2+} is able to displace the Na^+ and transiently bind to the chondroitin sulfate component of the matrix.

In another set of experiments using the IOI technique with fluorescently labeled lactoferrin, a molecule with $M_r = 80,000$, there was reason to think that this molecule transiently bound to the heparan sulfate component of the matrix [24]. The evidence was that when lactoferrin was complexed with heparin, for which lactoferrin had a higher affinity than with heparan sulfate, the larger complex diffused faster than the uncomplexed molecule. This suggested that when the lactoferrin was complexed it no longer transiently bound to the heparan groups in the extracellular matrix.

These results lend themselves to simulations using MCell, which incorporates bimolecular reactions. Because the experimental data may be fitted to the diffusion equation, it may be assumed that interaction of either Ca^{2+} or lactoferrin with components of the extracellular matrix should be represented by a fast-equilibrium reaction scheme (see Chap. 14 in [48]).

The development of this argument may be sketched as follows (see also [27]). Let C represent the concentration of the diffusing substance in IS, B, concentration of binding sites of the matrix and S, concentration of complex between substance and matrix, then the binding and unbinding processes can be described by a second-order (bimolecular) reaction scheme:

$$C + B \xrightarrow{k_f} S, \tag{6.13a}$$

$$S \xrightarrow{k_b} C + B, \tag{6.13b}$$

and k_f is the forward rate constant associated with Eq. (6.13a) and k_b the backward rate constant associated with Eq. (6.13b). Omitting the source, loss and flow terms, and following [48], the diffusion equation may be written as:

$$\frac{\partial C}{\partial t} = D^* \nabla^2 C - k_f C (B - S) + k_b S \tag{6.14}$$

and the kinetics described by:

$$\frac{\partial S}{\partial t} = k_f C (B - S) - k_b S. \tag{6.15}$$

Locally, the reaction process is assumed to be much faster than the diffusion process and there is equilibrium between the mobile and complexed molecules, so the derivative on the left hand side of Eq. (6.15) is zero; it is also assumed that $B \gg C$ so:

$$S = RC, \tag{6.16}$$

where R is the dimensionless parameter $R = k_f B/k_b$. Note that this may be written $R = B/K_D$ where $K_D = k_b/k_f$ is the equilibrium dissociation constant. This is essentially the Law of Mass Action applied to an appropriate local region.

Comparing Eq. (6.14) and Eq. (6.15) it is clear that the diffusion equation with the reaction process may be written as:

$$\frac{\partial C}{\partial t} = D^* \nabla^2 C - \frac{\partial S}{\partial t}, \tag{6.17}$$

then substituting for S in Eq. (6.17) using Eq. (6.16) results in

$$\frac{\partial C}{\partial t} = \frac{D^*}{1+R} \nabla^2 C, \tag{6.18}$$

and the basic diffusion equation is recovered with a new effective diffusion coefficient $D^*_{mat} = D^*/(1+R)$. There is a lack of consistency in this argument because the time derivative of S is regarded as zero in Eq. (6.15) but non-zero when the substitution is made in Eq. (6.17). The result may be derived more rigorously, taking into account appropriate time and length scales and the approach to local equilibrium [49, 50]. The result is also quoted by Crank (Chap. 14 in [48]) who simply assumes the validity of Eqs. (6.16) and (6.17).

The forgoing may be interpreted as saying that the final tortuosity is the product of the tortuosity arising from the geometry (λ_g) multiplied by the tortuosity arising from the interaction with the matrix (λ_m) (see [27] for some preliminary verification using MCell):

$$\lambda = \sqrt{\frac{D}{D^*}} \sqrt{1+R} = \lambda_g \times \lambda_m. \tag{6.19}$$

6.8   Conclusions

A molecule executing random walks in a structured environment will, over time, explore the entire connected space. If its progress can be tracked then the structure will be revealed. Modern single-particle tracking methods do just that but necessitate the observance of a great many trajectories to arrive at meaningful information [51]. A recent study with single-walled carbon nanotubes (SWCNT) hints that it may be possible to measure the nanoscale organization of the ECS with single-particle tracking [52]. In the 'point-source paradigm' approach, outlined here, a vast number of molecules are released from a single location and the resulting concentration sampled in space and time. This effectively reveals the structure, embodied in the two parameters, volume fraction (α) and tortuosity (λ), because there is a rigorous relationship through the diffusion equation between the microscopic behavior of a wandering molecule and the macroscopic concentration. The most informative parameter revealed by diffusion is the tortuosity, which measures the hindrance imposed on a diffusing molecule by the obstacles, dead-spaces, reversible binding reactions and other factors (Fig. 6.8). The volume fraction is a simpler parameter but often of great interest to a biologist. Deviations from a pure diffusion process may also demonstrate loss or clearance of molecules from the IS, which is also valuable information.

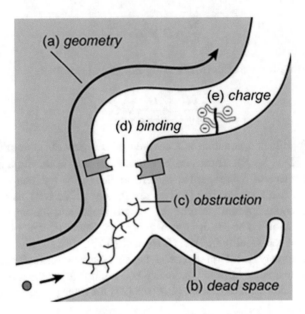

Fig. 6.8 Several factors influence the diffusion of a molecule in the IS: **a** geometry of the spaces, which imposes a delay on a diffusing molecule compared with its behavior in a free medium; **b** dead-space microdomain where molecules lose time exploring a dead-end (such a microdomain may resemble a "pocket" as shown, but it may also take the form of local enlargement of the IS or glial wrapping of a neuron); **c** obstruction by extracellular matrix molecules such as hyaluronan; **d** binding sites for the diffusing molecule either on cell membranes or extracellular matrix; and **e** fixed negative charges, also on the extracellular matrix, that may affect the diffusion of charged molecules. Reproduced from [13]

An underlying assumption involved in the interpretation of data and modeling the IS is that a molecule within the IS that is not in contact with the bounding cell walls or with the extracellular matrix, moves in an essentially free medium. A recent study [53] used Time-Resolved Fluorescence Anisotropy Imaging (TR-FAIM) to measure the local viscosity in the ECS and suggested that the diffusion coefficient of a small probe molecule (Alexa Fluor 360, $M_r = 349$) was reduced to 70% of that in a true free medium. Along with the SWCNT study mentioned above, this is another example of the increasing ability to resolve the IS on a nanometer scale.

Apart from revealing structure of the IS, these parameters arrived at through analysis of diffusion have great utility in designing techniques for drug delivery to the brain [54]. Unfortunately, getting drugs into the brain is difficult. The blood-brain barrier (BBB) around the penetrating blood vessels keeps out most substances unless the BBB has specific transporters (e.g. for glucose). Relatively small lipophilic molecules can cross the membranes of the cells forming the BBB but the factors that govern permeability are complex [55]. Methods of drug delivery such as convection-enhanced delivery, which is based on introducing a cannula into the brain and pressure injecting the drug, rely on a combination of induced bulk flow and drug diffusion and have some utility, especially when combined with magnetic resonance imaging (MRI—see also the introduction to MRI given in Sect. 12.5, notably Fig. 12.14) [56].

Studying the spreading of molecules in the interstitial (or extracellular) space of the brain is revealing much new information about this hitherto inaccessible region that suggests it is not just a 'space' but a complex and essential microenvironment.

References

1. C. Nicholson, Rep. Prog. Phys. **64**, 815 (2001)
2. J.J. Iliff et al., Sci. Trans. Med. **4**, 147ra111 (2012)
3. L. Xie et al., Science **342**, 373 (2013)
4. J. Hrabe, S. Hrabětová, K. Segeth, Biophys. J. **87**, 1606 (2004)
5. C.S. Patlak, J.D. Fenstermacher, Am. J. Physiol. **229**, 877 (1975)
6. C. Nicholson, Biophys. J. **68**, 1699 (1995)
7. N.J. Abbott, Neurochem. Int. **45**, 545 (2004)
8. M. Asgari, D. de Zelicourt, V. Kurtcuoglu, Sci. Rep. **5**, 15024 (2015)
9. H.F. Cserr, D.N. Cooper, P.K. Suri, C.S. Patlak, Am. J. Physiol. **240**, F319 (1981)
10. M. Asgari, D.D. Zelicourt, V. Kurtcuoglu, Sci. Rep. **6**, 38635 (2016)
11. B.-J. Jin, A.J. Smith, A.S. Verkman, J. Gen. Physiol. **148**, 489 (2016)
12. C. Nicholson, J.M. Phillips, J. Physiol. **321**, 225 (1981)
13. E. Syková, C. Nicholson, Physiol. Rev. **88**, 1277 (2008)
14. C. Nicholson, Brain Res. **333**, 325 (1985)
15. J.D. Fenstermacher, T. Kaye, Ann. New York Acad. Sci. **531**, 29 (1988)
16. I. Voříšek, E. Syková, J. Cerebral Blood Flow Metab. **17**, 191 (1997)
17. S. Hrabětová, C. Nicholson, Neurochem. Int. **45**, 467 (2004)
18. M. Anděrová, S. Kubinová, T. Mazel, A. Chvátal, C. Eliasson, M. Pekny, E. Syková, Glia **35**, 189 (2001)
19. M.E. Rice, Y.C. Okada, C. Nicholson, J. Neurophysiol. **70**, 2035 (1993)
20. S. Hrabětová, D. Masri, L. Tao, F. Xiao, C. Nicholson, J. Physiol. **587**, 4029 (2009)
21. R.G. Thorne, C. Nicholson, Proc. Natl. Acad. Sci. U.S.A. **103**, 5567 (2006)
22. R.G. Thorne, S. Hrabětová, C. Nicholson, J. Neurophysiol. **92**, 3471 (2004)
23. L. Tao, C. Nicholson, Neuroscience **75**, 839 (1996)
24. R.G. Thorne, A. Lakkaraju, E. Rodriguez-Boulan, C. Nicholson, Proc. Natl. Acad. Sci. U.S. A. **105**, 8416 (2008)
25. D.J. Wolak, M.E. Pizzo, R.G. Thorne, J. Control Release **197**, 78 (2015)
26. S. Hrabětová, C. Nicholson, in *Electrochemical Methods for Neuroscience*, ed. A.C. Michael, L.M. Borland (CRC Press, Taylor Francis Group, Boca Raton, 2007), p. 167
27. C. Nicholson, P. Kamali-Zare, L. Tao, Comput. Vis. Sci. **14**, 309 (2011)
28. J. Odackal, R. Colbourn, N.J. Odackal, L. Tao, C. Nicholson, S. Hrabětová, J. Vis. Exp. (125), e55755, (2017). doi:https://doi.org.10.3791/55755
29. A.D. Sherpa, F. Xiao, N. Joseph, C. Aoki, S. Hrabětová, Synapse **70**, 307 (2016)
30. C. Nicholson, J.A. Miyan, K.T. Potter, R. Williamson, N.J. Abbott, in *Cephalopod Neurobiology*, ed. by N.J. Abbott, R. Williamson, L. Maddock (Oxford University Press, Oxford, 1995), p. 383
31. A. Lehmenkühler, E. Syková, J. Svoboda, K. Zilles, C. Nicholson, Neuroscience **55**, 339 (1993)
32. A. Lehmenkühler, C. Nicholson, E.J. Speckmann, Brain Res. **561**, 292 (1991)
33. L. Tao, C. Nicholson, J. Theoret. Biol. **229**, 59 (2004)

34. J.R. Stiles, T.M. Bartol, in *Computational Neuroscience: Realistic Modeling for Experimentalists*, ed. by E. De Schutter (CRC Press, London, 2001), p. 87
35. S. Hrabětová, J. Hrabe, C. Nicholson, J. Neurosci. **23**, 8351 (2003)
36. A. Tao, L. Tao, C. Nicholson, J. Theoret. Biol. **234**, 525 (2005)
37. J.P. Kinney, J. Spacek, T.M. Bartol, C.L. Bajaj, K.M. Harris, T.J. Sejnowski, J. Comp. Neurol. **521**, 448 (2013)
38. F. Xiao, J. Hrabe, S. Hrabětová, Biophys. J. **108**, 2384 (2015)
39. C. Nicholson, Biophys. J. **108**, 2091 (2015)
40. C. Nicholson, L. Tao, Biophys. J. **65**, 2277 (1993)
41. L. Tao, C. Nicholson, J. Microsc. (Oxford) **178**, 267 (1995)
42. F. Xiao, C. Nicholson, J. Hrabe, S. Hrabětová, Biophys. J. **95**, 1382 (2008)
43. W.M. Deen, AIChE J. **33**, 1409 (1987)
44. B. Cragg, Trends Neurosci. **2**, 159 (1979)
45. M. Pallotto, P.V. Watkins, B. Fubara, J.H. Singer, K.L. Briggman, eLife **4**, e08206 (2015)
46. C. Nicholson, S. Hrabětová, Biophys. J. (2017). doi:https://doi.org.10.1016/j.bpj.2017.06.052
47. M. Stroh, W.R. Zipfel, R.M. Williams, W.W. Webb, W.M. Saltzman, Biophys. J. **85**, 581 (2003)
48. J. Crank, *The Mathematics of Diffusion*, 2nd edn. (Clarendon Press, Oxford, 1975)
49. A.S. Perelson, L.A. Segel, J. Math. Biol. **6**, 75 (1978)
50. B. Goldstein, C. DeLisi, J. Abate, J. Theoret. Biol. **52**, 317 (1975)
51. M.J. Saxton, Nat. Methods **5**, 671 (2008)
52. A.G. Godin, J.A. Varela, Z. Gao, N. Danne, J.P. Dupuis, B. Lounis, L. Groc, L. Cognet, Nat. Nano **12**, 238 (2017)
53. K. Zheng, T.P. Jensen, L.P. Savtchenko, J.A. Levitt, K. Suhling, D.A. Rusakov, Sci. Rep. **7**, 42022 (2017)
54. D.J. Wolak, R.G. Thorne, Mol. Pharm. **10**, 1492 (2013)
55. M.D. Habgood, D.J. Begley, N.J. Abbott, Cell. Mol. Neurobiol. **20**, 231 (2000)
56. R.R. Lonser, M. Sarntinoranont, P.F. Morrison, E.H. Oldfield, J. Neurosurg. **122**, 697 (2015)

Chapter 7
Turbulent Diffusion in the Atmosphere

Manfred Wendisch and Armin Raabe

7.1 Introduction

For good reasons, meteorology is often referred to as physics of the atmosphere. The motions of air parcels (called wind) and several other atmospheric processes, such as photon transport through the atmosphere, phase transitions during cloud evolution (water vapour diffusion growth of cloud droplets and ice crystals), and many others are of stochastic nature. They can often be considered and quantitatively be described as diffusion phenomena.

In this context, the term 'turbulent diffusion' is commonly used in meteorological applications. It refers to the fact that, in addition to the mean motion of air parcels (mean wind), they are subject to irregular (stochastic) fluctuating movements, which takes the air parcel both in wind direction and perpendicular to it. Since more than a century, meteorologists have attempted to quantify this problem. However, even today, using powerful computers and sophisticated measurement techniques, the atmospheric turbulent system is too complicated to allow a stringent description and prediction of turbulent atmospheric motions. This is one of the major reasons why weather forecast is still associated with significant uncertainties.

This chapter deals with an approach towards a physical description of the spreading of pollution in a turbulent atmosphere. We may witness this phenomenon quite commonly by, e.g., following the smoke of a chimney or the steam emerging from a power plant. It is remarkable that the equations to quantify atmospheric

M. Wendisch (✉) · A. Raabe
Faculty of Physics and Earth Sciences, Leipzig Institute for Meteorology,
University of Leipzig, Stephanstr. 3, 04103 Leipzig, Germany
e-mail: m.wendisch@uni-leipzig.de

A. Raabe
e-mail: raabe@uni-leipzig.de

© Springer International Publishing AG 2018
A. Bunde et al. (eds.), *Diffusive Spreading in Nature, Technology and Society*,
https://doi.org/10.1007/978-3-319-67798-9_7

turbulence reveal close similarities with the formalism provided by the Fick's diffusion laws, Eqs. (2.6) and (2.9).

7.2 Fick's Laws Applied for Turbulent Diffusion

The key problem is the prediction of the temporal changes of the concentration distribution c of a pollutant as a function of time and space. In our considerations we follow Reynolds [1] who was one of the first describing the mass transfer in a turbulent atmosphere.

The pollution is represented by its concentration $c(\vec{r}, t)$ (in kg m^{-3}), which is a function of space (represented by the position vector $\vec{r}$) and time (indicated by the symbol t). For the components of the position the two equivalent notations are applied in the following: x_i, with $i = 1, 2, 3$, and $\{x, y, z\}$. The components of the wind vector (air velocity) $\vec{v}(\vec{r}, t)$ will be denoted by the two following notations: v_i, with $i = 1, 2, 3$, and $\{u, v, w\}$.

Local changes of the concentration of a pollutant are caused by sources and sinks $P(\vec{r}, t)$ as well as by movements of air for non-uniformly distributed pollution:

$$\frac{\partial c}{\partial t} = P - \sum_{k=1}^{3} u_k(\vec{r}) \frac{\partial c}{\partial x_k}. \tag{7.1}$$

The latter influence is represented by the second term on the right-hand side of Eq. (7.1), from where flux in, e.g., $+x$ direction is seen to give rise to increasing local concentration ($\frac{\partial c}{\partial t} > 0$) for decaying concentration in x direction (i.e. for $\frac{\partial c}{\partial x} < 0$). This second term in Eq. (7.1) is the scalar product of the wind vector and the gradient of the pollutant concentration representing the advection/convection of the pollutant. For further treatment, Reynolds considered the stochastic character of turbulence by decomposing the physical variables

$$\begin{aligned} c &= \bar{c} + c' \\ u_k &= \bar{u}_k + u_k' \\ P &= \bar{P} + P' \end{aligned} \tag{7.2}$$

into their averaged values $\bar{c}, \bar{u}_k, \bar{P}$ and the deviations c', u_k', P' from the average due to turbulence.

With these relations, Eq. (7.1) is rearranged to:

$$\frac{\partial \bar{c}}{\partial t} = \bar{P} - \sum_{k=1}^{3} \bar{u}_k \frac{\partial \bar{c}}{\partial x_k} - \sum_{k=1}^{3} \frac{\partial \left(\overline{u_k' c'} \right)}{\partial x_k} \tag{7.3}$$

Derivation of Eq. (7.3) assumes the following premises:

(i) Though the time dependence of the respective quantities can be measured with high precision, it is still their average value one is generally interested in.

(ii) After inserting Eqs. (7.2) into Eq. (7.1), deviations from the average disappear due to the averaging $(\overline{c'}=0, \overline{u'_k}=0, \overline{P'}=0)$. This, however, does not hold for non-linear products such as $\overline{u'_k(\vec{r})\frac{\partial c'}{\partial x_k}}$ since the two factors are, mutually, correlated.

(iii) For attaining the third term on the right-hand side of Eq. (7.3) the equation of continuity (Eq. 2.8) has to be applied for velocities within an incompressible fluid, $\sum_{k=1}^{3}\frac{\overline{\partial u'_k}}{\partial x_k}=0$. This relation indicates that the wind divergence vanishes. This relation is equivalent with the reasonable requirement that there are neither sinks nor sources in the movement of air in the atmosphere.

The last term of Eq. (7.3) describes the temporal changes of the averaged field of concentration $\bar{c}$ by the divergence $\sum_{k=1}^{3}\frac{\partial\left(\overline{u'_k\cdot c'}\right)}{\partial x_k}$ of a quantity $\left(\overline{u'_k\cdot c'}\right)$, which is referred to as a turbulent flux (density) of the pollutant with physical unit of $(kg\ m^{-2}\ s^{-1})$. The turbulence causes a spreading of the various air parcels including the pollutant. Adapting the scheme of Fig. 2.2a to the present situation, the turbulent fluxes may immediately be assumed to be proportional to the spatial concentration gradient (see the second term on the right-hand side of Eq. (7.3)). In analogy with Fick's 1st law (Eq. 2.6) the following relation is obtained:

$$\overline{u'_k\cdot c'} = -K_{c,k}\frac{\partial\bar{c}}{\partial x_k} \tag{7.4}$$

with $K_{c,k}$ referred to as the turbulent diffusion coefficient in units of $m^2\ s^{-1}$.

In Sect. 7.3 the mechanisms of turbulent diffusion will be considered in more detail. It should already be noted, however, that the resulting diffusivities vary with both the height z and the space direction considered.

With Eq. (7.4), the explicit notation of Eq. (7.3) becomes:

$$\frac{\partial\bar{c}}{\partial t} = \bar{P} - \bar{u}\frac{\partial\bar{c}}{\partial x} - \bar{v}\frac{\partial\bar{c}}{\partial y} - \bar{w}\frac{\partial\bar{c}}{\partial z} + \frac{\partial}{\partial x}\left(K_{c,x}\frac{\partial\bar{c}}{\partial x}\right) + \frac{\partial}{\partial y}\left(K_{c,y}\frac{\partial\bar{c}}{\partial y}\right) + \frac{\partial}{\partial z}\left(K_{c,z}\frac{\partial\bar{c}}{\partial z}\right) \tag{7.5}$$

This equation represents the diffusion equation of an atmospheric constituent with concentration c in a turbulent atmosphere. It will be used, in the remainder of this chapter, to introduce into some diffusion-related atmospheric phenomena.

First, a coordinate system is chosen, which drifts with the mean wind vector, i.e. $\bar{u}=\bar{v}=\bar{w}=0$. An initial, instantaneous source emitting an amount Q^* (in kg) of pollution at position $\vec{r}_Q$ is assumed. It is represented by the initial condition

$\bar{c}(x, y, z, t = 0) = Q^* \delta(\vec{r} - \vec{r}_Q)$, with $\delta(\vec{r} - \vec{r}_Q)$ denoting the Dirac delta function (with $\int \delta(\vec{r} - \vec{r}_Q) \cdot dxdydz = 1$). The turbulent diffusion coefficients $K_{c,k}$ are assumed constant over the considered space scale so that $\frac{\partial}{\partial x} K_{c,x} = \frac{\partial}{\partial y} K_{c,y} = \frac{\partial}{\partial z} K_{c,z} = 0$. In this way, following e.g. Etling [2], Eq. (7.5) is transferred into the well-known diffusion equation, Fick's 2nd law, Eq. (2.9):

$$\frac{\partial \bar{c}}{\partial t} = K_{c,x} \frac{\partial^2 \bar{c}}{\partial x^2} + K_{c,y} \frac{\partial^2 \bar{c}}{\partial y^2} + K_{c,z} \frac{\partial^2 \bar{c}}{\partial z^2}, \qquad (7.6)$$

with the solution, see also Eq. (2.10):

$$\bar{c}(\vec{r}, t) = \frac{Q^*}{(4 \cdot \pi \cdot t)^{3/2} \cdot (K_{c,x} \cdot K_{c,y} \cdot K_{c,z})^{1/2}} \cdot \exp\left[-\frac{(x - x_q)^2}{4 \cdot K_{c,x} \cdot t} - \frac{(y - y_q)^2}{4 \cdot K_{c,y} \cdot t} - \frac{(z - z_q)^2}{4 \cdot K_{c,z} \cdot t} \right]$$

$$(7.7)$$

Equation (7.7) describes the distribution of the pollutant by turbulence over an enlarging volume. The range of pollution scales with the Einstein law of diffusion, Eq. (2.11), following the relation $\langle (x_k - x_{kq})^2 \rangle = \sqrt{2K_{c,k}t}$. The concentration at the position of the source is diluted in proportion with $t^{-3/2}$.

As a next step the more realistic scenario of a continuous chimney emission with a stationary point source is considered, where the resulting concentration has to obey the condition of stationarity $\frac{\partial \bar{c}}{\partial t} = 0$. Transport is assumed to occur in x direction only, so that $\bar{v} = \bar{w} = 0$. The flow in x direction is assumed to exceed turbulent diffusion ($|K_{c,x} \frac{\partial^2 \bar{c}}{\partial x^2}| \ll |\bar{u} \frac{\partial \bar{c}}{\partial x}|$) so that, starting with Eq. (7.5), one ends up with:

$$\bar{u} \frac{\partial \bar{c}}{\partial x} = K_{c,y} \frac{\partial^2 \bar{c}}{\partial y^2} + K_{c,z} \frac{\partial^2 \bar{c}}{\partial z^2}. \qquad (7.8)$$

The structure of Eq. (7.8) is easily reestablished from Eq. (7.6), considering that differentiation with respect to the x coordinate disappears on the right-hand side, and that differentiation with respect to time t is replaced by that with respect to x, with an additional factor $\bar{u}$ appearing on the left-hand side. Correspondingly, in the solution of Eq. (7.8),

$$\bar{c}(x, y, z) = \frac{\dot{Q}}{4 \cdot \pi \cdot x \cdot (K_{c,y} \cdot K_{c,z})^{1/2}} \cdot \exp\left[-\frac{\bar{u}}{4x} \cdot \left(\frac{y^2}{K_{c,y}} + \frac{(z - h)^2}{K_{c,z}} \right) \right] \qquad (7.9)$$

the structure of Eq. (7.7) is also maintained. Instead of the quantity z_q the height h is introduced, which could represent, e.g., the height of a smoke-emitting chimney. t is replaced by $x/\bar{u}$ which constitutes the time needed for covering the distance x at travelling speed $\bar{u}$. The value $\dot{Q}$ (in kg s^{-1}) quantifies the efficiency of the source of pollution, which is positioned at $x = 0$, $y = 0$ and $z = h$ (the height of the pollution

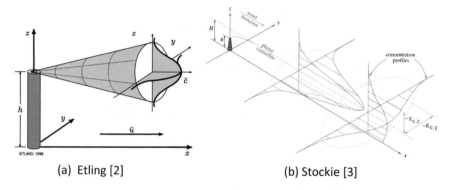

(a) Etling [2]   (b) Stockie [3]

Fig. 7.1 Spatial distribution of pollution originating from a continuous source of emission, predicted by Eq. (7.9) for turbulent diffusion ("Gaussian Plume"): Schematic overview (**a** after Etling [2]) and detailed representation (**b** after Stockie [3]), which also demonstrates the overshooting (H) of the emission height (h) and the contact of the plume with the Earth surface. This allows to describe the deposition of emitted substance in the surroundings of a continuous source

source). Equation (7.9) is widely used in applied meteorology to assess the concentrations of pollutants in the surroundings of a continuous point source. Two ways for illustrating its significance are shown in Fig. 7.1.

Figure 7.1a shows the simplest pattern where the turbulent diffusivities in horizontal and vertical directions are equal. A more detailed illustration is provided in (Fig. 7.1b). Here it is taken into account that the centerline of the plume does not start with the mouth of the chimney. Immediately after emission, smoke rather continues to be shifted upwards by buoyancy, till final temperature equilibration with the surroundings. Turbulent spreading differs notably in vertical and horizontal directions and is, moreover, a function of height z. It is, in particular, considered that the pollution will eventually reach the surface with all the negative consequences known from regions suffering under negligence of environmental protection.

Figure 7.2 illustrates some of the scenarios resulting from turbulence under the various atmospheric conditions. These scenarios are mainly controlled by the vertical gradient of the temperature in the atmosphere, which determines the atmospheric stratification (i.e. the temperature profile in vertical direction). These gradients are related to the change in temperature of a rising air parcel as compared to the respective ambient temperature of the environment. During ascent of an air parcel in the atmosphere, the parcel is adiabatically cooled (adiabatic refers to a process with negligible heat exchange between the air parcel and its surrounding air, which is well met for rising air parcels in the atmosphere due to the low thermal conductivity of air). The energy needed for the expansion of the air parcel during the ascent is consumed on the expense of the internal energy of this parcel, leading to a decrease in its temperature. This is the situation illustrated in Fig. 7.2 with the lines in red. The actual scenario depends on the stratification.

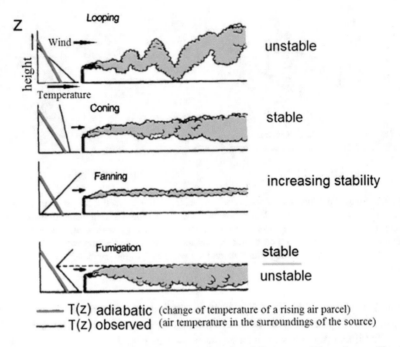

Fig. 7.2 Differently shaped plumes for visualizing the different conditions in stratification (vertical thermal structure of atmosphere) causing turbulent diffusion (schemes from Bierly and Hewson [13] see Hupfer and Kuttler [4])

In the case of the uppermost example shown in Fig. 7.2, the temperature change in the surrounding ambient air exceeds that of the rising air parcel. A rising air parcel will thus be surrounded by air, which (as compared to the air parcel) will become progressively colder during ascent. This results in a continuous increase of the buoyancy of the rising air parcel, just as in a continuous decrease of buoyancy for downward movement, with the effect of an essentially unlimited movement of the air parcel both up- and downwards.

This unrestricted movement in either direction is, obviously, inhibited in the opposite case shown further below. Now, with the temperature gradient in atmosphere below that of the rising air, the plume is stabilized in vertical direction. Fume confinement to a certain range becomes more stringent with further decreasing of the atmospheric temperature gradient (third representation), resulting in fan-like distributed plumes.

Temperature profiles in the atmosphere may exhibit a vertical inversion as exemplified in the bottom of Fig. 7.2. Such situations may result from different winds (shear) in different heights. As an example, the wind may carry warmer air on top of colder air. Under these conditions, smoke is kept below a certain altitude. Such situations are well known for enabling beautiful views over the tops of the mountains, with a sea of fog in the valleys in between.

7.3 Quantification of Turbulent Diffusion

So far the stochastic character of movement in the atmosphere was discussed without worrying about its origin. To establish a quantitative correlation between the fluctuating velocities in air, the law of continuity is applied to the density of the mechanic momentum flux in space:

$$\rho \cdot \vec{\mathbf{v}} = \bar{\rho} \cdot \left(\vec{\mathbf{v}} + \vec{\mathbf{v}}' \right) \left[\frac{\text{kg} \frac{\text{m}}{\text{s}}}{\text{m}^3} \right], \tag{7.10}$$

rather than to the spatial density of particles, i.e. concentration c, as considered in Eq. (7.1). By replacing the concentration c by the momentum flux and assuming incompressibility $(\rho = \bar{\rho} = const.)$, analogous reasoning (see Eq. (7.3) and items (i) to (iii) following this equation) leads to:

$$\rho \frac{\partial \bar{u}}{\partial t} = \bar{P}_u - \rho \cdot \bar{u} \cdot \frac{\partial \bar{u}}{\partial x} - \rho \cdot \bar{w} \cdot \frac{\partial \bar{u}}{\partial z} - \rho \cdot \frac{\partial \left(\overline{u' \cdot u'} \right)}{\partial x} - \rho \cdot \frac{\partial \left(\overline{w' \cdot u'} \right)}{\partial z} \tag{7.11}$$

To simplify the situation, a two-dimensional wind $\vec{v}(u, 0, w)$ with a horizontal component in x- and a vertical component in z-direction is assumed, which varies with altitude z. We furthermore consider the idealized conditions of horizontal homogeneity and of stationarity. The latter condition implies that the averaged momentum flux $\rho \bar{u}$ does not change with time (the wind has a constant velocity). There should be neither sources nor sinks and, finally, it is taken into account that the averaged vertical velocity $\bar{w}$ near the Earth surface must be zero. Then Eq. (7.11) is easily seen to reduce to:

$$\rho \cdot \frac{\partial \left(\overline{w' \cdot u'} \right)}{\partial z} = 0 \tag{7.12}$$

This means that under such conditions the quantity $\tau = \rho \cdot \overline{w'u'}$, referred to as the vertical flux of turbulent momentum, remains invariant with both time (due to the required stationarity) and altitude z.

Since the vertical gradient $\frac{\partial \bar{u}}{\partial z}$ is easily recognized as the driving force for the generation of turbulent momentum fluxes one may note (see also Boussinesq [5])

$$\tau = \rho \cdot \left| \overline{w'u'} \right| = \rho \cdot K_{u,z} \cdot \frac{\partial \bar{u}}{\partial z} \tag{7.13}$$

where the factor of proportionality $K_{u,z}$ has the dimension of a diffusivity and may, moreover, be understood to assume the role of the turbulent diffusion coefficient, similar that introduced with Eq. (7.4).

Following Prandtl [6], velocity fluctuations may be understood as being caused by the ascent or decline of air masses over a certain length l referred to as the Prandtl mixing length. The magnitude u' of velocity fluctuation may thus be estimated as

$$u' = \overline{u}(z+l) - \overline{u}(z) = \overline{u}(z) + l \cdot \frac{\partial \overline{u}}{\partial z} + \cdots - \overline{u}(z) \approx l \cdot \frac{\partial \overline{u}}{\partial z} \qquad (7.14)$$

where, in the second equation, we have made use of a Taylor expansion. With the approach given by Eq. (7.14), fluctuations in velocity are seen to become the larger, the larger the gradient in mean velocity is. Taking into account that, as a consequence of continuity, $u' = w'$, Eq. (7.13) is transformed into:

$$\tau = \rho \cdot \overline{|w'u'|} = \rho \cdot l^2 \cdot \left(\frac{\partial \overline{u}}{\partial z}\right)^2 \qquad (7.15)$$

By comparison with Eq. (7.13) the expression for the turbulent diffusivity is obtained:

$$K_u = l^2 \cdot \frac{\partial \overline{u}}{\partial z} \qquad (7.16)$$

Since turbulent mixing is commonly diminished approaching the surface, Prandtl implied, as a first-order estimate, a direct proportionality between the mixing length l and the height z above the surface. In flow-dynamic experiments v. Karman [7] determined a value of 0.4 for the factor of proportionality. The factor is referred to as the v. Karman constant κ:

$$l = \kappa \cdot z \qquad (7.17)$$

It is common to abbreviate the ratio $\tau/\rho = u_*^2$ $[\text{m}^2/\text{s}^2]$. The parameter u_* is the so-called friction or shear stress velocity:

$$u_* = \sqrt{\frac{\tau}{\rho}} = \sqrt{\overline{|w'u'|}} = l \cdot \frac{\partial \overline{u}}{\partial z} = \kappa \cdot z \cdot \frac{\partial \overline{u}}{\partial z}. \qquad (7.18)$$

By inserting u_* into Eq. (7.16), the turbulent diffusivity appears in the form:

$$K_u = l \cdot u_* = \kappa \cdot z \cdot u_*. \qquad (7.19)$$

By comparing Eq. (7.19) with the corresponding gas diffusion, as provided by conventional gas-kinetic theory, the positions of the mean free path and of the mean thermal velocity are represented by the mixing length and the shear stress velocity (representing a measure of the velocity fluctuations).

Assuming that the dependence of the mean velocity $\bar{u}(z)$ on height z above the ground exceeds that of the shear stress velocity and the v. Karman constant, Eq. (7.18) can be transferred into the following integral relation:

$$\frac{u_*}{\kappa} \int\limits_{z_0}^{z} \frac{\mathrm{d}z}{z} = \int\limits_{\bar{u}=0}^{\bar{u}(z)} \mathrm{d}\bar{u}, \qquad (7.20)$$

which is solved yielding:

$$\bar{u}(z) = \frac{u_*}{\kappa} \cdot \ln\left(\frac{z}{z_0}\right). \qquad (7.21)$$

z_0 is referred to as the aerodynamic roughness of the ground; it denotes the height above the ground (i.e. above $z = 0$) where the mean velocity of wind may be considered to be zero. It is a function of surface texture and may vary with the conditions, such as the roughness over sea surface, which increases with increasing wind velocity (as a consequence of increasing turbulences of the water waves), while the opposite could be true with a meadow of high grass when—at stormy weather—the leaves/grass haulms are bent to the ground, which decreases surface roughness.

Equation (7.21) constitutes a so-called logarithmic wind profile (Prandtl [6]) under near neutral stratification. Extensions of this theory to stable and unstable stratification conditions (see discussion of Fig. 7.2) are described using the Monin-Obukhov-similarity theory (Monin and Obukhov [12]).

As a prerequisite of the dependence given by Eq. (7.21), u_* (and thus the flux of mechanic momentum) is assumed constant over the considered layer in the atmosphere. Via Eq. (7.21), a measurement of the increase of wind velocity, $\bar{u}(z)$, with increasing height z provides a direct measure of both u_* and z_0. Furthermore, by inserting the resulting value of u_* into Eq. (7.19), the turbulent diffusivity is derived, which increases linearly with height z.

Figure 7.3 illustrates an example of such measurements and the subsequent data analysis, which comply with the formalism developed above. The measurement results confirm, in particular, the predicted logarithmic wind profile. The aerodynamic roughness z_0 of the ground is determined to be 1 mm, based on the two measured profiles. The surface of the sandy ground in the considered example is essentially unaffected by the considered wind velocities. The turbulent diffusivities increase with increasing wind velocity (comparing the two profiles P2 with P1) and with increasing height z. The latter dependence is in close agreement with the prediction given by Eq. (7.19). The turbulent diffusivities, moreover, exceed the gas diffusivities (which are, in the atmosphere, of the order of 10^{-5} m^2 s^{-1}) by several orders of magnitude. As a consequence of the turbulences, atmospheric mixing is thus seen to occur at much higher rates than it would be expected due to mere gas phase diffusion (Table 7.1).

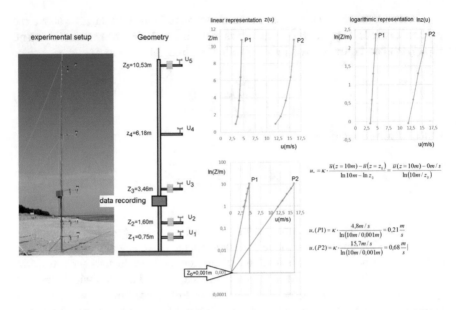

Fig. 7.3 Measurement of wind profiles over sandy ground. Wind velocity is measured by five sensors (U_1–U_5) at five different altitudes (z_1–z_5). At two times of measurement (P1, P2), the recorded gradients in wind velocities are exploited for determining the shear stress velocity u_* and the aerodynamic roughness z_0. All data are summarized in Table 7.1, jointly with the coefficients of turbulent diffusion determined via Eq. (7.19)

Table 7.1 Primary data of the measurement of wind velocity illustrated by Fig. 7.3 and values of shear stress velocity u_* and turbulent diffusivity K_u derived from these data on the basis of Eqs. (7.18) and (7.19) at different altitudes z (with the aerodynamic roughness z_0 in both cases determined to be 1 mm)

	U(z = 10 m)	u_*	z:	0.5 m	1 m	2 m	4 m	10 m	20 m
P1	4.8 m/s	0.21 m/s	K_u (m²/s)	0.042	0.084	0.168	0.336	0.84	1.68
P2	15.7 m/s	0.68 m/s	K_u (m²/s)	0.136	0.272	0.544	1.088	2.72	5.44

7.4 Conclusions

The formalism presented in this chapter to quantify stochastic mass transfer in the atmosphere by turbulent diffusion is based on the mixing theory introduced by Prandtl almost a century ago. It implies the definition of a mixing path length. Reasonable agreement between observations and theory was attained by assuming that the mixing path length is proportional to the height above ground, with a factor of proportionality (referred to as the v. Karman constant) $\kappa = 0.4$. There are numerous attempts to determine the v. Karman constant on the basis of theoretical estimates, which contributes to a stringent theory of turbulent diffusion. Baumert [8], e.g., succeeded in developing a self-consistent system of differential equations for describing turbulent diffusion in the atmosphere, yielding a value of $\kappa = (2\pi)^{-1/2} = 0.399$, in remarkable agreement with the observations. However, irrespective of a time period of more than a century between Reynolds [1], Prandtl [6], Monin and Obukhov [12] and Baumert [8], even today a generally accepted theory to describe atmospheric turbulence is still missing.

Other atmospheric turbulence models are based on approaches where turbulent fluxes are described by the superposition of turbulent eddies of different size. Conceptions of this type were promoted by Richardson [9] and Kolmogorov [10]; they conclude with spectral turbulent diffusivities, which take into account that a turbulent flow consists of different whirls of different size—a so called spectrum of turbulent eddies. Models of spectral turbulent diffusion are nowadays exploited for the analysis and interpretation of turbulence measurements of high spatial and temporal resolution. They are, thus, an important constituent of weather prediction models considering turbulence in the atmosphere (large eddy simulation, Sagaut [14]).

By the conceptions of Reynolds/Prandtl and of Richardson/Kolmogorov, the same phenomenon is described from different perspectives. A unified view on turbulent diffusion might be based on either of them. The development of such an all-embracing view, however (see, e.g., Kraus [11]), will continue to remain a challenging task for both observational measurement and theoretical modelling for the next decades.

References

1. O. Reynolds, Philos. Trans. R. Soc. Lond. A **186**, 123–164 (1894)
2. D. Etling, *Theoretische Meteorologie, eine Einführung* (Vieweg, 1996)
3. J.M. Stockie, SIAM Rev. **53**(2), 349–372 (2011)
4. P. Hupfer, W. Kuttler (eds.), *Witterung und Klima*, 12th edn. (Teubner Vlg, Wiesbaden, 2006)
5. L. Boussinesq, *Theorie de l'ecoulement tourbillonnant* (Paris, 1897)
6. L. Prandtl, Z. angew. Math. u. Mech. **5**, 136–139 (1925)
7. Th. v. Karman, Nachr. Akad. Wiss. Göttingen, Math.-phys. Kl. 58–76 (1930)
8. H.Z. Baumert, Universal equations and constants of turbulent motion. Phys. Scr. T155 014001, 12 pp (2013). https://doi.org/10.1088/0031-8949/2013/T155/014001
9. L.F. Richardson, *Weather Prediction by Numerical Process* (Cambridge University Press, 1922)
10. A.N. Kolmogorov, Dissipation of energy in locally isotropic turbulence. Dokl. Akad. Nauk SSSR **32**, 8–16 (1941)
11. H. Kraus, *Grundlagen der Grenzschicht-Meteorologie* (Springer, 2008)
12. A.S. Monin, A.M. Obukhov, Basic laws of turbulent mixing in the surface layer of the atmosphere. Tr. Akad. Nauk. SSSR Geophiz. Inst. **24**, 163–187 (1954)
13. E.W. Bierly, E.W. Hewson, Some restrictive meteorological conditions to be considered in design of stacks. J. Appl. Meteorol. **1**, 383–390 (1962)
14. P. Sagaut, *Large Eddy Simulation for Incompressible Flows—An Introduction*, 3rd edn. (Springer, 2006)

Chapter 8
Hot Brownian Motion

Klaus Kroy and Frank Cichos

8.1 Introduction

Brownian motion, as characterized by Albert Einstein in 1905 [1], is the thermal motion of suspended particles that are small enough to jiggle perceptibly, but large enough to be visible in the microscope. This sort of "motion from heat" is not forbidden by the second law of thermodynamics, as incorrectly suggested by Wilhelm Röntgen in a letter to Einstein, but rather reveals its atomistic origin. Indeed, Jean Perrin received the Nobel prize in 1926 "for proving atoms real", on this basis, so that even the "Energetiker" group around Wilhelm Ostwald no longer openly denied their existence. And some 35 years later, Richard Feynman started his famous lectures [2] with the words: "If, in some cataclysm, all of scientific knowledge were to be destroyed, and only one sentence passed on to the next generation of creatures, what statement would contain the most information in the fewest words? I believe it is [...] that all things are made of atoms—little particles that move around in perpetual motion, attracting each other when they are a little distance apart, but repelling upon being squeezed into one another".

In the following, we give a non-technical introduction[1] to recent developments that extend Einstein's work and the notion of Brownian motion to conditions very far from equilibrium. In particular, we consider colloidal particles in non-isothermal solvents, i.e., under conditions arising whenever either the particles themselves,

---

[1]For further introductory reading see Refs. [3–6]. The names of our collaborators (partly funded by the Deutsche Forschungsgemeinschaft and the Humboldt foundation), who did much of the original work reviewed here, can be found in the references, at the end.

K. Kroy (✉)
Institute of Theoretical Physics, Leipzig University, Leipzig, Germany
e-mail: klaus.kroy@uni-leipzig.de

F. Cichos
Peter Debye Institute for Soft Matter Physics, Leipzig University, Leipzig, Germany
e-mail: cichos@physik.uni-leipzig.de

DOI: 10.1007/978-3-319-67798-9_8, © Springer International Publishing Switzerland 2015

selected container walls, or any other embedded (nano-)structures are locally heated above the ambient temperature. Such "hot" conditions invalidate, in one stroke, all of our conventional statistical mechanics tools, such as equilibrium ensembles and Boltzmann factors. The latter are the main tools used to leapfrog over the forbiddingly complicated microscopic dynamics of many-body systems, yet to still achieve quantitative control on an atomistic basis. But they generally cease to work far from equilibrium. So it seems as if hot Brownian motion destroys all the nice concepts that Brownian motion once helped to establish. It is not quite that bad, in the end. Below, we explain how to deal with such complicated non-isothermal situations, even if they involve directed autonomous particle motion, viz., active swimming.

Although there is no lack of motivation for studying the physics of swimmers [7], hot Brownian motion provides compelling intrinsic reasons to do so. Namely, in absence of special symmetries, any nanoparticle in a non-isothermal solvent is automatically a "hot Brownian swimmer" and thereby a realization of a most promising swimmer design in terms of potential (scalable, biocompatible, sustainable, steerable, ...) applications [8]. Vice versa, all so-called active Brownian particles or microswimmers (heated or not) are "hot" in the sense that their diffusion is strongly enhanced by their non-equilibrium self-propulsion [9]. Swimming moreover gives rise to unconventional inelastic interactions with other particles and container walls. These may cause unusual effects akin to those observed in agitated granular gases [10] or even cell colonies [11], and they seriously undermine apparently unimpeachable thermodynamic notions such as pressure and surface tension [12]. In general, hot Brownian particles and swimmers are thus not easily treated along the same lines as conventional colloids. And a world made of such objects as its "atoms" may seem infinitely more complicated than our actual physical world. Yet, it actually is a very good toy model for living matter, which relies on a most delicate interplay between nonequilibrium ("active") and equilibrium ("passive") Brownian motion to fulfill its complex tasks (see e.g., [5, 13–15]). Such models should thus bring us a good step closer to linking the notions of Brownian motion and diffusion to fluctuation phenomena studied in other disciplines, some of which are addressed in this volume.

8.2 Brownian Motion

To understand hot Brownian motion, it is useful to first recall a few facts about ordinary equilibrium Brownian motion. A qualitative understanding on the level of Einstein's pioneering work [1] will suffice for the remainder. Einstein looks at small particles suspended at low concentration in an isothermal solvent, which could for example be pollen grains in water or aerosols in still air. Assuming the particles themselves to be made out of a somewhat denser material than the surrounding solvent, gravity always pulls them downwards. As a consequence, if you follow a particle over time, you will always find it drifting downwards, on average. Yet, random Brownian fluctuations prevent the particle ensemble from settling completely to the ground. They give rise to an osmotic pressure p proportional to temperature T and particle number concentration c, very much as for a dilute gas.[2] This pressure thus increases near the ground such as to exactly balance the gravitational force density. Replicating the conventional arguments for the atmospheric pressure and density as a function of altitude, one finds the very same result, namely that pressure p and particle concentration c both decrease exponentially as a function of height h,

$$p(h) \propto c(h) = c(h')e^{-[U(h)-U(h')]/k_B T}. \tag{8.1}$$

The potential energy difference $U(h) - U(h')$ between heights h and h' could explicitly be written as particle mass times gravitational acceleration times height difference. Boltzmann's constant k_B (historically the gas constant over Loschmidt's number) in the exponent actually made this "Brownian barometer equation" so exciting for Einstein and Perrin, since it implicitly refers to the number of atoms per mole and therefore bears witness of the atomic structure of all matter.

Now, in a second ingenious step, Einstein suggests to revisit the classical argument leading to Eq. (8.1) from a different perspective. Consider, so he says, the same situation not in terms of a static force balance (gravity versus osmotic pressure) but in terms of the corresponding balancing particle fluxes. According to Stokes' law, gravity excites a downward drift flux $-c\nabla U/\zeta$ (concentration $\times$ velocity), inversely proportional to the friction ζ of the particles with the solvent. This is the flux witnessed by an observer concentrating on the average motion of an individual particle. According to Fick's law, Brownian fluctuations give rise to an opposing diffusion flux $-D\nabla c \overset{(8.1)}{=} Dc\nabla U/k_B T$, quantified in strength by the diffusion coefficient D, and directed from high to low solute concentrations c. As for the static forces, the dynamic fluxes have to be precisely balanced everywhere, in equilibrium. This condition of *detailed balance* allows Einstein to infer a most remarkable relation, first obtained by William Sutherland, namely

$$D = k_B T/\zeta. \tag{8.2}$$

It constitutes a universal link between a measure of the fluctuations (the diffusivity) and a dissipative transport coefficient (the friction) of a dilute solute, and can therefore be called the mother of all fluctuation-dissipation relations. Again, the remarkable thing for Einstein and Perrin was that two transport coefficients appearing in two mesoscopic equations, namely the diffusion equation for the solutes and Stokes' equation for a particle dragged through a fluid, are somehow linked to Loschmidt's number, and hence to the notion of atoms. In the joints between some unsuspicious smooth macroscopic continuum equations lurked, much to the dismay of Ostwald, the grotesque face of the atomic world. In this sense, Eq. (8.2) is comparable to another famous equation put forward by Einstein at about the same time, namely $E = \hbar\omega$, which relates the two classical notions of energy and frequency via the microscopic Planck constant. In a more practical reading, Eq. (8.2) allows a

---

[2]The analogy might seem compelling, but the opponents of the atomistic world view would have objected to the application of thermodynamic notions to colloidal particles.

Brownian particle to be used as a thermometer, if its friction coefficient ζ is known, or as a rheometer for the fluid viscosity η, if the functional form of $\zeta(\eta)$ is known.

Finally, towards the end of his paper [1], Einstein pushes his analysis of Brownian motion one step further, proposing a microscopic model (in modern language "a random walk") to argue that the dynamics of a Brownian particle of mass m is indeed diffusive at late times $t \gg m/\zeta$ (in modern language "in the Markov limit"), where inertial effects have been damped out. Thereby, he underpins the interpretation of Eq. (8.2) as a fluctuation-dissipation relation by demonstrating that D indeed characterizes the particle fluctuations. He first shows that the particle concentration, which he identifies with the probability for finding a single particle released at the origin at position $\mathbf{r}$ after time t, obeys the diffusion equation (see also the notations in Eqs. (2.9) and (2.12) for diffusion in one and two dimensions)

$$\frac{\partial}{\partial t}c(\mathbf{r}, t) = D\nabla^2 c(\mathbf{r}, t). \tag{8.3}$$

As an important consequence, the particle velocity, which is the central observable in Newtonian mechanics, turns out to be ill-defined (formally divergent) for the commonly accessible times $t \gg m/\zeta$ (Fig. 8.1, left panel). It is therefore not a good idea to try and measure a Brownian particle's velocity, as experimentalists commonly did in many futile attempts throughout the late 19th century. Instead, as Einstein finds, the mean-square displacement and the diffusivity,

$$\Delta \mathbf{r}^2(t) \equiv \int d\mathbf{r} \, \mathbf{r}^2 c(\mathbf{r}, t) \quad \text{and} \quad D \equiv \frac{1}{6}\frac{\partial \Delta \mathbf{r}^2(t)}{\partial t}, \tag{8.4}$$

are well-behaved (good observables) over a wide range of time scales.

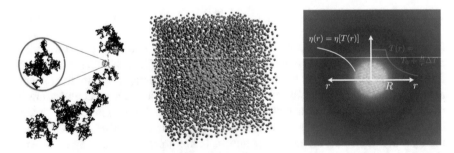

Fig. 8.1 Principle of hot Brownian motion. *Left*: the trajectory of a hot Brownian particle (at late times $t \gg m/\zeta$) is a "diffusive" fractal (see, e.g., Sect. 11.4 for a more general introduction to fractals), as for ordinary Brownian motion, just traversed faster; it is nowhere differentiable, hence the velocity is ill-defined. *Center*: snapshot of an atomistic non-equilibrium molecular dynamics simulation featuring a hot nanoparticle in a Lennard–Jones fluid (particles color-coded for their kinetic energy or "molecular temperature"). *Right*: coarse-grained co-moving molecular temperature and viscosity fields, $T(r)$ and $\eta(r)$, around a uniformly heated spherical nanoparticle of radius R (courtesy of M. Selmke)

8.3 Hot Brownian Motion

Hot Brownian motion is simply the Brownian motion of colloidal particles that are hotter than their solvent. More generally, one may speak of non-isothermal Brownian motion, if any spatially varying temperature profile is somehow maintained in the solvent around the particle. The recent advances in measuring, manipulating, and theoretically characterizing the motion of hot Brownian things heavily exploit that "Brownian" means "small enough to jiggle perceptibly, but large enough to be visible in the microscope". Even more importantly, they exploit that it means "*large enough to admit some systematic coarse-graining*", or, briefly, that Brownian particles belong to a mesoscopic "middle-world" [5].

This crucial property allows some universal (i.e., independent of microscopic details) and practically useful (as opposed to merely formal) exact mathematical statements about Brownian motion to be formulated, without tinkering with atoms. And it is the main reason why Einstein's paper from 1905 has started an unfinished "slower revolution" [3] that makes it very popular[3] and relevant till today [4]. As a consequence, the theoretical arguments put forward by Einstein in 1905 only need very little amendment to generalize the laws of Brownian motion from the important (but conceptually very special) class of equilibrium conditions to situations very far from equilibrium. Moreover, much like their equilibrium counterparts, the resulting predictions are still of a universal character, although less than in equilibrium. And they may also still serve as a paradigm for fluctuations in many other, apparently unrelated mesoscopic devices, e.g., in electrical engineering and nanophotonics [16, 17], and even for living matter [6, 18]. The ability to coarse-grain is the key to make progress in all these directions, and, quite generally, our most important tool to bring an infinitely complicated world within reach of analysis and comprehension [19].

The heat emanating from a hot particle rapidly diffuses into the surrounding solvent and thereby establishes a comoving temperature gradient around the comparatively slowly moving particle, as sketched in Fig. 8.1 (center and right). This invalidates the conventional discussion of isothermal Brownian motion, which predicts the particle dynamics directly from equilibrium thermodynamics, i.e., on a coarse-grained level, without ever referring to the complicated dynamics of interacting atoms. We cannot directly copy this elegant trick, here, since the fluid viscosity and thermal fluctuations vary spatially in the vicinity of the particle, so that it is not a priori clear which temperature or viscosity should be relevant for the particle motion, and whether we can apply conventional thermodynamic arguments, at all. How can we then know how much the translational and rotational Brownian motion are enhanced by heating? Does Eq. (8.2) still hold? And what is then meant by T and ζ?

Universal answers (independent of molecular details) to these questions are provided by the theory of hot Brownian motion [20, 21]. In order to avoid the complexity of the general results, we focus on the Markov limit ($t \gg m/\zeta$), here. In other words, as Einstein, we consider only late times, where we can neglect memory effects aris-

---

[3] At the time of writing, Google Scholar lists more than 6000 citations.

ing from the slow solvent hydrodynamics [22, 23]. It can be shown that the non-equilibrium effects can then actually be subsumed into a small number of effective transport coefficients, chiefly an effective reduced friction coefficient ζ_{HBM} and an effective Brownian temperature T_{HBM}. These two quantities together determine the effective diffusivity D_{HBM} according to a generalized Einstein relation [20, 24]

$$D_{HBM} = k_B T_{HBM}/\zeta_{HBM}. \tag{8.5}$$

In the Markov limit, hot Brownian motion can thus be mapped onto equilibrium Brownian motion in a solvent with an effective temperature and viscosity. This is reminiscent of a classical trick in thermodynamics, where a nonequilibrium process is replaced by an equivalent equilibrium process effecting the same state change, except that there is no state change, here. The generalized Einstein relation implies, in particular, that a hot Brownian particle can still be employed as a thermometer or as a rheometer, in the Markov limit, if one is aware of the fact that it measures an effective temperature or viscosity, respectively. In general, the effective quantities are complicated functions of the local "molecular" temperature and viscosity fields, $T(\mathbf{r})$ and $\eta(\mathbf{r})$, throughout the whole fluid, but they can be calculated to good precision, for many practical purposes.

For example, to estimate the effective friction of a hot sphere, one needs to generalize the classical calculation by Stokes for the friction coefficient ζ to a radially varying temperature $T(r)$ and viscosity $\eta(r)$. If the equation of state $\eta(T)$ is known, explicit predictions for the effective translational (t) and rotational (r) friction coefficients $\zeta_{HBM}^{t,r}$ of a hot sphere can then be computed [24–26].

It is conceptually (and also practically) more interesting to understand the effective temperature that characterizes the thermal agitation of a hot Brownian particle. By virtue of the Brownian scale separation, a local thermal equilibrium can still be assumed to hold, even when the molecular temperature $T(\mathbf{r})$ varies appreciably over distances comparable to the particle diameter but long compared to the molecule size. This allows a consistent linear theory of non-isothermal fluctuating hydrodynamics to be constructed, based on the non-isothermal Stokes equation for the solvent dynamics [21]. The theory determines how the local solvent fluctuations dictated by the spatially varying molecular temperature field $T(\mathbf{r})$ throughout the whole solvent volume are propagated by hydrodynamic interactions to the Brownian particle and contribute to its apparent Brownian temperature T_{HBM}. For a hot Brownian sphere,[4] the result can be written as the weighted average [24]

$$T_{HBM} = \frac{\int \phi(\mathbf{r})T(\mathbf{r})\,d\mathbf{r}}{\int \phi(\mathbf{r})\,d\mathbf{r}}. \tag{8.6}$$

The weight $\phi(\mathbf{r})$ is the so-called dissipation function—essentially the product of the viscosity and the squared velocity gradient. Due to the different flow fields for

---

[4]The general expression for a non-spherical particle in an arbitrary temperature field is slightly more complex than Eq. (8.6), but its basic structure is the same [20].

translational (t) and rotational (r) motion of the particle, this prescription leads to different effective temperatures for translation and rotation, namely [24, 25]

$$T_{\text{HBM}}^{\text{t}} \approx T_0 \left(1 + \frac{5}{12}\Delta T\right), \qquad T_{\text{HBM}}^{\text{r}} \approx T_0 \left(1 + \frac{3}{4}\Delta T\right). \qquad (8.7)$$

Here ΔT is the difference between the solvent temperature at the particle surface and the ambient temperature T_0 (cf. Fig. 8.1), and higher order terms in ΔT are usually small in actual applications. The effective temperatures can in turn be related to effective diffusivities $D_{\text{HBM}}^{\text{t,r}}$ for translation and rotation, via Eq. (8.5).

In hot Brownian thermometry, one thus measures different effective temperatures for different degrees of freedom, which can provide hints at the spatial structure of the molecular temperature field $T(\mathbf{r})$. Note that these effective temperatures are not merely postulated, as in some other areas of non-equilibrium statistical mechanics, but can systematically be calculated from the underlying non-isothermal fluctuating hydrodynamic theory [21]. Along the same lines, it is even possible to deal with Brownian memory effects, and to show that the effective temperature Eq. (8.6) is merely the low-frequency limit of an equivalent formula in which all terms are frequency dependent [20]. The corresponding *temperature spectrum* replaces T_{HBM} outside the Markov limit. One then finds that the translational and rotational *velocities* are systematically hotter than the corresponding position- and orientation-coordinate degrees of freedom, respectively. All this eventually leads to the fancy notion of hot Brownian thermospectrometry [27], which means that one can, in principle, indirectly infer the molecular temperature field $T(\mathbf{r})$ in a non-isothermal solvent from observations of the Brownian fluctuations of a suspended particle, over a large frequency range.

So far, a number of predictions of the theory of hot Brownian motion could be validated experimentally and in numerical simulations [24, 28, 30, 31]. We specifically mention the experimental verification of the translational effective temperature from Eq. (8.7). Interestingly, the experiment can take advantage of the solvent heating due to the hot particle to achieve a highly sensitive and background-free detection of its Brownian motion, as described in Sect. 8.5. Figure 8.2 (left panel) provides a parameter-free comparison of the average diffusion time τ_D of a Brownian particle heated within a laser focus, which has been obtained by this method, with the prediction from $T_{\text{HBM}}^{\text{t}}$ in Eq. (8.7). An even more comprehensive analysis of the various temperatures, i.e., the conventional local molecular solvent temperature and the effective temperatures characterizing the Brownian dynamics of various degrees of freedom (rotational, translational positions and velocities) of the particle is in principle possible in atomistic simulations (Fig. 8.2, right panel). But the high compressibility of the Lennard–Jones solvent causes technical difficulties that currently still impede the validation of several aspects of the theory by computer simulations.

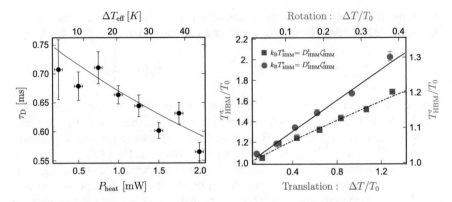

Fig. 8.2 Practice of hot Brownian motion (plots adapted from Ref. [8], with kind permission of The European Physical Journal (EPJ)). *Left*: parameter-free experimental test of Eq. (8.7) employing the Twin-PhoCS method (Sect. 8.5) to measure the average time τ_D for crossing the laser focus [28]. *Right*: non-equilibrium molecular dynamics simulations [29] deduce the effective temperature T_{HBM} from the particle diffusion via Eq. (8.5); lines indicate the predictions for the rotational (solid) and translational (dot-dashed) effective Brownian temperatures from Eq. (8.7)

8.4 Hot Brownian Swimming

The study of microswimmers has a long history dating back to the 17th century, when they were first studied under the microscope by the Dutch draper Antoni van Leeuwenhoek. Only much later, starting with systematic investigations as those by Robert Brown and Adolphe Brogniard in the early 19th century, researchers slowly became aware of the interference of Brownian motion with micro-scale swimming, and much of the pioneering work was devoted to disentangling both effects. So the study of so-called "animalcules" and their self-propulsion predated that of molecules and their thermal motion, and what started as an investigation of the former eventually furnished proof of the existence of the latter.

As pointed out in the introduction, there are compelling intrinsic reasons to retrace the historic path backwards and extend the (by now established) analysis of Brownian motion to swimming. Non-isothermal conditions automatically turn all thermally asymmetric particles into hot swimmers. An archetypal example for a hot microswimmer is provided by a Janus sphere half covered with gold [32], as illustrated in Fig. 8.3 (left). Such a particle heats up asymmetrically and therefore excites an asymmetric temperature gradient in the surrounding solvent, when illuminated by a green laser. The temperature gradient, in turn, excites a boundary-layer flow along the particle surface [8, 33], which gives rise to a net drift motion of the particle along its symmetry axis. The same mechanism is responsible for thermophoretic motion in an external temperature gradient, but here the gradient is caused by the hot particle itself. This is why one also speaks of self-phoresis [34]. A very convenient coarse-grained description condenses the boundary-layer flow into a simple slip boundary condition for the hydrodynamic solvent velocity on the particle surface. This acknowledges that phoresis, as any kind of swimming, is a force-free type of motion or self-propulsion, and it allows for a good analytical control over mutual hydrodynamic interactions of swimmers and related interesting behavior near walls and obstacles [35].

Swimming thus contributes a systematic "ballistic" drift along the instantaneous particle axis $\hat{\mathbf{u}}(t)$ to the random Brownian motion. As any systematic drift, it is outpaced by the diffusive translational Brownian motion at short times but prevails over it at late times. However, the particle axis $\hat{\mathbf{u}}(t)$ is itself subject to *rotational* Brownian motion, which randomizes the motion again, at very late times, so that it becomes once more diffusive. The particle dynamics for $t \gg m/\zeta$ is therefore well-captured by the two Langevin equations

$$\frac{d\mathbf{r}}{dt} = v_p \hat{\mathbf{u}} + \sqrt{2D_{HBM}^t}\,\xi_t\,, \qquad \frac{d\hat{\mathbf{u}}}{dt} = \sqrt{2D_{HBM}^r}\,\xi_r \times \hat{\mathbf{u}}\,, \qquad (8.8)$$

for the position $\mathbf{r}(t)$ and orientation $\hat{\mathbf{u}}(t)$ of the swimmer. Here $\xi_t(t)$ and $\xi_r(t)$ represent the translational and rotational thermal noise. More precisely, the values at any time are independently drawn from Gaussian distributions of vanishing mean and unit variance. Langevin equations are simply an alternative formalism to describe diffusive dynamics. In fact, the first Eq. (8.8) is equivalent to a diffusion equation with a drift given by the swim velocity of magnitude v_p and direction $\hat{\mathbf{u}}$. Similarly, the second equation is equivalent to a diffusion equation for the particle axis, which is a unit vector and thus diffuses on a unit sphere.

In the Markov limit, hot Brownian motion thus merely enters the discussion of the swimmer's motion through the effective translational and rotational diffusivities, D_{HBM}^t and D_{HBM}^r. In unbiased hot Brownian motion (for a homogeneous hot sphere), one usually only notices the former. But for the swimmer, D_{HBM}^r is more easily noticeable via the randomization of the swim direction. Accordingly, Eq. (8.8) predicts a crossover in the mean-square displacement

$$\Delta\mathbf{r}^2(t) = 6D_{HBM}^t t + \frac{2v_p^2}{(D_{HBM}^r)^2}\left[D_{HBM}^r t + e^{-D_{HBM}^r t} - 1\right] \sim \begin{cases} v_p^2 t^2 \\ 2v_p^2 t/D_{HBM}^r \end{cases} \qquad (8.9)$$

The motion is "ballistic" at short times $t \ll 1/D_{HBM}^r$ (upper line) and diffusive at late times $t \gg 1/D_{HBM}^r$ (lower line), corresponding to distances shorter/longer than the persistence length v_p/D_{HBM}^r. The particle trajectories thus resemble conformations of semiflexible polymers. An experimental confirmation is provided in Fig. 8.3 (right). The late-time diffusivity $v_p^2/3D_{HBM}^r \gg D_{HBM}^t$ is strongly enhanced by the propulsion and therefore increases strongly upon heating.

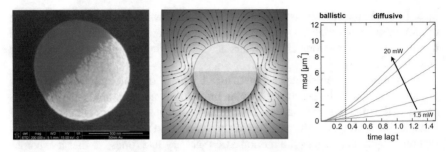

Fig. 8.3 Principle of hot Brownian swimming. *Left*: electron microscopy image of a 1 μm polystyrene Janus particle covered with a 50 nm gold film on one hemisphere. *Center*: the streamlines around a realistically modeled (force-free) hot Janus swimmer with a heated gold cap, calculated in the boundary-layer approximation, exhibit a pronounced near-field structure that is responsible for its directed self-propulsion (adapted from Ref. [8], with kind permission of The European Physical Journal (EPJ)). *Right*: the trajectories of hot Brownian swimmers are persistent random walks, resembling semiflexible-polymer conformations; rotational hot Brownian motion limits their persistence, giving rise to a crossover from a "ballistic" to a diffusive form of the mean-square displacement (msd), Eq. (8.9), at lag times $t \simeq 1/D^r_{HBM}$ (near dotted line)

8.5   Detecting and Steering Hot Brownian Particles

A well controlled and well defined model system for studies of hot Brownian motion is provided by a gold nanoparticle suspended in water. Gold nanoparticles strongly absorb visible light due to a surface plasmon resonance, a collective excitation of their conduction band electrons. It is this effect, which gives gold suspensions their distinct reddish color. The excited electrons thermalize within femtoseconds and release their energy within picoseconds to the metal ions, from where it spreads further via thermal vibrations. The heat conductivity of gold exceeds that of typical solvents by more than two orders of magnitude, so that a gold sphere that has a radius R of some tens or hundreds of nanometers can be represented as an essentially isothermal nano heat source in an infinite heat bath of ambient temperature T_0. If the temperature increment of the solvent at the particle surface is denoted by ΔT, Fourier's law[5] yields the stationary temperature profile

$$T(r) = T_0 + \Delta T R / r. \tag{8.10}$$

This solution is practically attained within hundreds of nanoseconds. As heat diffusion is several orders of magnitude faster than Brownian diffusion, the temperature profile around a hot Brownian particle can be treated as a comoving field. The associated radial heat flow is responsible for the nonequilibrium character of hot Brownian motion. But it also allows for a very efficient detection of its motion by optical microscopy, which would otherwise be difficult to achieve because of the faint optical contrast. The scattered intensity scales with the square of the volume for small

---

[5]Joseph Fourier assumed heat to diffuse, an idea adapted to particles by Adolf Fick in 1855.

scattering sources, so that non-fluorescent nanometer-sized objects are optically very hard to detect. The absorbed heat scales linearly in the particle volume and moreover modifies the density and refractive index $n(r)$ of the surrounding medium,

$$n(r) = n_0 + (\partial n/\partial T)\,\Delta TR/r \qquad (8.11)$$

The long-ranged temperature profile of Eq. (8.10) thus gives rise to an associated long-ranged "mirage" or "thermal lens" around a small heat source. The refractive index changes are quite small $(-\partial n/\partial T \simeq 10^{-3} - 10^{-4}\,\mathrm{K}^{-1}$, with n commonly decreasing upon thermal expansion) but can be probed very sensitively by optical means. The deflection of a laser beam by such a thermal lens is the principle behind the methods of *photothermal microscopy* and *photothermal spectroscopy* (PhoCS). Such photothermal techniques have first been developed in the 1960s [36, 37] but have only more recently been refined to achieve single-particle and single-molecule sensitivity [38–41].

Light scattering from the thermal lens created by a point-like heat source has a perfect microscopic analogue that is very familiar from atomic physics and quantum mechanics, namely Rutherford or Coulomb scattering. As such it sparked much of our current understanding of the inner structure of atoms [43]. In its photothermal variant, the dielectric permittivity profile $\epsilon(\mathbf{r}) = n^2(\mathbf{r})$ plays the role of the Coulomb potential. Since the second terms on the right-hand sides of Eqs. (8.10) and (8.11) are small compared to the first (n_0), $\epsilon(\mathbf{r})$ also decays with the inverse distance from the heat source. The alternative mathematical descriptions of the phenomenon by Fermat's principle (ray-optics) and by Helmholtz's equation (wave optics) find their perfect analogues in the classical (as adopted by Rutherford) and the full quantum mechanical (using Schrödinger's equation) treatments of the Coulomb scattering problem. Photons are deflected by the photonic potential $\epsilon(\mathbf{r})$ in the very same manner as Rutherford's α-particles were deflected by Coulomb interaction with the atomic nuclei.

The left panel of Fig. 8.4 shows a macroscopic setup for a demonstration experiment of photonic Rutherford scattering. A green laser beam entering the cube from the left heats a small metal sphere embedded in an acrylamide block. The emerging refractive index profile around the metal sphere is probed by a red laser beam passing at a distance b from the scattering center, corresponding to the impact parameter in Rutherford's analysis of his experiment. The beam is deflected by an angle θ related to the scattering parameter by $\cot(\theta/2) \propto b$. Photonic Rutherford scattering thus probes the refractive index profile in the same way as Rutherford probed Coulomb's law (and possible deviations from it) [42], except that there is no sizable backscattering, as the refractive index changes relative to the vacuum value are typically very weak.

The role of the mirage as a diverging lens is easily demonstrated in the macroscopic experiment, as well, as seen from Fig. 8.4 (center), which shows a photo of the distorted image of a rectangular grid positioned behind the acrylamide block. In the same manner, a tightly focused probe laser beam, which does not itself excite surface plasmons, is defocussed by the thermal lens around a hot nanoparticle. This

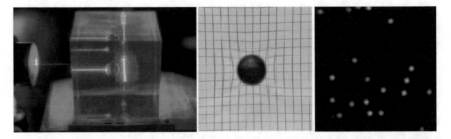

Fig. 8.4 Principle of photothermal detection (Partly adapted from Ref. [42]). *Left*: a macroscopic experiment demonstrating the photothermal version of Rutherford scattering: a metal sphere of 1 cm diameter, embedded in an acrylamide cube with edge length of 10 cm, is heated by a green laser beam from the left; the emerging "mirage" is detected by a red laser. *Center*: it distorts a rectangular grid photographed through the acrylamide block (courtesy of M. Selmke and M. Braun). *Right*: by the same principle, photothermal scanning microscopy images tiny gold nanoparticles (10 nm) embedded in a polymer via their defocusing of a probe beam

defocussing is the actual signal measured in photothermal microscopy by a lock-in detection scheme [42, 44]. The long-range character of the refractive-index profile allows very small absorbers such as quantum dots [45] or even single molecules [40] to be detected with high contrast.

The time-scale separation between heat diffusion and Brownian motion allows a perfect detection of the latter by photothermal microscopy. In particular, one can record and analyze the fluctuations of the photothermal signal caused by particles traversing the focal volume of a photothermal microscope. Each particle crossing the focal volume causes a photothermal burst much like the fluorescence bursts in fluorescence correlation spectroscopy (FCS) [46]. The length of these bursts corresponds to the time spent by the hot Brownian particles within the focal volume (see Fig. 8.5 left). Its average determines the decay of the autocorrelation function of the photothermal signal. Based on this principle, photothermal correlation spectroscopy (PhoCS) was independently developed and applied to single gold nanoparticle diffusion by diverse groups [47–49]. A refined version, called twin-PhoCS [28], exploits the peculiarities of the lensing mechanism in a more sophisticated way. It is based on the insight that a lens placed exactly in the focal plane of the focussed probe beam leaves the divergence of the probe beam unchanged, whereas a positioning slightly below or above the focal plane changes the divergence of the probe beam in opposite directions. In other words, the effective photothermal detection volume splits up into two sharply separated lobes giving a positive/negative photothermal signal S_+/S_- corresponding to diminished/enhanced beam divergence, as exemplified by Fig. 8.5. (The twin-focus splitting naturally occurs in axial direction, but can be created in the focal plane, as well [44].) Importantly, the statistics of the sign changes of the photothermal signal are independent of the precise size and shape of the two lobes, allowing for considerably improved quantitative control. As an additional benefit, drift components in the particle motion, as e.g., due to swimming or radiation pressure, are readily revealed by an analysis of the sign changes. Figure 8.5 displays the

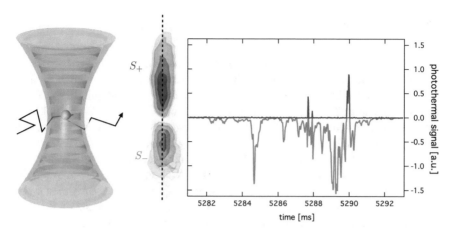

Fig. 8.5 Principle of twin-PhoCS. *Left*: a particle diffusing through the focal volume of a photothermal microscope and heated by an intensity-modulated laser (typically at 200 kHz), generates a photothermal lens probed by a second laser. *Center*: the lensing mechanism creates a splitting of the focal volume into two parts ("twin-focus"). *Right*: time trace of the photothermal signal from a hot gold nanoparticle; negative/positive signals S_-/S_+ correspond to enhanced/diminished probe-beam divergence by its surrounding thermal lens

signal from a single particle entering the lower lobe and briefly visiting the upper lobe before eventually escaping. A detailed analysis of such photothermal microscopy time traces by auto-correlation and cross-correlation methods allows for very precise tests of the theory of hot Brownian motion, as shown in Fig. 8.2. What requires more experimental effort is to test the more complex dynamics at short times, and to test the rich theoretical predictions for hot Brownian motion beyond the Markov limit, alluded to in Sect. 8.3. Even ordinary Brownian motion in isothermal fluids is not easily studied on such short timescales. It has only recently become possible to push the resolution down to a few nanoseconds, so that a quantitative experimental characterization of its complex hydrodynamic memory and inertial effects could be achieved for the first time [23, 50].

Completely new experimental opportunities arise for the self-propelling (i.e., asymmetric) hot particles, discussed in the preceding section. Single-particle tracking can be used to detect the position and orientation of Janus particles. Advanced real-time detection and feedback techniques put the modern experimenter in the position of a Maxwell demon, who literally knows the positions and orientations of all his "atoms". This knowledge can be exploited to switch the laser light that fuels the swimmers' engines on and off at the right time, in order to impose complex swimming patterns and force-free particle steering onto hot Brownian particles by so-called *photon-nudging* techniques [51, 52].

8.6 Exact Symmetry of Hot Brownian Swimming

As recalled in Sect. 8.2, Einstein invoked the condition of detailed balance for the thermodynamic forces and fluxes in order to arrive at the Brownian barometer formula and the fluctuation-dissipation relation, Eqs. (8.1) and (8.2), respectively. Thereby, he imposed strong symmetries on the motion of the Brownian particle. One may ask what survives of all this in situations far from equilibrium, when the mesoscopic physics is still stationary, but the underlying dynamics is not time-symmetric any more, so that detailed balance is broken. (Time symmetry may for instance be broken by persistently shearing a colloidal suspension or, more severely, by shining a laser on it to turn the suspended particles into self-propelled hot Brownian swimmers.) The answer to this question is provided by the so-called fluctuation theorem, which is a generalization of the second law of thermodynamics and the fluctuation-dissipation theorem, contained in Eqs. (8.2) and (8.12). Like them, it comes in several closely related formulations [53]. Its essence is that the entropy production provides an objective metric for the "distance from equilibrium" (the degree of time-reversal symmetry breaking) of nonequilibrium processes. However, "hard to compute, and even harder to measure experimentally, it has been little studied in active systems" [54]. Luckily, for a hot Brownian swimmer in the Markov limit, a fully quantitative analytical formulation of the fluctuation theorem can be worked out. Moreover, with dedicated experimental techniques, such as described in Sect. 8.5, the theory can be tested with high precision, as demonstrated in Ref. [31], and outlined in the remainder.

The key to generalizing the equilibrium detailed-balance relation to a nonequilibrium fluctuation theorem is its reformulation in terms of the mesoscopic thermodynamic energy, heat, and entropy changes associated with fluctuating particle paths and their probabilities [55]. Consider the probabilities $P(h \to h')$ and $P(h' \leftarrow h)$ for particle paths starting at height $h > h'$ and ending at height h' and their time-reversed paths, respectively. In equilibrium, they must be identical:

$$1 = \frac{P(h \to h')}{P(h \leftarrow h')} = \frac{P(h \to h'|h)c(h)}{P(h \leftarrow h'|h')c(h')} \qquad \text{(detailed balance)}. \qquad (8.12)$$

This amounts to the time-reversal symmetry of all particle currents between any two positions h and h'. The second equality provides an alternative formulation employing the conditional probabilities $P(h \to h'|h)$ and $P(h \leftarrow h'|h')$ for the particle to go from height h to h' and *vice versa*, given that it started in either of these two heights. Now, one notices that the ratio of the probabilities for the starting positions is nothing but the ratio $c(h)/c(h') = e^{[U(h')-U(h)]/k_B T}$ of the local particle concentrations. And that, since no external work δW is performed on or by the suspended particle, energy conservation[6] allows the potential energy difference to be rephrased as the heat $\Delta Q^{(\text{rev})} = -\Delta Q_R^{(\text{rev})} = U(h') - U(h) < 0$ transferred from the particle to the heat reservoir R (i.e., the solvent) when falling from h to h'. For equilibrium

---

[6]We count incoming energies as positive in the first law of thermodynamics: $dU = \delta Q + \delta W$.

Brownian motion, this heat exchange is fully reversible and therefore amounts to the reversible entropy reduction/increase

$$\Delta S^{(\text{rev})} = -\Delta S_R^{(\text{rev})} = [U(h') - U(h)]/T = k_B \ln[c(h)/c(h')], \qquad (8.13)$$

for the particle and its solvent, respectively. Accordingly, Eq. (8.12) implies

$$\frac{P(h \to h'|h)}{P(h \leftarrow h'|h')} = e^{\Delta Q_R^{(\text{rev})}/k_B T} \qquad \text{(in equilibrium)}. \qquad (8.14)$$

Out of equilibrium, detailed balance is broken, so that the ratio in Eq. (8.12) will no longer be unity for all h, h'. While a stationary probability ratio, such as Eq. (8.1), will still obtain under stationary conditions, the detailed pairwise symmetry of the individual particle trajectories between heights h and h' will be replaced by more complicated flux patterns. How is Eq. (8.14) then to be modified? If the barometer distribution in Eq. (8.1) is unchanged and the solvent is isothermal, the answer is obviously that the reversible entropy change has to be replaced by the total entropy change, including the spontaneous entropy production that breaks the mesoscopic reversibility or time-reversal invariance. In other words, the total heat ΔQ_R added to the solvent contains some reversible contribution $\Delta Q_R^{(\text{rev})}$ and some "irreversible" or dissipative contribution $\Delta Q_R^{(\text{irr})}$. Thereby, we obtain $P(h \to h'|h)/P(h \leftarrow h'|h') = e^{\Delta Q_R/k_B T}$ and from this the steady-state *fluctuation theorem*

$$\frac{P(h \to h')}{P(h \leftarrow h')} = e^{\Delta Q_R^{(\text{irr})}/k_B T} \Rightarrow \langle e^{-\Delta Q_R^{(\text{irr})}/k_B T} \rangle = 1. \qquad (8.15)$$

The last formulation follows by reshuffling the terms, such as to isolate the forward path probability $P(h \to h')$ and its associated dissipation $\Delta Q_R^{(\text{irr})}$ on the same side of the equality. One then sums over all possible forward paths with the appropriate boundary conditions and exploits the normalization of $P(h \leftarrow h')$. Note that the heat $\Delta Q_R^{(\text{irr})}$ added irreversibly to the bath is nothing but the dissipated heat, also briefly called "*the dissipation*". The corresponding net entropy production associated with the particle path is $\Delta Q_R^{(\text{irr})}/T$. On average non-negative by virtue of the second law, which is incidentally recovered from Eq. (8.15) by a first order Taylor expansion, it fluctuates as a function of the path taken by the particle, such as to conform with Eq. (8.15).

To eventually make Eq. (8.15) useful for a hot Brownian swimmer, we appeal to the above mapping of the corresponding non-isothermal Brownian motion to an equivalent equilibrium problem. In the Markov limit, it allows us to take over all of the above formulas for a hot Brownian particle in a non-isothermal solvent by simply writing T_{HBM} in place of the equilibrium bath temperature T (and similarly for the friction, if we want to explicitly write out the dissipated heat). With this modification, Eq. (8.15) expresses an exact symmetry in the motion of a hot Brownian swimmer. It reveals what is left over of the perfect symmetry imposed onto the space

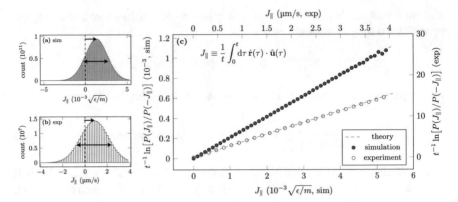

Fig. 8.6 Fluctuation theorem for a hot Brownian swimmer (adapted from Ref. [31]). *Left*: statistics of longitudinal particle currents $J_\parallel$ (formula in panel **c**) measured in simulation (**a**) and experiment (**b**); average propulsion is along ($\parallel$) the swimmer's axis $\hat{\mathbf{u}}$, but Brownian fluctuations ($\leftrightarrow$) can displace it against the mean drift current $\langle J_\parallel \rangle$ ($\rightarrow$). *Right*: test of the fluctuation theorem, Eq. (8.15), using histograms (**a**), **b** as proxies for path probabilities $P(J_\parallel)$; the logarithm of their ratio $P(J_\parallel)/P(-J_\parallel)$ for forward/backward motion is linear in $J_\parallel t$, irrespective of the (different) conditions prevailing in experiment/simulation

of path probabilities by the strict time-reversal invariance of normal diffusion, when the particle becomes an active swimmer. In Ref. [31], we have tested this prediction with high precision. Illuminating a large region of space by green laser light, which is strongly absorbed by the gold cap of our Janus particles, we achieved a good degree of stationarity for the swimming. As the data points lying on a straight line in Fig. 8.6 demonstrate, the logarithm of the probability ratio $P(h \rightarrow h'|h)/P(h \leftarrow h'|h')$ is a linear function of the Markovian particle velocity along its axis (here expressed in terms of the longitudinal particle current $J_\parallel$). To show that this is exactly what is predicted by Eq. (8.15), the dissipation rate $\dot{Q}_R^{(\mathrm{irr})} = v_\mathrm{p}\zeta_\mathrm{HBM}J_\parallel$ was explicitly calculated from the analytically known path weight for Eq. (8.8), in Ref. [31]. The result is however easily deduced directly from thermodynamic arguments, as it has the plainly obvious form "particle current $J_\parallel$ times thermophoretic force $\zeta_\mathrm{HBM}v_\mathrm{p}$" (both along the particle axis $\hat{\mathbf{u}}$). Although the "thermophoretic force" enters in the same way as an external driving force, it should of course not be mistaken for one, since the free swimmer moves by self-propulsion, entirely without the action of external forces. Also note that the dissipation rate $\dot{Q}_R^{(\mathrm{irr})}$ and its associated "hot Brownian entropy production rate" $\dot{Q}_R^{(\mathrm{irr})}/k_B T_\mathrm{HBM}$ are virtual quantities, referring to a virtual equilibrium bath defined by the theory of hot Brownian motion, while the actual bath is non-isothermal and bears a substantial housekeeping heat flux (from the hot swimmer into the cool ambient solvent), not included in $\dot{Q}_R^{(\mathrm{irr})}$.

8.7 Conclusions

We have discussed hot Brownian particles and swimmers, two examples for Brownian motion very far from equilibrium. Thanks to the strong scale separation between the Brownian particles and their atoms, substantial theoretical progress could be made along the lines first laid out by Einstein, yielding exact analytical predictions for the hot Brownian dynamics by systematic coarse graining. Wherever these predictions were tested so far, they were found to be in excellent agreement with experimental observations and simulation data. In fact, the basic ideas are readily generalized to situations where the colloidal particles are cooled with respect to the solvent (so-called "cold Brownian motion" [56]) or dissolved in gases instead of liquids, and even to situations in ultra-high vacuum (the so-called Knudsen regime) [57, 58]. There remains the experimental challenge to explore the more intriguing features of hot Brownian motion at shorter times, where hydrodynamic memory and inertia come into play and give rise to the frequency-dependent noise temperature [20, 27, 31], alluded to in Sect. 8.3.

Good progress has already been made with another modification of equilibrium Brownian motion that is hard to avoid under general non-isothermal conditions, namely self-propulsion. We have mentioned the steering of hot swimmers by Maxwell-demon type methods summarily known as photon nudging. These methods could in the future be harnessed for studying micron-sized artificial swarms of active particles that mimic the interactions and behavior of schools of fish and flocks of birds, thereby creating a microscopic laboratory for controlled studies of some most amazing large-scale biological phenomena.

As we have discussed, such swimming (or flying) motion is subject to exact symmetries that arise from the time-reversal symmetry breaking caused by the net entropy production associated with the swimmer's directed motion (not with the operation of its engine). It is worthwhile to point out that the same mechanism also applies to other broken symmetries, in particular spatial ones [59], as was also— to the best of our knowledge for the first time—experimentally demonstrated in Ref. [31]. This observation hints at a quite general underlying pattern governing the nonequilibrium dynamics of any driven mesoscopic degrees of freedom. Indeed, the symmetry breaking associated with the violation of detailed balance can be formulated in a language familiar from quantum field theory. In this language, the net entropy production takes the role of a gauge field [60, 61]. If integrated around a loop in state space, reversible entropy changes will add up to zero, and only the dissipative heat or irreversible entropy production will contribute, so that the subscript (irr) can be omitted from expressions like those in Eq. (8.15), e.g.,

$$\langle e^{-\oint dt\, \dot{Q}_R/k_B T} \rangle = 1. \tag{8.16}$$

Although the entropy of a state is only defined up to an additive gauge constant, and although the reversible entropy change $\Delta S^{(rev)}$ during a state change will change sign upon time (or path) reversal, the dissipation and the associated net entropy production are both gauge invariant and independent of symmetry transformations, such as time reversal, hence "objective". The closed line integrals over the entropy change in the heat bath thus play a role akin to Berry's phase in quantum mechanics or holonomies (essentially, how much the sum of angles in a triangle deviates from 180°) in differential geometry. The symmetry breaking due to thermodynamic irreversibility can thereby be quantified objectively, in an analogous way as Wilson loops can help to detect space-time curvature in relativistic quantum field theory. The symmetry relation (8.16) that constrains nonequilibrium processes is in this context also interpreted as the consequence of a partial breaking of a supersymmetry that comprises time-reversal invariance [62, 63]. How far this structure can be extended to short-time hot Brownian motion is currently an open question [64, 65].

References

1. A. Einstein, Ann. Phys. **322**, 549 (1905)
2. R. Feynman, R. Leighton, M. Sands, *The Feynman Lectures of Physics*, vol. 1 (Addison-Wesley, Reading, MA, 1963)
3. M. Haw, Phys. World **18**(1), 19 (2005)
4. E. Frey, K. Kroy, Ann. Phys. (Leipzig) **14**, 20 (2005)
5. M. Haw, *Middle World: The Restless Heart of Matter and Life* (Macmillan, New York, 2006)
6. K. Kroy, Physik J. **15**, 20 (2016). (in German)
7. G. Gompper et al. (eds.), Microswimmers. From single particle motion to collective behavior. Eur. Phys. J. Spec. Top. **225**, 11–12 (2016)
8. K. Kroy, D. Chakraborty, F. Cichos, Eur. Phys. J. Spec. Top. **225**, 2207 (2016)
9. J. Palacci, C. Cottin-Bizonne, C. Ybert, L. Bocquet, Phys. Rev. Lett. **105**, 088304 (2010)
10. N.V. Brilliantov, T. Pöschel, *Kinetic Theory of Granular Gases* (Oxford University Press, Oxford, 2004)
11. B. Smeets et al., Proc. Natl. Acad. Sci (USA) **113**, 14621 (2016)
12. A.P. Solon et al., Nat. Phys. **11**, 673 (2015)
13. H. Morowitz, E. Smith, Complexity **13**, 51 (2007)
14. P. Sartori, S. Pigolotti, Phys. Rev. X **5**, 041039 (2015)
15. C. Battle et al., Science **352**, 604 (2016)
16. M. Skolnik, *Radar Handbook* (McGraw-Hill, New York, 1970)
17. T.A. Milligan, *Modern Antenna Design* (Wiley, USA, 2005)
18. H. Turlier et al., Nat. Phys. **12**, 513 (2016)
19. E. Smith, Rep. Prog. Phys. **74**, 046601 (2001)
20. G. Falasco, M.V. Gnann, D. Rings, K. Kroy, Phys. Rev. E **90**, 032131 (2014)
21. G. Falasco, K. Kroy, Phys. Rev. E **93**, 032150 (2016)
22. T. Li, S. Kheifets, D. Medellin, M.G. Raizen, Science **328**, 1673 (2010)
23. T. Franosch et al., Nature **478**, 85 (2011)
24. D. Chakraborty et al., Europhys. Lett. **96**, 60009 (2011)
25. D. Rings, D. Chakraborty, K. Kroy, New J. Phys. **14**, 053012 (2012)
26. N. Oppenheimer, S. Navardi, H.A. Stone, Phys. Rev. Fluids **1**, 014001 (2016)
27. G. Falasco, R. Pfaller, M. Gnann, K. Kroy, arXiv:1406.2116 (unpublished)
28. M. Selmke, R. Schachoff, M. Braun, F. Cichos, RSC Adv. **3**, 394 (2013)
29. R. Schachoff et al., Differ. Fund. **23**, 1 (2015)
30. D. Rings et al., Phys. Rev. Lett. **105**, 090604 (2010)
31. G. Falasco et al., Phys. Rev. E **94**, 030602(R) (2016)
32. H.-R. Jiang, N. Yoshinaga, M. Sano, Phys. Rev. Lett. **105**, 268302 (2010)

33. T. Bickel, A. Majee, A. Würger, Phys. Rev. E **88**, 012301 (2013)
34. W.C.K. Poon, in *Physics of Complex Colloids*, Vol. 184 of *Proceedings of the International School of Physics "Enrico Fermi"*, ed. by F.S.C. Bechinger, P. Ziherl (IOS, SIF, Amsterdam, Bologna, 2013), p. 317
35. I. Llopis, I. Pagonabarraga, J. Non-Newton, Fluid Mech. **165**, 946 (2010)
36. W.B. Jackson, N.M. Amer, A.C. Boccara, D. Fournier, Appl. Opt. **20**, 1333 (1981)
37. M. Harada, K. Iwamotok, T. Kitamori, T. Sawada, Anal. Chem. **65**, 2938 (1993)
38. D. Boyer et al., Science **297**, 1160 (2002)
39. S. Berciaud, L. Cognet, G. Blab, B. Lounis, Phys. Rev. Lett. **93**, 257402 (2004)
40. A. Gaiduk, M. Yorulmaz, P.V. Ruijgrok, M. Orrit, Science **330**, 353 (2010)
41. M. Selmke, M. Braun, F. Cichos, ACS Nano **6**, 2741 (2012)
42. M. Selmke, F. Cichos, Am. J. Phys. **81**, 405 (2013)
43. E. Rutherford, Philos. Mag. Ser. (1911)
44. M. Selmke, F. Cichos, Phys. Rev. Lett. **110**, 103901 (2013)
45. L.C.B.L.S. Berciaud, Nano Lett. **5**, 2160 (2005)
46. D. Magde, E. Elson, W.W. Webb, Phys. Rev. Lett. (1972)
47. P.M.R. Paulo et al., J. Phys. Chem. C **113**, 11451 (2009)
48. V. Octeau et al., ACS Nano **3**, 345 (2009)
49. R. Radünz, D. Rings, K. Kroy, J. Phys. Chem. A (2009)
50. S. Kheifets et al., Science **343**, 1493 (2014)
51. B. Qian et al., Chem. Sci. **4**, 1420 (2013)
52. A.P. Bregulla, H. Yang, F. Cichos, ACS Nano **8**, 6542 (2014)
53. U. Seifert, Rep. Prog. Phys. **75**, 126001 (2012)
54. E. Fodor et al., Phys. Rev. Lett. **117**, 038103 (2016)
55. K. Sekimoto, *Stochastic Energetics*, Vol. 799 of *Lecture Notes in Physics* (Springer, Berlin, Heidelberg, 2010)
56. P.B. Roder et al., Proc. Natl. Acad. Sci. (USA) **112**, 15024 (2015)
57. J. Millen, T. Deesuwan, P. Barker, J. Anders, Nat. Nanotechnol. **9**, 425 (2014)
58. K. Kroy, Nat. Nanotechnol. **9**, 415 (2014)
59. P.I. Hurtado, C. Pérez-Espigares, J.J. Pozo, P.L. Garrido, Proc. Natl. Acad. Sci. (USA) **108**, 7704 (2011)
60. H. Feng, J. Wang, J. Chem. Phys. **135**, 234511 (2011)
61. M. Polettini, Europhys. Lett. **97**, 30003 (2012)
62. K. Mallick, M. Moshe, H. Orland, J. Phys. A: Math. Theoret. **44**, 095002 (2011)
63. C. Aron, G. Biroli, L.F. Cugliandolo, J. Stat. Mech.: Theoret. Exp. **2010**, P11018 (2010)
64. L. Joly, S. Merabia, J.-L. Barrat, Europhys. Lett. **94**, 50007 (2011)
65. A. Argun et al., Phys. Rev. E **94**, 062150 (2016)

Chapter 9
On Phase Transitions in Biased Diffusion of Interacting Particles

Philipp Maass, Marcel Dierl and Matthias Wolff

9.1 Introduction

When traveling on a highway, we can make the unpleasant experience of getting trapped in a jam like the one shown in Fig. 9.1. Or we see that a queue of cars comes to a standstill and gets a bit later again into motion, without any apparent reason for this strange behavior. This behavior may remind us of a liquid freezing and melting, i.e. a phase transition between its liquid and solid states. In the case of the liquid, we know that the transition is commonly initiated by the variation of the temperature or pressure, or, more generally speaking, by the variation of parameters that control the state of the system. In this chapter we will see that jamming phenomena like that of vehicles on a road can be understood quite similarly as a phase transition between two non-equilibrium steady states (NESS) under variation of control parameters.

As a typical feature of the situation under consideration we recognize that there is a bias in motion and some sort of interaction. Cars are moving in one direction and they change their speed in relation to the cars in front of them. This combination of biased motion and interaction is not restricted to vehicular traffic, but can appear under quite different conditions [1]. Amazingly, it in particular arises on a molecular level in a number of biophysical processes. A prominent example is the biased motion of motor proteins, like kinesin and dynein, along filamentary tracks in eucaryotic cells, the so-called microtubules [2], as illustrated in Fig. 9.1. Dyneins

P. Maass (✉) · M. Wolff
Fachbereich Physik, Universität Osnabrück, Barbarastraße 7,
49076 Osnabrück, Germany
e-mail: maass@uos.de

M. Wolff
e-mail: mawolff@uos.de

M. Dierl
Physikalisch-Technische Bundesanstalt, Abbestraße 2-12,
10587 Berlin, Germany
e-mail: Marcel.Dierl@ptb.de

© Springer International Publishing AG 2018
A. Bunde et al. (eds.), *Diffusive Spreading in Nature, Technology and Society*,
https://doi.org/10.1007/978-3-319-67798-9_9

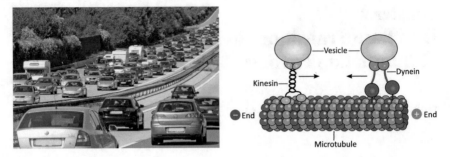

Fig. 9.1 Left: Cars moving along a highway with a jam in one direction (photography: Th. Rein-hardt/pixelio.de). Right: Illustration of the directed motion of motor proteins along a microtubule. Dynesin motors move in the retrograde direction toward the minus ends of microtubules whereas kinesin motors transport cargo in the anterograde direction toward the plus ends

are important for propelling sperm, bacteria and other cells, and kinesins support intracellular transport [3]. Other translocating motor proteins are myosins that move along actin filaments in the cytoskeleton and are responsible for muscle contraction [4]. Interactions between these motors become important, when many of them move simultaneously along a single filament. Similarly as for a traffic jam of cars, it may be possible that a domain wall appears, which separates two phases of protein motors on the microtubule, one, where the densities of motors is low (like in a gas), and one, where the density of motors is high (like in a liquid) [5, 6]. In fact, an experimental realization of such state of coexisting phases has been realized *in vitro* with a fluores-cently labeled single-headed kinesin motor [7]. Further examples of biased motion of interacting particles in biophysical applications include biopolymerization [8] and the directed motion of ions through channels in cell membranes [9–11].

The origin of phase transitions in equilibrium systems, as the liquid-gas or liquid-solid transition of a fluid, can be traced back to an interplay of energy and entropy, when the system settles down to its equilibrium state of minimum free energy. Can certain generic mechanisms be identified also for the occurrence of phase transi-tions in biased diffusion systems out of equilibrium? And is it really conceivable that analogous underlying mechanisms lead to jamming of cars and molecular motors? For answering such questions, physicists usually search for some kind of "minimal model", where the basic features, that means a biased transport and the presence of interactions here, are introduced in a most simple way. It has turned out that already a slight modification of the simple jump model of Fig. 9.2 helps in understanding and predicting the phase transitions. This is referred to as the asymmetric simple exclusion process (ASEP) and corresponds to a driven lattice gas, where a biased stochastic hopping of particles between nearest-neighbor sites on a one-dimensional lattice is considered. Originally the ASEP has been introduced to describe protein synthesis by ribosomes [8] and it now appears as a basic building block in various applications. In addition to the vehicular traffic and the biophysical processes men-

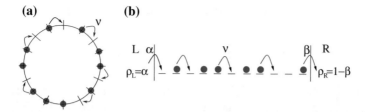

Fig. 9.2   Illustration of the totally asymmetric simple exclusion process (TASEP) in one dimension for **a** a ring system (periodic boundary conditions) and **b** an open channel coupled to two particle reservoirs L and R to the left and right, from and to which particles are injected and ejected with rates α and β. If $\alpha, \beta \leq \nu$ one can assign the particle densities $\rho_L = \alpha/\nu$ and $\rho_R = 1 - \beta/\nu$ to the left and right reservoirs, respectively (see Eq. (9.5) and text for details)

tioned above, this includes, for example, charge transfer in photovoltaic devices [12, 13] and surface growth phenomena [14, 15].

Moreover, the ASEP is of fundamental interest also for studying general properties of NESS [16–19], including fluctuation theorems for thermodynamic quantities defined on a microscopic level (cf. Chap. 8). Generalizations of the ASEP have given rise recently to the discovery of an intriguing set of universality classes in nonlinear hydrodynamics [20]. In this set, which includes ordinary diffusions as one case, the temporal spreading of particles, or density fluctuations, is characterized by dynamical exponents that agree with the Kepler ratios 1/1, 2/1, 3/2, 5/3, 8/5, ..., a sequence that converges towards the golden mean $(1 + \sqrt{5})/2 \cong 1.618$. The Kepler ratios follow from taking ratios of consecutive numbers in the famous Fibonacci sequence 1, 1, 2, 3, 5, 8, ..., where an element is constructed by taking the sum of the two preceding ones.

In the following we will discuss how phase transitions in driven diffusion systems arise and will review recent work on the implications of repulsive nearest-neighbor interactions in driven lattice gases, with a focus on open systems coupled to particle reservoirs. Results for these models form a basis to explain the occurrence of recently discovered phase transitions in more complicated periodically driven systems [21, 22] and to understand their consequences for Brownian pumps and motors, which provide useful models for molecular machines [3].

Though starting with the most simple assumptions for the elementary steps of motion, the methods applied will be shown to lead to remarkable insights enlightening the peculiarities of such systems. It is noteworthy that there is no need for the application of any advanced mathematical calculus. Probability considerations turn out to be, as a rule, completely sufficient. The way of reasoning will, however, sometimes deviate from conventional routes.

9.2   The Asymmetric Simple Exclusion Process (ASEP)

9.2.1   ASEP Along a Ring

Let us first consider the ASEP for particles that are driven along a ring of sites by performing jumps to vacant nearest-neighbor sites, as illustrated in Fig. 9.2a. A particle occupying a site hinders the other particles to occupy the same site, implying that each site can be occupied by at most one particle. In general, jumps occur in both clockwise and counter-clockwise direction with rates v_+ and v_-, where $v \equiv (v_+ - v_-) > 0$ for a bias in clockwise direction. In this very simple setup, the ASEP already includes the key features of biased motion and interaction. In Fig. 9.2a, we have, for simplicity, considered the extreme case of the totally asymmetric simple exclusion process (TASEP), where particles jump solely in clockwise direction ($v_- = 0$). Actually, as we will see below, what is decisive is the presence of a bias, while its strength $(v_+ - v_-)/(v_+ + v_-)$ is not essential for the occurrence of phase transitions. The reason is that the functional form of the particle current on the density matters, but not the magnitude of the current.

To derive how the current depends on the particle density, let us specify the possible particle configurations $C = \{n_i\}$ on the ring by introducing the site occupation numbers n_i, where $n_i = 1$ if site i is occupied and zero otherwise. In the NESS, the current must be constant along the ring. We thus need to consider the current between the sites i and $(i + 1)$ only. It can be decomposed into a partial current in clockwise and a partial current in counter-clockwise direction. A particle is able to jump in clockwise direction from site i to $(i + 1)$, if site i is occupied and site $(i + 1)$ empty. For the corresponding joint probability of the two occupation numbers we write $p_2(n_i = 1, n_{i+1} = 0)$. Because the rate for a jump from i to $(i + 1)$ is v_+, the partial current in clockwise direction is $p_2(n_i = 1, n_{i+1} = 0) v_+$. Analogously, the partial current in counter-clockwise direction is $p_2(n_i = 0, n_{i+1} = 1) v_-$. Hence we can write for the bulk current in the stationary state

$$
\begin{aligned}
j_{\text{B}} &= p_2(n_i = 1, n_{i+1} = 0) v_+ - p_2(n_i = 0, n_{i+1} = 1) v_- \\
&= \langle n_i(1 - n_{i+1}) \rangle v_+ - \langle n_{i+1}(1 - n_i) \rangle v_- .
\end{aligned} \tag{9.1}
$$

In going from the first to the second line we have used that $\langle n_i(1 - n_{i+1}) \rangle = \sum_{n_i=0,1} \sum_{n_{i+1}=0,1} p_2(n_i, n_{i+1}) n_i (1 - n_{i+1}) = p_2(n_i = 1, n_{i+1} = 0)$, and analogously $\langle n_{i+1}(1 - n_i) \rangle = p_2(n_i = 0, n_{i+1} = 1)$. This identification of joint probabilities for specific sets of occupation numbers with averages over products of occupation numbers is always possible because the n_i are either zero or one. Here and in the following $\langle \ldots \rangle$ refers to an average in the NESS.

Intuitively, one may expect that in the NESS none of the particle configurations on the ring is preferred over the other. This is indeed the case and can be rationalized as follows, where, for the sake of simplicity, we consider a totally asymmetric simple exclusion process (TASEP). Under stationary-state conditions, the probability $P(C)$ for all configurations C must not vary with time. When introducing the transition rate $W(C \rightarrow C')$ for a configuration C to change into another C' by one particle jump, this implies

$$P(C) \sum_{C''} W(C \rightarrow C'') = \sum_{C'} P(C')W(C' \rightarrow C), \tag{9.2}$$

because constancy of $P(C)$ requires that the total rate $\sum_{C'} P(C')W(C' \rightarrow C)$ of jumps transferring configurations C' into C (gain terms) must be balanced by the total rate $P(C) \sum_{C''} W(C \rightarrow C'')$ of jumps transferring C into configurations C'' (loss terms).

We now show that Eq. (9.2) indeed holds if all probabilities $P(C)$ are equal. In this case $P(C) = P(C')$, and Eq. (9.2) simplifies to [19]

$$\sum_{C''} W(C \rightarrow C'') = \sum_{C'} W(C' \rightarrow C). \tag{9.3}$$

Now, for proving the validity of this equation, we first realize that any configuration C may be thought to be built up by chains ("clusters") of particles occupying nearest-neighbor sites, including the case where such a cluster consists of only a single particle. The number of clusters in a configuration C is denoted as $N_{cl}(C)$. For the configuration C shown in Fig. 9.2a, for example, $N_{cl}(C) = 6$. There are six new configurations C'' which can emerge from C by jumps of particles from the clockwise ends of the clusters in the configuration C. Because these jumps have the same rate v, we find $\sum_{C''} W(C \rightarrow C'') = 6v$ for the configuration in Fig. 9.2a, and $\sum_{C''} W(C \rightarrow C'') = vN_{cl}(C)$ in general. Correspondingly, there are as well six different configurations C', which could have generated the configuration C in Fig. 9.2a via jumps of particles that joined the counter-clockwise ends of the clusters in C. This also gives $\sum_{C'} W(C' \rightarrow C) = 6v$, or $\sum_{C'} W(C' \rightarrow C) = vN_{cl}(C)$ in general.

It is straightforward to generalize this line of reasoning to the ASEP with jumps to both clockwise and counter-clockwise direction. We thus showed that if all configurations C are equally probable, the ASEP is in a steady state. One may expect also that the ASEP actually approaches this steady state in the long-time limit. This indeed can be proven by resorting to the complete mathematical description of the ASEP dynamics, which is given by a master equation [23].

Because all configurations C are equally probable, we can replace the averages over products of occupation numbers by the product of their averages. Applied to the products appearing in Eq. (9.1), this yields $\langle n_i(1 - n_{i+1}) \rangle = \langle n_{i+1}(1 - n_i) \rangle = \rho(1 - \rho)$ where $\rho = \langle n_i \rangle$ is the mean occupation number or particle number density. Hence we obtain

$$j_{\mathrm{B}}(\rho) = (v_+ - v_-)\rho(1 - \rho) = v\rho(1 - \rho) \tag{9.4}$$

for the bulk current as a function of the density.

Strictly speaking, Eq. (9.4) is exactly true only if we neglect the constraint and associated correlation implied by the fixed particle number N_p on the ring. The joint probability of finding a site occupied and a neighboring (or any other) site empty is given by the probability of finding one site occupied times the conditional probability of finding a site to be empty if one site is occupied. The probability of finding one site occupied is N_p/N, where N is the number of sites. The probability that the $(N_p - 1)$ other particles leave one of the remaining $(N - 1)$ sites empty is $[1 - (N_p - 1)/(N - 1)]$. Hence, in this exact treatment one obtains $\langle n_i(1 - n_{i+1})\rangle = \langle n_{i+1}(1 - n_i)\rangle = (N_p/N)[1 - (N_p - 1)/(N - 1)]$. In the "thermodynamic limit" ($N \to \infty$, N_p/N fixed), this reduces to Eq. (9.4). Let us note that the replacement of averages over products of occupation numbers by the corresponding products of their averages, is commonly referred to as the "mean-field approximation".

In summary, we can say then that the mean-field expression for the current, in which correlations between occupation numbers are factorized, is exact in a bulk system and gives the parabola (9.4) plotted in Fig. 9.3a for the bulk current-density relation in the NESS. The current vanishes in the limit of zero and complete occupation of the ring since then one of the two factors ρ or $(1 - \rho)$ in Eq. (9.4) becomes zero. It attains its maximum at half filling, when the two factors are equal to each other. Let us remark that the current is symmetric with respect to a transformation $\rho \to (1 - \rho)$, i.e. $j_{\mathrm{B}}(\rho) = j_{\mathrm{B}}(1 - \rho)$. This is a consequence of particle-hole symmetry, which means that one can equivalently view the process as driving "mutually excluding" vacancies with concentration $(1 - \rho)$ in counter-clockwise direction with effective rate v.

9.2.2 ASEP Coupled to Particle Reservoirs

Cars or motor proteins are not driven along a ring, but they are entering and leaving a road or filament from some region. This motivates the study of an open ASEP along a channel of sites, where the particles are injected from a reservoir L coupled to the channel at its left boundary, and ejected to a reservoir R at its right boundary, as illustrated in Fig. 9.2b. If the site next to the left reservoir (left boundary site) is empty, injection takes place with a rate α, and if the site next to the right reservoir (right boundary site) is occupied, ejection takes place with a rate β. After some transient time, a NESS evolves that depends on the injection and ejection rates, which act as control parameters. Here and in the following we will always consider corresponding stationary states.

The open ASEP coupled to reservoirs is much more interesting than the closed ASEP along a ring, because it leads to phase transitions between NESS, where the

(a)   **(b)**

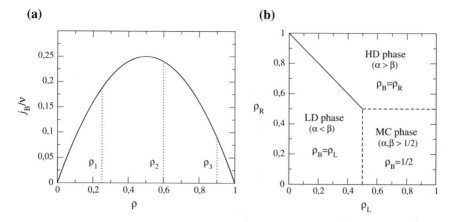

Fig. 9.3 a Bulk current as a function of density in the ASEP according to Eq. (9.4). **b** Phase diagram of the ASEP with the low-density (LD), maximum current (MC), and high-density (HD) phases. The solid line separating the LD and HD phases marks a first-order transition (jump in ρ_{B} when crossing the line), and the dashed lines separating the MC phase from the LD and HD phases mark second-order transitions (jump in the derivative of ρ_{B} as a function of the varying control parameters when crossing the line). To demonstrate the application of the minimum and maximum current principles given in Eq. (9.6), three densities $\rho_1 = 0.25$, $\rho_2 = 0.6$, and $\rho_3 = 0.9$ are indicated by the dotted lines in (a). If $(\rho_{\text{L}}, \rho_{\text{R}}) = (\rho_1, \rho_2)$ or $(\rho_{\text{L}}, \rho_{\text{R}}) = (\rho_1, \rho_3)$, the minimum current principle applies and the ASEP is in the LD phase with $\rho_{\text{B}} = \rho_1$ or in the HD phase with $\rho_{\text{B}} = \rho_3$, respectively. For $(\rho_{\text{L}}, \rho_{\text{R}}) = (\rho_2, \rho_1)$ or $(\rho_{\text{L}}, \rho_{\text{R}}) = (\rho_3, \rho_1)$, the maximum current principle applies and the ASEP is in the MC phase with $\rho_{\text{B}} = 1/2$

particle density can change abruptly with the variation of α and β, in analogy to the abrupt change of the density of a liquid upon crystallization when varying the temperature. Nevertheless, as will be discussed in connection with Eq. (9.6) below, the result derived for the bulk current in Eq. (9.4) turns out to be useful for understanding the origin of the phase transitions and for deriving the corresponding phase diagram.

For values not exceeding v, the injection and ejection rates can be associated with particle number densities ρ_{L} and ρ_{R} in the left and right reservoirs as follows:

$$\rho_{\text{L}} = \alpha/v, \qquad\qquad\qquad (9.5a)$$
$$\rho_{\text{R}} = 1 - \beta/v. \qquad\qquad\qquad (9.5b)$$

While these relations may be intuitively clear, it is instructive for our later analysis to give a reasoning. To this end, let us note that the rate for filling of an empty site in the bulk is given by ρv, i.e. by the probability ρ of finding the site next to the empty site to be occupied times the rate v for a jump. Consider now the corresponding rate $v\rho_{\text{L}}$ for the filling of the left boundary site from a reservoir site with mean occupation number ρ_{L}. By setting this rate equal to the injection rate α, we obtain $\alpha = v\rho_{\text{L}}$, or $\rho_{\text{L}} = \alpha/v$, in agreement with Eq. (9.5a). Analogously, the rate of emptying an occupied site in the bulk is $v(1 - \rho)$, and we can consider the corresponding rate $v(1 - \rho_{\text{R}})$ of emptying the right boundary site, where $(1 - \rho_R)$ refers to the mean hole occupation number in the respective reservoir. Setting this rate equal to the ejection rate β gives $\beta = \nu(1 - \rho_R)$, or $\rho_R = 1 - \beta/\nu$ in accordance with Eq. (9.5b). This method of associating injection and ejection rates with reservoir densities by resorting to the dynamics in the bulk, will be referred to as the "bulk-adapted" way. Because the reservoir densities in Eq. (9.5) cannot exceed one, the association is limited to the regime $\alpha, \beta \leq \nu$ here.

In contrast to the ring, the probabilities of the particle configurations C in the open ASEP (or TASEP) are no longer all the same. Recalling the line of reasoning for Eq. (9.3) based on the clusters, this can be understood from the fact that near the boundaries, the rates for feeding and decay of clusters become different, as a consequence of the differences in the values of α and β. It was, however, exactly the equality of the feeding and decaying rates from which, with Eq. (9.3), all configurations could be concluded to be equally probable. The difference in the feeding and decaying rate therefore implies that the distribution $P(C)$ becomes non-uniform.

While in equilibrium systems, the probabilities $P_{eq}(C)$ of configurations are given by Boltzmann weights, no general concept is yet available for predicting the $P(C)$ in NESS, as discussed also in Chap. 8 in connection with colloidal particles in non-isothermal solvents. In statistical physics it is thus of fundamental importance to find examples of non-trivial NESS, where the $P(C)$ can be exactly derived. The ASEP indeed constitutes one of these examples. For the reader interested in this challenging topic, we mention that $P(C)$ can be expressed by a matrix product form [16, 19], and that it can be calculated also from recursion relations [24, 25] or the Bethe ansatz [16, 18].

Here our focus is on the occurrence of phase transitions of the bulk density ρ_B in the channel's interior as a function of the controlling reservoir densities, which were first reported in Ref. [26]. The corresponding phase diagram is displayed in Fig. 9.3b. There appears a low-density (LD) phase for $0 \leq \rho_L < \min(1/2, 1 - \rho_R)$ where $\rho_B = \rho_L$, a high-density (HD) phase for $\max(1/2, 1 - \rho_L) < \rho_R \leq 1$, where $\rho_B = \rho_R$, and a maximum current (MC) phase for $\rho_L > 1/2 \wedge \rho_R < 1/2$, where $\rho_B = 1/2$ irrespective of ρ_L and ρ_R. Note that j_B from Eq. (9.4) assumes its maximum at $\rho_B = 1/2$.

How can we understand the origin of these different phases and the transitions between them? For answering this question, it is helpful to reformulate the conditions for the occurrence of the different phases, given above in terms of the densities ρ_L and ρ_R, in terms of the rates α and β with the help of Eqs. (9.5). The condition $\rho_R < (1 - \rho_L)$ for the LD phase corresponds to $\alpha < \beta$, the condition $\rho_R > (1 - \rho_L)$ for the HD phase to $\alpha > \beta$, and for the MC phase to occur, both α and β must be larger than $1/2$ (see also Fig. 9.3b). Thus, with respect to the LD and HD phases, a "hand-waving" argument would be to view the open ASEP as an assembly line, and to argue that the throughput of goods (particles) is governed by the slowest worker. This would give us an idea why the LD phase is realized for $\alpha < \beta$ $[\rho_R < (1 - \rho_L)]$ and the HD phase for $\beta < \alpha$ $[\rho_R > (1 - \rho_L)]$ in Fig. 9.3b, but it cannot help us to understand why there is the MC phase if α and β are both larger than $1/2$.

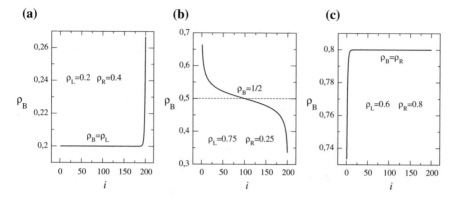

(a)   **(b)**   **(c)**

Fig. 9.4 Representative density profiles for **a** the LD, **b** the MC, and **c** the HD phase of the open ASEP with $N = 200$ sites. The profiles were calculated from the exact recursion relations given in [25]. In the MC phase, a bulk regime with nearly constant $\rho_B = 1/2$ is not seen, because of the slow power law decay $\sim 1/\sqrt{r}$ of the profile towards the bulk regime (strictly valid in the asymptotic limit of infinite N). The bulk value $\rho_B = 1/2$ can nevertheless be read off from the saddle point in the profile

A better approach for understanding the phase diagram is to look at density profiles in the different phases, for which representative examples are shown in Fig. 9.4 for a channel with $N = 200$ sites. These profiles were calculated from exact recursion relations formerly derived in [25]. Looking at these profiles, we see that there are bent parts because of boundary conditions to be fulfilled with respect to the reservoirs (see below). In this situation the particle flux is no longer solely generated by the bias. In addition to the bias-induced drift (bulk) current j_B, there is a diffusion current caused by the concentration gradient, which, as in the situation of diffusion considered, e.g., in Chap. 2, is directed towards decreasing concentration. The total flux j_{tot} is the sum of the drift and diffusion currents, and in the NESS, j_{tot} must be the same everywhere along the channel.

In the LD phase (Fig. 9.4a), the profile is flat except for a small region close to the right boundary, and the bulk density in the flat part matches the reservoir density ρ_L. In the region close to the right boundary, the density rapidly increases, but there is no matching of the density ρ_N at the site $i = N$ next to the right reservoir with ρ_R, as one might first guess. In the example in Fig. 9.4a, we see that $\rho_N \cong 0.27$, while $\rho_R = 0.4$. What determines ρ_N is the requirement of constant total current. Because j_{tot} is constant, it can be calculated from the flat regime $\rho_i = \rho_L$, where the diffusion current vanishes. Accordingly, $j_{tot} = j_B(\rho_L) = v\rho_L(1 - \rho_L)$ (cf. Eq. (9.4)). The ejection current from site $i = N$ is simply given by $\beta\rho_N = v(1 - \rho_R)\rho_N$ and this must equal j_{tot} also, giving $v(1 - \rho_R)\rho_N = v\rho_L(1 - \rho_L)$, or $\rho_N = \rho_L(1 - \rho_L)/(1 - \rho_R)$. For the parameters $\rho_L = 0.2$ and $\rho_R = 0.4$ in Fig. 9.4a this yields $\rho_N = 4/15 \cong 0.27$ in agreement with the data.

Similarly, in the HD phase (Fig. 9.4c), the profile is flat except for a small region close to the left boundary, and the bulk density in the flat part matches the reservoir density ρ_R. When approaching the left boundary, the density rapidly decreases. Its value ρ_1 at the site $i = 1$ next to the left reservoir follows from equating the total current $j_{tot} = v\rho_R(1 - \rho_R)$ with the injection current $\alpha(1 - \rho_1) = v\rho_L(1 - \rho_1)$. This gives $\rho_1 = 1 - \rho_R(1 - \rho_R)/\rho_L$. For the parameters $\rho_L = 0.6$ and $\rho_R = 0.8$ in Fig. 9.4c we obtain $\rho_1 = 11/15 \cong 0.73$, in agreement with the data.

In the MC phase (Fig. 9.4b), the current is at its maximum, $j_{tot} = v/4$. The boundary conditions at the left boundary $[v\rho_L(1 - \rho_1) = v/4]$ and right boundary $[v(1 - \rho_R)\rho_N) = v/4]$ then give $\rho_1 = 1 - 1/(4\rho_L)$ and $\rho_N = 1/[4(1 - \rho_R)]$. The corresponding values $\rho_1 = 2/3 \simeq 0.67$ and $\rho_N = 1/3 \simeq 0.33$ for the parameters in Fig. 9.4b again agree with the data.

It can be further shown [25] that the bent profile parts in the LD and HD phases decay exponentially towards the bulk value ρ_B (with possible power law corrections in sub-phases), while in the MC phase the profile decays very slowly as $1/\sqrt{r}$ towards ρ_B, meaning that $|\rho_{N/2 \pm r} - \rho_B| \sim r^{-1/2}$. These different behaviors are clearly reflected by the profiles shown in Fig. 9.4.

How can this insight into the behavior of the density profiles help us to understand the occurrence of the phase transitions? Despite of the missing matching with the reservoir densities for the bent parts of the profiles, we can infer from our evaluation that the density increases from the left to the right side of the channel for $\rho_L < \rho_R$, while it decreases for $\rho_L > \rho_R$. The diffusion current thus is negative (flowing from right to left) for $\rho_L < \rho_R$ and positive (flowing from left to right) for $\rho_R < \rho_L$, and it is zero in the flat region of the interior channel part (which extends to the left or right boundary in the LD and HD phases, respectively).

When the density starts to deviate from its constant value ρ_B in the interior part, where $j_{tot} = j_B$, the concentration gradients give rise to a diffusion flux. This additional flux must be compensated by a change of the drift current j_B to keep j_{tot} constant everywhere along the channel. For $\rho_L < \rho_R$ the diffusion current must be compensated by an increase of j_B, and for $\rho_L > \rho_R$ it must be compensated by a decrease of j_B. This implies that for $\rho_L < \rho_R, j_B = j_B(\rho)$ must assume its minimal value in the interval $\rho_1 \le \rho \le \rho_N$, while for $\rho_R < \rho_L$, it must assume its maximal value in the interval $\rho_N \le \rho \le \rho_1$. The density corresponding to this minimal or maximal j_B is the bulk value ρ_B appearing in the channel's interior. It is important to emphasize that this reasoning requires monotonically varying profiles.

With the known ρ_1 and ρ_N as a function of ρ_L and ρ_R from above, one can check that ρ_1 and ρ_N can actually be replaced by ρ_L and ρ_R in the reasoning, so as if the profiles would monotonically decrease or increase between ρ_L and ρ_R. The reasoning allows us to formulate the following rules for the determination of the phase diagram [26, 27]:

$$\rho_B = \begin{cases} \text{argmin}_{\rho_L \le \rho \le \rho_R} \{j_B(\rho)\}, & \rho_L \le \rho_R, \\ \\ \text{argmax}_{\rho_R \le \rho \le \rho_L} \{j_B(\rho)\}, & \rho_R \le \rho_L. \end{cases} \tag{9.6}$$

Here, $\text{argmin}_{\rho_L \le \rho \le \rho_R} \{j_B(\rho)\}$ and $\text{argmax}_{\rho_R \le \rho \le \rho_L} \{j_B(\rho)\}$ return that values of the (function argument) ρ, where, in the indicated interval, $j_B(\rho)$ assumes its minimum and maximum, respectively. The rules in Eq. (9.6) are referred to as the minimum and maximum current principles. Only j_B as a function of ρ is needed for their application. The application is demonstrated in Fig. 9.3 for three densities marked by the dotted lines in Fig. 9.3a, which serve as possible values for the reservoir densities (see the description in the figure caption for details).

Note that for $\rho_L > \rho_R$ the particle density decreases in the direction of the bias, implying that the diffusion current adds positively to the drift current. As a consequence, the maximum current principle applies. In the MC phase, where $\rho_L > 1/2 > \rho_R$, the value $\rho = 1/2$ at which j_B has its maximum, always lies in the interval $[\rho_R, \rho_L]$. Accordingly, the current in this phase attains the largest possible value $v/4$ and the bulk density in this phase is $\rho_B = 1/2$ irrespective of the reservoir densities.

For a more detailed characterization of the situation displayed in Fig. 9.3b and the conditions giving rise to switches between different phases we recollect that, in thermodynamics, one distinguishes between phase transitions of first and second order. Well known examples of first-order phase transitions are the freezing-melting and boiling-condensing transitions of matter upon a variation of, e.g., the temperature. First-order phase transitions are characterized by a discontinuity of the quantity characterizing the different phases, commonly referred to as the "order parameter". Second-order phase transitions are characterized by a discontinuity of the derivative of the order parameter with respect to a control parameter (e.g., the temperature), while the respective quantity itself exhibits no discontinuity.

In our case, the bulk density ρ_B is the order parameter characterizing the different phases in Fig. 9.3b, and ρ_L and ρ_R are the control parameters. Transitions to the maximum current phase are of second order, which means that ρ_B continuously varies when passing the corresponding transition lines, while the derivative of ρ_B with respect to ρ_L and ρ_R exhibits a discontinuity. For example, when increasing ρ_L for fixed $\rho_R = 0.2$, the phase diagram in Fig. 9.3b tells us that $\rho_B = \rho_L$ for $\rho_L \leq 1/2$, and that $\rho_B = 1/2$ for $\rho_L \geq 1/2$. Hence $d\rho_B/d\rho_L$ jumps from one to zero when passing the transition line at $\rho_L = 1/2$. By contrast, transitions between the low- and high-density phase are of first order, meaning that there is a discontinuity in ρ_B when passing the transition line: ρ_B jumps from $\rho_B = \rho_L$ to $\rho_B = \rho_R$ when crossing the line on a path from the LD to the HD phase (e.g., for $\rho_R = 0.7$ and ρ_L increasing).

Let us dwell for a minute to consider the situation when the reservoir densities have values corresponding to points on the first-order transition line, i.e. $\rho_L = 1 - \rho_R$, $\rho_L \in [0, 1/2[$. In this case two phases with different ρ_B, namely $\rho_B = \rho_L$ and $\rho_B = \rho_R$ can coexist. This may remind us of the situation of a liquid in equilibrium with its vapor phase within a closed vessel. They are separated from each other by an interface, i.e. the surface of the liquid. Thus, in analogy, we may in the ASEP as well expect the occurrence of interfaces or "domain walls", i.e. of boundaries between different phases. Indeed these domain walls separating different phases appear for reservoir densities on the first-order transition line. An example of a corresponding density profile with a domain wall is shown in Fig. 9.5a. This profile was obtained from a kinetic Monte Carlo (KMC) simulation, where the fluctuating occupation numbers n_i in the stationary state were averaged over a suitable time window. The

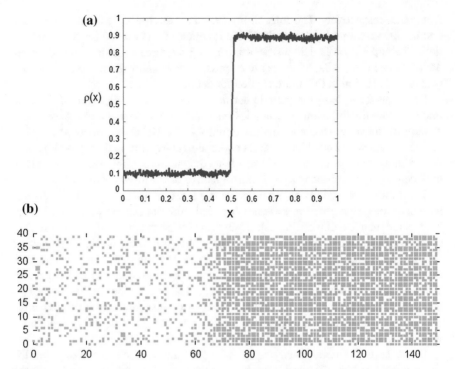

Fig. 9.5 a Snaphot of the time-dependent density profile for reservoir densities $\rho_{\mathrm{L}} = 0.1$ and $\rho_{\mathrm{R}} = 0.9$ on the first-order transition line between the LD and HD phases (see Fig. 9.3b). The profile was obtained from a KMC simulation of a TASEP with $N = 200$ sites by an averaging of occupation numbers in a time window τ with $v^{-1} \ll \tau \ll N^2/D_{\mathrm{w}}$. It is plotted here as a function of the quasi-continuous variable $x = i/N$ (i: lattice site number). **b** Snapshot of the occupation numbers in a two-dimensional TASEP with $N \times N_{\perp} = 150 \times 40$ sites for reservoir densities $\rho_{\mathrm{L}} = 0.2$ and $\rho_{\mathrm{R}} = 0.8$ on the first-order transition line. The configuration was obtained from a KMC simulation with periodic boundary conditions in the direction perpendicular to the bias (see text for further details)

domain walls in the ASEP resemble shock fronts in Burger's turbulence [28] and traffic jam models.

Obviously, there is no reason why the sudden step in the mean site occupancy (density) should occur at a certain position as, for example, the one shown in Fig. 9.5a around the site $0.5N \simeq 100$. In fact, the domain wall position is not fixed but fluctuates in time. These fluctuations have been analyzed in detail using a Boltzmann-Langevin approach [29]. Except for very small times, the domain wall position follows a random walk (see Chap. 2) process with a diffusion constant D_{w}. Hence, by exploiting the Einstein relation (Eq. (2.5)), the typical time for the domain wall position to visit all sites of the channel with approximately equal probability is of the order N^2/D_{w}. This implies that after an averaging of the occupation numbers over time scales larger than N^2/D_{w}, the corresponding profile in Fig. 9.5a would appear as a linearly increasing function from ρ_{L} to ρ_{R}.

The phase transitions are not a peculiar feature of one-dimensional systems. Figure 9.5b shows a snapshot of the site occupancies (particle configuration) from a KMC simulation of a TASEP in two dimensions. As in the one-dimensional channel sketched in Fig. 9.2b, the bias was applied in the x-direction from left to right, and particles were injected from a reservoir L at the left system boundary and ejected to a reservoir R at the right boundary. In the orthogonal y-direction no bias was applied, i.e. the rates for jump upwards (positive y-direction) and downwards (negative y-direction) were equal. Periodic boundary condition with respect to the y-direction were used, meaning that a particle attempting to leave the upper boundary by a jump in positive y-direction, is inserted at the opposing site at the lower boundary, if that site is empty. This amounts to a TASEP on an open channel forming a torus. By taking averages over the mean occupation numbers of sites along the y-direction, it is easy to show that the phase diagram in Fig. 9.3b is valid also for this two-dimensional system. In the example shown in Fig. 9.5b, the reservoir densities were chosen as $\rho_L = 0.2$ and $\rho_R = 0.8$, corresponding to a point on the transition line as in Fig. 9.5a. In Fig. 9.5b, however, we can visualize the transition directly in the snapshot of the particle configuration because the domain wall is much slower fluctuating. One can show that the diffusion coefficient D_w of its position decreases as $1/N_\perp$ with the width (number of sites) in the y-direction [30].

9.3 Driven Lattice Gases with Repulsive Interactions

What happens if particle interactions beyond site exclusions are present in the driven lattice gas? This question was first addressed in a specific model [31] and later studied in a more general context for repulsive interactions $V > 0$ between nearest-neighbor particles [32–34]. In this case, a particle configuration $C = \{n_i\}$ has an energy

$$E(C) = V \sum_i n_i n_{i+1} . \tag{9.7}$$

It was shown that one can again restrict the analysis to the extreme case of unidirectional jumps in order to capture the essential features with respect to phase transition between NESS [34].

In the presence of the nearest-neighbor interactions, the rates Γ_i for a jump from site i to a vacant neighboring site $(i + 1)$ depend on the occupation numbers left to the initial and right to the target site, as illustrated in Fig. 9.6a, i.e. $\Gamma_i = \Gamma(n_{i-1}, n_{i+2})$. For example, if $n_{i-1} = 1$ and $n_{i+2} = 0$, a particle on site i is pushed by its neighboring particle on site $(i - 1)$, leading to an increased jump rate. To the contrary, if $n_{i-1} = 0$ and $n_{i+2} = 1$, a particle on site i has to jump against the repulsive interaction with the particle at site $(i + 2)$, leading to a decreased jump rate. Considering the four different possibilities of the occupation of the sites $(i - 1)$ and $(i + 2)$, and introducing the hole occupation numbers $\tilde{n}_i = 1 - n_i$ for shorter notation, the current flowing from site i to site $(i + 1)$ can, in generalization of Eq. (9.1), be written as

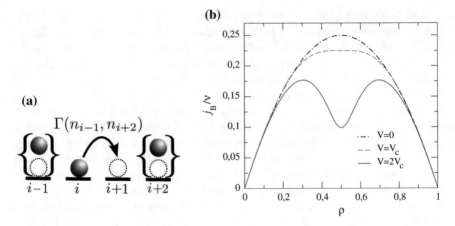

Fig. 9.6 **a** Illustration of the dependence of the jump rates on the occupation numbers in a one-dimensional driven lattice gas with nearest-neighbor interactions. **b** Current-density relation for three different strengths $V = 0$, $V = V_c$, and $V = 2V_c$ of the repulsive nearest-neighbor interaction in case where the jump rates are given by the Glauber rates in Eq. (9.9); V_c is the critical value, where the current-density relation develops a double-hump structure

$$
\begin{aligned}
j_i &= \sum_{l=0,1} \sum_{m=0,1} p_4(n_{i-1} = l, n_i = 1, n_{i+1} = 0, n_{i+2} = m)\Gamma(l,m) \\
&= \langle \tilde{n}_{i-1} n_i \tilde{n}_{i+1} \tilde{n}_{i+2} \rangle \, \Gamma(0,0) + \langle n_{i-1} n_i \tilde{n}_{i+1} \tilde{n}_{i+2} \rangle \, \Gamma(1,0) \\
&\quad + \langle \tilde{n}_{i-1} n_i \tilde{n}_{i+1} n_{i+2} \rangle \, \Gamma(0,1) + \langle n_{i-1} n_i \tilde{n}_{i+1} n_{i+2} \rangle \, \Gamma(1,1) .
\end{aligned}
\tag{9.8}
$$

Here $p_4(n_{i-1} = l, n_i = 1, n_{i+1} = 0, n_{i+2} = m)$ is the joint probability of finding site i to be occupied, site $(i + 1)$ to be empty and the sites $(i − 1)$ and $(i + 2)$ to have occupations according to the values l and m, respectively. As in Eq. (9.1), we have replaced these joint probabilities in Eq. (9.8) with the corresponding averages over products of occupation numbers in the NESS.

The system is considered to be in contact with a heat reservoir at temperature T and we use the thermal energy $k_B T$, where k_B is the Boltzmann constant, as our energy unit ($k_B T = 1$, $V/k_B T = V$). The Glauber rates [35]

$$
\Gamma(n_{i-1}, n_{i+2}) = \frac{2\nu}{\exp[(n_{i+2} - n_{i-1})V] + 1}
\tag{9.9}
$$

are taken as jump rates, which, in the absence of the interaction ($V = 0$), reduce to $\Gamma(n_{i-1}, n_{i+2}) = \nu$.

9.3.1   Current-Density Relation in the Bulk

In a bulk (ring) system correlations and densities are translationally invariant (modulo N) in the NESS. To derive the relation between the density and the current, we need to express the correlations between occupation numbers in Eq. (9.8) by the density $\rho = \langle n_i \rangle$. As mentioned above, this in general is a difficult task, because there are no universal laws providing the probabilities for the particle configurations C (or microstates) in NESS. However, we can take advantage here of a remarkable fact that is valid in one dimension, namely that for rates satisfying the relations

$$\Gamma(0, 1) = \Gamma(1, 0) \, e^{-V} \,, \tag{9.10a}$$

$$\Gamma(0, 0) + \Gamma(1, 1) - \Gamma(0, 1) - \Gamma(1, 0) = 0 \,, \tag{9.10b}$$

the probability distribution for the configurations in the NESS becomes the equilibrium Boltzmann distribution [34, 36, 37], i.e. $P(C) \propto \exp[-E(C)]$ with $E(C)$ from Eq. (9.7). Since the Glauber rates in Eq. (9.9) satisfy Eqs. (9.10a) and (9.10b), the correlations between occupation numbers in Eq. (9.8) equal the equilibrium correlations in the corresponding one-dimensional Ising model, which can be calculated by various means, as, for example, the transfer matrix technique [38]. As a result, one finds [34]

$$j_{\mathrm{B}}(\rho) = 2v \left[\left(\rho - C^{(1)} \right)^2 \frac{2f - 1}{2\rho(1 - \rho)} + \left(\rho - C^{(1)} \right) (1 - f) \right] , \tag{9.11}$$

where $f = 1/[\exp(V) + 1] = \Gamma(0, 1)/2v$, and

$$
\begin{aligned}
C^{(1)} &= \langle n_i n_{i+1} \rangle_{\mathrm{eq}} \\
&= \frac{1}{2(1 - e^{-V})} \left[2\rho(1 - e^{-V}) - 1 + \sqrt{1 - 4\rho(1 - \rho)(1 - e^{-V})} \right]
\end{aligned} \tag{9.12}
$$

is the average of the product of two occupation numbers at neighboring sites in equilibrium.

Figure 9.6b shows the current-density relation for three different interaction strengths V. Because the bulk dynamics is particle-hole symmetric, $j_{\mathrm{B}}(\rho) = j_{\mathrm{B}}(1 - \rho)$. For $V \to 0$, $j_{\mathrm{B}}(\rho)$ approaches the parabola $j_{\mathrm{B}} = v\rho(1 - \rho)$ from Eq. (9.4) for particles with site exclusion only. For large V, a minimum in the current must occur at half filling $\rho = 1/2$, because the preferred particle configurations then correspond to staggered arrangements of occupied sites and holes (antiferromagnetic ordering). In such configurations, jumps have a very small rate $\propto \exp(-V)$ that vanishes in the limit $V \to \infty$. Moreover, because $j_{\mathrm{B}}(\rho) \to 0$ for $\rho \to 0$ (or $\rho \to 1$) and $j_{\mathrm{B}}(\rho) = j_{\mathrm{B}}(1 - \rho)$, the appearance of a minimum at $\rho = 1/2$ must go along with the appearance of two maxima at densities $\rho_{1,2}^\star$ with $\rho_2^\star = 1 - \rho_1^\star$. An analysis of Eqs. (9.11) and (9.12) yields that the corresponding double hump structure in the current occurs for V exceeding the critical value

$$V_c = 2 \ln 3 \cong 2.20, \tag{9.13}$$

and that the two maxima for $V > V_c$ occur at

$$\rho^*_{1,2}(V) = \frac{1}{2} \mp \sqrt{\frac{3}{4} - \frac{1}{2}\sqrt{\frac{2e^V}{e^V - 1}}} . \tag{9.14}$$

In Fig. 9.6b, the current-density relation is displayed for $V = 0$, $V = V_c$, and $V = 2V_c$.

9.3.2 Phase Diagram for Bulk-Adapted Couplings

How do the interactions influence the phase transitions between NESS if the driven lattice gas is brought into contact with two particle reservoirs L and R at the left and right boundary? One answer to this question is readily provided by applying the minimum and maximum current principles, Eq. (9.6), to the current-density relation (9.11). In the strong interaction regime $V > V_c$, this yields a phase diagram with in total seven phases, where two of them are "left-boundary phase" with $\rho_B = \rho_L$, two are "right-boundary phase" with $\rho_B = \rho_R$, two are maximum current phases with either $\rho_B = \rho^*_1$ or $\rho_B = \rho^*_2$, and one is a minimum current phase with $\rho_B = 1/2$. The diagram for $V = 2V_c$, corresponding to the curve with the double hump in Fig. 9.6b, is shown in Fig. 9.7a. Solid lines separating phases in this figure mark transitions of first order, and dashed lines indicate transitions of second order. As a consequence of the particle-hole symmetry in the bulk dynamics, the phase diagram, as the one in Fig. 9.3b, exhibits a symmetry with respect to the off-diagonal.

However, as discussed in Sect. 9.2.2, application of the minimum and maximum current principles requires the density profiles to be monotonically varying. In general this condition is not fulfilled for the system-reservoir couplings in the presence of interactions. Analogous to the case of equilibrium systems, density oscillations typically occur at the system boundaries [32, 33] due to the effects of modified interactions close to the boundaries, as, e.g., caused by missing neighbors. The modified interactions change the rates involved in the injection and ejection of particles compared to the rates for jumps in the bulk, as illustrated in Fig. 9.8. At the left boundary, these rates are the injection rate $\Gamma_L(n_2)$ that, for nearest-neighbor interaction should depend on the occupation at site $i = 2$ and the rate $\Gamma_1(n_3)$ that depends only on the occupation at site $i = 3$ due to the missing neighbor to the left. Analogously, at the right boundary the modified rates are $\Gamma_{N-1}(n_{N-2})$ and $\Gamma_R(n_{N-1})$. A physical choice of these rates is discussed in the next Sect. 9.3.3. It leads to another phase diagram with ρ_L and ρ_R as control parameters, as exemplified in Fig. 9.7b.

An interesting question is whether it is possible to define the rates at the system boundaries in such a manner that the phase diagram predicted by the minimum and maximum current principles becomes valid. This is indeed possible as will be shown now. The guiding principle is to adapt the dynamics near the reservoirs with density

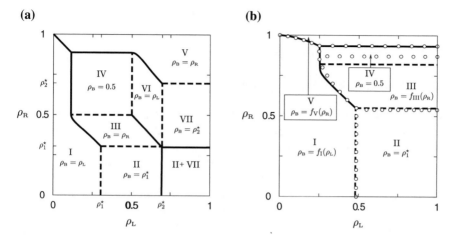

Fig. 9.7 NESS phase diagrams of driven lattice gases with nearest-neighbor interactions for **a** bulk-adapted couplings and **b** equilibrated-bath couplings to the particle reservoirs. The interaction strength is $V = 2V_c$. Solid lines indicate first-order transitions and dashed lines second-order transitions. The shaded area in **a** marks a region, where the two maximum current phases coexist. The lines in **b** refer to analytical calculations based on the time-dependent density functional theory of lattice fluids [34] and the circles mark the results from kinetic Monte Carlo simulations

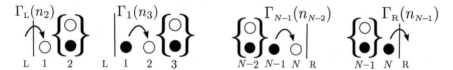

Fig. 9.8 Illustration of the rates involved in the injection and ejection of particles in the driven lattice gas with nearest-neighbor interactions and unidirectional bias

ρ_{LR} to the bulk dynamics of a system with density $\rho = \rho_{LR}$. To be specific, let us consider the left reservoir with density ρ_L and the injection rate $\Gamma_L(n_2)$ to a vacant site $n_1 = 0$, where the occupation n_2 corresponds to the site right of the target site of a jump. To adapt this rate to the bulk dynamics, we consider a bulk system of density $\rho = \rho_L$ and a jump in the system's interior from a site i to a vacant site $(i + 1)$ ($n_{i+1} = 0$) with neighbor occupation n_{i+2}. Such a jump can occur only if the site i is occupied. Its rate is either $\Gamma(0, n_{i+2})$ for $n_{i-1} = 0$ or $\Gamma(1, n_{i+2})$ for $n_{i-1} = 1$. For given occupancies $(n_{i+1} = 0, n_{i+2})$, the occupancies $(n_{i-1} = 0, n_i = 1)$ and $(n_{i-1} = 1, n_i = 1)$ occur with the conditional probabilities $p_{2|2}(n_{i-1} = 0, n_i = 1 | n_{i+1} = 0, n_{i+2}; \rho_L, V)$ and $p_{2|2}(n_{i-1} = 1, n_i = 1 | n_{i+1} = 0, n_{i+2}; \rho_L, V)$, respectively, where the explicit designation of ρ_L and V reminds us that the conditional probabilities are for a bulk system with density $\rho = \rho_L$ and particle interactions V. Because these probabilities are translationally invariant in the bulk, we can write them simply as $p_{2|2}(01|0n_2; \rho_L, V)$ and $p_{2|2}(11|0n_2; \rho_L, V)$, where we set $n_{i+2} = n_2$. Accordingly, we obtain

$$\Gamma_L(n_2) = p_{2|2}(01|0n_2; \rho_L, V)\Gamma(0, n_2) + p_{2|2}(11|0n_2; \rho_L, V)\Gamma(1, n_2).\qquad(9.15a)$$

With the same type of reasoning we find

$$\Gamma_1(n_3) = p_{1|3}(0|10n_3; \rho_L, V)\Gamma(0, n_3) + p_{1|3}(1|10n_3; \rho_L, V)\Gamma(1, n_3),\qquad(9.15b)$$

where $p_{1|3}(m_1|m_2\,m_3\,m_4)$ is the probability in a bulk system of finding an occupation m_1 at site $(i + 1)$ for given occupations $(m_2\,m_3\,m_4)$ at sites $(i + 2)$, $(i + 3)$, and $(i + 4)$. Analogously, bulk-adapted rates $\Gamma_R(n_{N-1})$ and $\Gamma_{N-1}(n_{N-2})$ can be defined.

The specification of the bulk-adapted rates here extends the preliminary considerations in Sect. 9.2.2. Its underlying concept can be generalized to even more complicated situations including interactions of longer range and periodically driven diffusion in time-dependent potentials [21]. In all these cases, KMC simulations with the bulk-adapted rates have shown that phase diagrams of NESS agree with the predictions of the minimum and maximum current principles.

Since the bulk-adapted couplings to the reservoirs are in general not physical (except for the simplest case of the ASEP for $\alpha, \beta \leq v$), one may ask, why these couplings should be useful. The reason is as follows. Having a bulk region of constant (or nearly constant) density ρ_B in the system's interior, a slightly enlarged region may be defined, where the density profile approaches the bulk region monotonically. The densities at the boundaries of this enlarged region can be used as effective densities [31]. Application of the minimum and maximum current principles with respect to these effective densities as "boundary densities" then will consistently provide the correct NESS phase with the order parameter ρ_B. This reasoning shows that all *possible* phases are predicted by applying the minimum and maximum current principles to the bulk current-density relation. The bulk-adapted specification of the rates involved in the injection and ejection of particles then provides a means to actually generate all these possible NESS phases in a systematic and controlled manner.

We conclude this section by noting that a study of the density profiles in the various phases of the diagram shown in Fig. 9.7a yielded characteristic features that agree with that found in the phases of the ASEP for $V = 0$ [39]. In the boundary phases, the profiles decay exponentially towards the bulk value ρ_B, with possible power-law corrections, and in the minimum or maximum current phases the decay follows power laws.

A new phenomenon occurs in the parameter region $\rho_L > \rho_2^\star \wedge \rho_R < \rho_1^\star$ marked by the shaded area in Fig. 9.7a. In this region, the maximum current phases with $\rho_B = \rho_1^\star$ and $\rho_B = \rho_2^\star$ coexist. Similar to the density profile for the case of the two coexisting boundary LD and HD phases in Fig. 9.5a, a domain wall appears in the system, but its mean position appears to be pinned at a location that depends on ρ_L and ρ_R. Also the width of the fluctuations around the mean position depends on the value of the reservoir densities. A detailed analysis of this coexistence of two self-organized phases, in particular of the mean positions and fluctuations of the domain wall is lacking yet and shall be pursued in the future [40].

9.3.3   Phase Diagram for Equilibrated-Bath Couplings

As discussed above, in the NESS of driven lattice gases with interactions, there typically appear oscillations in the density profiles in regions close to the particle reservoirs, if the rates at the system boundaries are different from the particular bulk-adapted ones. For example, when the particle reservoirs are represented by equilibrated ideal Fermi gases with chemical potentials μ_L and μ_R, giving the reservoir densities

$$\rho_{L,R} = \frac{1}{\exp(-\mu_{L,R}) + 1}, \tag{9.16}$$

a natural form of the rates at the boundaries is [34]

$$\Gamma_L(n_2) = \rho_L \frac{2v}{\exp(n_2 V - \mu_L) + 1}, \tag{9.17a}$$

$$\Gamma_R(n_{N-1}) = (1 - \rho_R) \frac{2v}{\exp(-n_{N-1} V + \mu_R) + 1}, \tag{9.17b}$$

$$\Gamma_1(n_3) = \frac{2v}{\exp(n_3 V) + 1} = \Gamma(0, n_3), \tag{9.17c}$$

$$\Gamma_{N-1}(n_{N-2}) = \frac{2v}{\exp(-n_{N-2} V) + 1} = \Gamma(n_{N-2}, 0). \tag{9.17d}$$

The rates in Eqs. (9.17c) and (9.17d) have the same form as the bulk rates given in Eq. (9.9), where the missing neighbors are accounted for by vanishing occupation numbers. The injection and ejection rates in Eqs. (9.17a) and (9.17b) correspond to Eq. (9.9), if to the particles in the reservoirs are assigned the "site energies" $\mu_{L,R}$. The additional factors ρ_L and $(1 - \rho_R)$ in Eqs. (9.17a) and Eq. (9.17b) take into account the filling of the baths. Overall, Eq.(9.17) gives transition rates that resemble forms resulting from Fermi's golden rule [41]. In the following we will refer to these system-reservoir couplings as the "equilibrated-bath rates".

   The phase diagram for equilibrated-bath rates looks quite different from that for the bulk-adapted rates. As an example we show in Fig. 9.7b the diagram for the same interaction strength $V = 2V_c$ as in Fig. 9.7a. Compared to Fig. 9.7a, five instead of seven phases appear, where one is a minimum current phase with $\rho_B = 1/2$, one is a maximum current phase with $\rho_B = \rho_1^\star$, one is a left-boundary phase, and two are right-boundary phases. Because of the mismatch between boundary and bulk dynamics for the equilibrated-bath couplings, the bulk density ρ_B in the boundary phases no longer equals one of the reservoir densities, but rather is a function of either ρ_L and ρ_R, as indicated in Fig. 9.7b. Note also that the diagram in Fig. 9.7b is not symmetric with respect to the off-diagonal, which seems to violate the particle-hole symmetry in the system.

   Although the phase diagram in Fig. 9.7b looks quite different from that in Fig. 9.7a, there is in fact a general connection between all phase diagrams plotted in dependence of the control parameters ρ_L and ρ_R. This is because from a theoretical point of view, the NESS phases are in fact controlled by the complete set of rates Γ_η ($\eta = L, R, 1$ or $N-1$, cf. Fig. 9.8) governing the dynamics at the boundaries. In this higher dimensional space of theoretical control parameters Γ_η, the phase diagram is unique, and for nearest-neighbor interactions exhibits a symmetry reflecting the particle-hole symmetry in the overall dynamics [34].

In applications, as, e.g., a molecular wire attached to some metal electrodes, the system-reservoir couplings will be given by some setup, and the experimentalist most likely will be able to influence the boundary dynamics in a controlled manner by changing ρ_L or ρ_R (or μ_L and μ_R). One thus can regard ρ_L and ρ_R as the experimental control variables. The connection between the unique phase diagram in the Γ_η-space and the non-unique diagrams in the (ρ_L, ρ_R)-space is given by the functional dependence of the rates Γ_η on the reservoir densities. For example, in the case of the bulk-adapted couplings we have $\Gamma_L = h_L^{ba}(\rho_L)$ and $\Gamma_1 = h_1^{ba}(\rho_L)$ with functions $h_L^{ba}(\rho_L)$ and $h_1^{ba}(\rho_L)$ defined by Eqs. (9.15) (note the conditional probabilities in these equations are fixed once ρ_L and V are given), while in the case of the equilibrated-bath couplings, we have $\Gamma_L = h_L^{eb}(\rho_L)$ and $\Gamma_1 = h_1^{eb}(\rho_L)$ with functions $h_L^{eb}(\rho_L)$ and $h_1^{eb}(\rho_L)$ defined by Eqs. (9.16) and (9.17). Analogously, Γ_R and Γ_{N-1} are parameterized by ρ_R. This means that reservoir densities (ρ_L, ρ_R) correspond to unique points in the Γ_η-space, but these points are in general different for different system-reservoir couplings. The distinct phase diagrams in Fig. 9.7a, b are thus originating from different projections of submanifolds in the Γ_η-space onto the (ρ_L, ρ_R)-plane. For the bulk-adapted couplings, all phases in the Γ_η-space appear in this projection.

The foregoing discussion explains why the phase diagrams in dependence of ρ_L and ρ_R can be quite different for different system-reservoir couplings. It remains to develop a theory, which can predict the (unique) phase diagram in the Γ_η-space. To tackle this problem, a kinetic theory was developed [32, 34] based on the time-dependent density functional theory of lattice fluids [42, 43] and exact results for density functionals in one dimension [44]. The key assumption in this theory is that relations between correlation functions and densities in the NESS can be represented on a local scale by the corresponding relations in equilibrium systems (for details, see Ref. [34]). Representative results of this theory are marked by the lines in Fig. 9.7b and agree well with the results from kinetic Monte Carlo simulations that are represented by the circles.

9.4 Conclusions

Non-equilibrium steady states generated by driven diffusion show a rich variety of self-organized structures. In this chapter we have reviewed recent findings on the phase structure of NESS in driven lattice gases coupled to particle reservoirs. After giving an introduction to the phase transitions between NESS in the asymmetric simple exclusion process, we discussed the physics of these transitions if repulsive nearest-neighbor interactions between the particles beyond site exclusions are included in the process.

Our key findings can be summarizes as follows: Due to the effect of modified particle interactions at the system-reservoir boundaries, the minimum and maximum current principles are no longer sufficient to predict phase diagrams of NESS based solely on the current-density relation in the bulk. These principles are nevertheless useful to predict all possible NESS phases that can appear in the driven diffusion system. A systematic procedure exists to define bulk-adapted system-reservoir couplings such that the phase diagram predicted by the minimum and maximum current becomes valid. In this way one can generate all possible NESS phases and investigate their properties. In applications there will be other system-reservoir couplings present and the phase diagrams for these couplings, plotted as a function of the experimentally controllable parameters, generally differ from that predicted by the minimum and maximum current principles. From a theoretical point of view, one can consider these differing phase diagrams as projections of submanifolds in one unique phase diagram in a high-dimensional space onto a lower dimensional space of the experimentally controllable variables. The high-dimensional space is spanned by all parameters needed for the complete specification of the system-reservoir couplings. To calculate the phase diagram for the repulsive nearest-neighbor interactions, good results were obtained by using a kinetic approach based on time-dependent density functional theory of lattice fluids.

These findings and the underlying methods of analysis were recently shown to be relevant also for understanding the occurrence of phase transitions in periodically driven NESS [21, 22]. This in particular has important consequences for the physics of Brownian motors that have found widespread applications ranging from cellular ion pumps to quantum ratchets [45, 46]. For Brownian motors operating on interacting particles, it was argued that the occurrence of phase transition between NESS is a generic feature irrespective of their type. The knowledge of the phase structure is important to determine control parameters for optimal motor efficiencies, and extremal current phases can be utilized to make the motor-generated particle flow robust against fluctuations of the control parameters. Further developments include the application to multiple-phase structures in traffic flows [47], to the linear-response behavior of NESS [48], as well as future studies on the connection between the non-equilibrium physics of classical driven diffusion systems and the equilibrium quantum physics of strongly correlated spinless fermions [17, 49, 50].

References

1. A. Schadschneider, D. Chowdhury, K. Nishinari, *Stochastic Transport in Complex Systems: From Molecules to Vehicles*, 3rd edn. (Elsevier Science, Amsterdam, 2010)
2. J. Howard, *Mechanics of Motor Proteins and the Cytoskeleton* (Sinauer Associates, Sunderland, 2001)
3. A.B. Kolomeisky, J. Phys. Condens. Matter **25**, 463101 (2013)

4. A.F. Huxley, in Progress, in *Biophysics and Biophysical Chemistry*, vol. 7, ed. by J.A.V. Butler, B. Katz (Pergamon Press, New York, 1957), pp. 255–538
5. R. Lipowsky, S. Klumpp, T.M. Nieuwenhuizen, Phys. Rev. Lett. **87**, 108101 (2001)
6. E. Frey, K. Kroy, Ann. Phys. **14**, 20 (2005)
7. K. Nishinari, Y. Okada, A. Schadschneider, D. Chowdhury, Phys. Rev. Lett. **95**, 118101 (2005)
8. C.T. MacDonald, J.H. Gibbs, A.C. Pipkin, Biopolymers **6**, 1 (1968)
9. T. Chou, D. Lohse, Phys. Rev. Lett. **82**, 3552 (1999)
10. B. Hille, *Ionic Channels of Excitable Membranes*, 3rd edn. (Sinauer Associates, Sunderland, 2001)
11. P. Graf, M.G. Kurnikova, R.D. Coalson, A. Nitzan, J. Phys. Chem. B **108**, 2006 (2004)
12. K.O. Sylvester-Hvid, S. Rettrup, M.A. Ratner, J. Phys. Chem. B **108**, 4296 (2004)
13. M. Einax, M. Dierl, A. Nitzan, J. Phys. Chem. C **115**, 21396 (2011)
14. T. Halpin-Healy, Y.C. Zhang, Phys. Rep. **254**, 215 (1995)
15. J. Krug, Adv. Phys. **46**, 139 (1997)
16. B. Derrida, Phys. Rep. **301**, 65 (1998)
17. G.M. Schütz, in *Phase Transitions and Critical Phenomena*, ed. by C. Domb, J.L. Lebowitz , vol. 19 (Academic Press, San Diego, 2001)
18. O. Golinelli, K. Mallick, J. Phys. A **39**, 12679 (2006)
19. R.A. Blythe, M.R. Evans, J. Phys. A **40**, R333 (2007)
20. V. Popkov, A. Schadschneider, J. Schmidt, G.M. Schütz, Proc. Nat. Acad. Sci. **112**, 12645 (2015)
21. M. Dierl, W. Dieterich, M. Einax, P. Maass, Phys. Rev. Lett. **112**, 150601 (2014)
22. M. Dierl, P. Maass. Preprint (2016)
23. J.F. Gouyet, M. Plapp, W. Dieterich, P. Maass, Adv. Phys. **52**, 523 (2003)
24. B. Derrida, E. Domany, D. Mukamel, J. Stat. Phys. **69**, 667 (1992)
25. G.M. Schütz, E. Domany, J. Stat. Phys. **72**, 277 (1993)
26. J. Krug, Phys. Rev. Lett. **67**, 1882 (1991)
27. V. Popkov, G.M. Schütz, Europhys. Lett. **48**, 257 (1999)
28. J. Bec, K. Khanin, Phys. Rep. **447**, 1 (2007)
29. P. Pierobon, A. Parmeggiani, F. von Oppen, E. Frey, Phys. Rev. E **72**, 036123 (2005)
30. G.D. Lam, Collective effects and phase transitions in simple Brownian motors. Master thesis, University of Osnabrück, 2016
31. J.S. Hager, J. Krug, V. Popkov, G.M. Schütz, Phys. Rev. E **63**, 056110 (2001)
32. M. Dierl, P. Maass, M. Einax, Europhys. Lett. **93**, 50003 (2011)
33. M. Dierl, P. Maass, M. Einax, Phys. Rev. Lett. **108**, 060603 (2012)
34. M. Dierl, M. Einax, P. Maass, Phys. Rev. E **87**, 062126 (2013)
35. R.J. Glauber, J. Math. Phys. **4**, 294 (1963)
36. H. Singer, I. Peschel, Z. Phys. B **39**, 333 (1980)
37. S. Katz, J.L. Lebowitz, H. Spohn, J. Stat. Phys. **34**, 497 (1984)
38. M. Takahashi, *Thermodynamics of One-Dimensional Solvable Models* (Cambridge University Press, 1999)
39. M. Wolff. Untersuchung stationärer Dichteprofile in getriebenen Diffusionssystemen (in German). Bachelor thesis, Universität Osnabrück, 2014
40. M. Wolff, M. Dierl, P. Maass (To be published)
41. A. Nitzan, *Chemical Dynamics in Condensed Phases* (Oxford University Press, Oxford, 2006)
42. D. Reinel, W. Dieterich, J. Chem. Phys. **104**, 5234 (1996)
43. S. Heinrichs, W. Dieterich, P. Maass, H.L. Frisch, J. Stat. Phys. **114**, 1115 (2004)
44. J. Buschle, P. Maass, W. Dieterich, J. Phys. A **33**, L41 (2000)
45. P. Reimann, Phys. Rep. **361**, 57 (2002)
46. E.R. Kay, D.A. Leigh, F. Zerbetto, Angew. Chem. Int. Ed. **46**, 72 (2007)
47. M.E. Foulaadvand, P. Maass, Phys. Rev. E **94**, 012304 (2016)
48. U. Seifert, T. Speck, EPL **89**, 10007 (2010)
49. F.C. Alcaraz, M. Droz, M. Henkel, V. Rittenberg, Ann. Phys. **230**, 250 (1994)
50. R. Stinchcombe, Adv. Phys. **50**, 431 (2001)

Part III
Technology

Chapter 10
Diffusive Spreading of Molecules in Nanoporous Materials

Christian Chmelik, Jürgen Caro, Dieter Freude, Jürgen Haase, Rustem Valiullin and Jörg Kärger

10.1 Introduction

Materials with pore diameters in the range of 1–100 nm are referred to as "nanoporous" [1]. They are found in nature and may be fabricated artificially with both inorganic and organic frameworks. Their ability to interact with molecules and ions on their large inner surface offers ideal prospects for their application in matter upgrading, including catalysis, separation, purification and ion exchange (see Fig. 10.1) [2, 3]. The purposeful design of such materials has given rise to tremendous productivity enhancement. This is in particular true with zeolites, an inorganic nanoporous material distinguished by its regular pore structure with extensions in the range of molecular sizes. The annual benefit worldwide by their

C. Chmelik · D. Freude · J. Haase · R. Valiullin · J. Kärger (✉)
Faculty of Physics and Earth Sciences, Leipzig University, Linnéstr. 5,
04103 Leipzig, Germany
e-mail: kaerger@physik.uni-leipzig.de

C. Chmelik
e-mail: chmelik@physik.uni-leipzig.de

D. Freude
e-mail: freude@uni-leipzig.de

J. Haase
e-mail: j.haase@uni-leipzig.de

R. Valiullin
e-mail: valiullin@uni-leipzig.de

J. Caro
Institute of Physical Chemistry and Electrochemistry, Leibniz University Hanover,
Callinstr. 3A, 30167 Hanover, Germany
e-mail: caro@pci.uni-hannover.de

© Springer International Publishing AG 2018
A. Bunde et al. (eds.), *Diffusive Spreading in Nature, Technology and Society*,
https://doi.org/10.1007/978-3-319-67798-9_10

Fig. 10.1 Nanoporous materials (bottom right), available as crystallites (bottom middle), often in compressed form (bottom left), are key elements in refineries and other chemical plants for matter upgrading (top). The top picture was cut from an image by Walter Siegmund, licensed under CC BY 2.5 [5]. Bottom pictures reproduced with permission from Ref. [6], copyright (2013) Chemiewerk Bad Köstritz GmbH and Wiley-VCH Verlag GmbH & Co. KGaA

exploitation in only petroleum refining has, e.g., been estimated to at least 10 billion US dollars [4].

The gain in value-added products by the use of such materials can clearly never be higher than allowed by the diffusion rate of the involved molecules. The intracrystalline diffusivity (i.e. the rate of molecular migration within the individual particles of the material) does thus become a key number for the efficiency of the given process.

Simultaneously, however, within the context of the book, guest molecules in nanoporous materials offer ideal opportunities for illustrating and quantitating spreading phenomena. In what follows we shall be able to refer to quite a number of items which have been mentioned already in Chap. 2 on introducing into the theoretical foundation of spreading phenomena: Completely different from the situation with human societies or ecological systems, molecules in nanoporous materials offer the unique opportunity of observing spreading phenomena under "initial" and "boundary" conditions which are largely controlled by the investigator. This includes, in particular, the option of repeating experiments under essentially identical starting conditions as well as a thoughtful variation of the spreading conditions. Such variations may quite easily be achieved by changes in temperature or molecular concentration ("loading"), with the latter caused by a variation of the partial pressure of the guest molecules in the surrounding atmosphere. Similarly, also variations in the type of guest molecules (by considering, e.g., molecules of different diameter, chain length or polarity) and of the host material (in particular by surface modification) leads to a variation in the spreading conditions. The option of such variations facilitates the search for fundamental laws and, eventually, their final proof.

As a consequence of their thermal energy, atoms and molecules are subject to a continuous irregular movement, referred to as diffusion, in all states of matter. Thus, in addition to nanoporous host-guest systems, the book does as well deal with diffusion and spreading phenomena in, e.g., solution and suspension (Chap. 8), solids (Chap. 13), biological systems (Chaps. 5 and 6) and our atmosphere (Chap. 7). As a common feature of condensed matter, it is often the structure of the system itself which is affected by mass transfer. Nanoporous host-guest systems are distinguished also in this respect since the host framework generally turns out to remain, in a very good approach, unaffected by amount and nature of the guest molecules.

All what has been said so far necessitates, as a matter of course, the possibility to obtain reliable and unambiguous information on mass transfer in such systems. Diffusion measurements with molecules in nanoporous materials have been conventionally performed by a (macroscopic) recording of mass gain or release upon pressure variation in the surrounding atmosphere. Statements on diffusion had to be based therefore on model assumption. The more recent application of microscopic measuring techniques has shown that these assumption have been unjustified in numerous cases, giving rise to a paradigm shift in our understanding of molecular spreading and diffusion in nanoporous materials [1, 7]. Sections 10.2 and 10.3 introduce into the fundamentals of these novel techniques. The examples given in the further sections of this chapter have in particular been chosen for exemplifying some of the fundamental laws of spreading and diffusion which have been introduced in Chap. 2. They are, moreover, thought to introduce into the more specific cases considered in the Chap. 11 which deals, among others, with different strategies for technology improvement by transport enhancement in nanoporous materials.

10.2   Monitoring Molecular Spreading by Pulsed Field Gradient NMR

Here, we confine ourselves to a short introduction into the principles of diffusion measurement by the pulsed field gradient (PFG) technique of NMR (also known as pulsed gradient spin-echo (PGSE) NMR, NMR diffusometry and q-space imaging) and refer, for a more extensive treatise, to Chap. 12. For illustrating the principle of measurement it is sufficient to adopt the classical view on nuclear magnetic resonance (NMR) [1, 7, 8]. At exactly the same result one would also arrive by a rigorous treatment on considering the expectation values of the quantum mechanical operators corresponding to the relevant physical quantities. Most atoms, notably hydrogen which is mainly considered in PFG NMR diffusion studies, possess a nuclear spin. It bears both a magnetic dipolar moment (like the needle of a compass) and a mechanic momentum (like a gyroscope). Thus, similarly as a spinning gyroscope under the influence of gravity, nuclear spins perform a rotational ("processional") motion as soon as they are placed in a magnetic field. The rotational frequency is given by the relation

$$\omega = \gamma B. \tag{10.1}$$

B and γ denote, respectively, the intensity of the magnetic field and a factor of proportionality characteristic of the given nucleus, referred to as the gyromagnetic ratio. The superposition of many rotating nuclear spins gives rise to a rotating macroscopic magnetization. This rotating magnetization induces a voltage in a surrounding coil, known as the physical principle of an electric generator. It is recorded as the NMR signal.

In PFG NMR diffusion measurements, a constant magnetic field B_0 is superimposed, over a short time interval δ, by an additional, inhomogeneous field

$$B_{add} = gx, \tag{10.2}$$

the "field gradient pulses". By combining Eqs. (10.1) and (10.2), the resonance frequency is seen to become, during this time interval, a well-defined function of the spatial coordinate x

$$\omega(x) = \gamma(B_0 + gx) = \gamma B_0 + \gamma gx. \tag{10.3}$$

With the second equation, the rotational frequencies of the spins are seen to vary with their location x within the sample, with values above and below the average value γB_0 if the zero point ($x = 0$) of the space scale is placed in the center of the sample. These differences lead to a spreading of the direction of the individual spins. The macroscopic magnetization resulting as a superposition of the magnetizations of the individual spins does, correspondingly, decrease. It vanishes totally when, eventually, the spins show into all directions.

In combination with an appropriately chosen sequence of high-frequency pulses (with the frequency as given by Eq. (10.1)), however, one is able to re-establish the macroscopic magnetization by applying an identical second field gradient pulse. This re-establishment is caused by refocusing the individual spins. As a prerequisite of complete refocusing, each spin has to remain fixed in space. Otherwise there remains a phase difference $\gamma gx\delta$, with x now denoting the difference in the positions which the given nuclear spin (and, hence, the atom/molecule, to which it belongs) occupies during the first and second gradient pulses. The thus displaced spin does contribute to the macroscopic magnetization with only its projection on the mean direction of all spins, i.e. with $\cos(\gamma gx\delta)$. The attenuation of the NMR signal, i.e. the ratio between the signals with and without gradient pulses applied, is thus immediately seen to be given by the relation

$$\frac{S(g\delta, t)}{S(0)} = \int\limits_{-\infty}^{\infty} P(x, t)\cos(\gamma gx\delta)dx. \tag{10.4}$$

$P(x, t)$ denotes the probability that an arbitrarily selected molecule within the sample is shifted (during t, given by the separation between the two gradient pulses) over a distance x (in the direction of the field gradient applied). We recollect that exactly this probability has been considered already in Fig. 2.4 of Chap. 2 where, for homogenous systems, it has been calculated to be (Eq. 2.10)

$$P(x, t) = \frac{1}{\sqrt{4\pi Dt}}\exp\left(-\frac{x^2}{4Dt}\right). \tag{10.5}$$

Inserting Eq. (10.5) into Eq. (10.4) yields

$$\frac{S(g\delta, t)}{S(0)} = \exp\left(-\gamma^2 g^2 \delta^2 Dt\right). \tag{10.6}$$

The diffusivity D is thus seen to immediately result from the signal attenuation in PFG NMR experiments.

In the context of PFG NMR, $P(x, t)$ is referred to as the mean propagator [8–10]. With Eq. (10.4) it is seen to be nothing else than the Fourier transform of the PFG NMR signal attenuation. In this way, it becomes directly experimentally accessible. Figure 10.2 illustrates the various facets of information on molecular spreading thus attainable with beds of nanoporous particles.

Though nanoporous materials, i.e. holes in a framework, are not homogeneous in the strict sense of the word, Fig. 10.2a is seen to reveal the pattern as introduced already with Fig. 2.4 in Chap. 2 for spreading in homogeneous systems. This, however, is an immediate consequence of the given experimental conditions ensuring that the molecular displacements are large enough in comparison with the pore diameters and sufficiently small in comparison with the extension of the nanoporous particles/crystals [12].

It is no problem, therefore, to discuss mass transfer in terms of concentrations and fluxes (as appearing in Eqs. 2.6–2.13) by considering unit volumes and areas which notably exceed the pore sizes, but are small enough in comparison with the crystal sizes. The situation becomes completely different in Fig. 10.2b, where molecular displacements are confined to the crystal sizes. Here, the PFG NMR is seen to provide information on structural parameters (like the size of the crystals/ compartments which the diffusants are confined to), operating not unlike a microscope. Hence, this type of analysis has become popular under the term dynamic imaging [8, 10].

At sufficiently high temperatures, in Fig. 10.2d, the guest molecules are shown to leave, during the covered observation times, the individual crystals. Now, a sufficiently large amount of them is able to get, from the lower level of potential

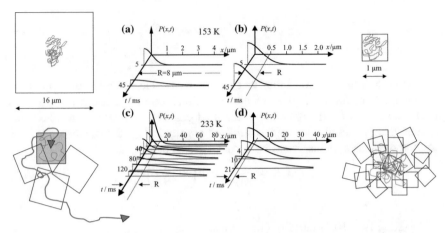

Fig. 10.2 Propagator representation of PFG NMR for visualizing diffusive molecular spreading (ethane) in beds of nanoporous crystals (zeolites of type NaCaA) of two different sizes (radius R). The plots show the increase in spreading with increasing time. Due to symmetry, only one half of the distribution curves are shown (after [9]). Ethane concentration in the gas phase between the crystals is negligibly small in comparison with the intracrystalline concentration. Gas and adsorbed phase are in equilibrium (measurement with fused sample tubes). Reproduced with permission from Ref. [11], copyright (2010) The Royal Society of Chemistry

energy within the nanoporous crystals, to the higher one in the surrounding atmosphere. Correspondingly, one is following molecular spreading over the whole batch of crystals rather than within only the interior of each individual crystals as considered in Fig. 10.2a, b. With the larger crystals considered in Fig. 10.2c, one is even able to distinguish between two constituents of the distribution curves, namely a narrow one referring to those molecules which, during the observation time, did not exchange their positions between different crystals, and a broad one, referring to the other ones. Their fraction (given by the area under the broad constituent) is, expectedly, seen to increase with increasing time.

10.3 Recording the Evolution of Concentration Profiles by Microimaging

Knowledge of the molecular propagation probabilities as accessible by PFG NMR does not automatically allow the prediction of the evolution of molecular distributions within nanoporous materials. This type of information has most recently become available by the introduction of microimaging by interference microscopy (IFM) and IR microscopy (IRM) [1, 7, 13]. Figure 10.3 introduces their measuring principles.

IRM is based on the operation of a set of detectors arranged in a plane ("focal plane array detector") allowing the determination of the IR signal from areas of

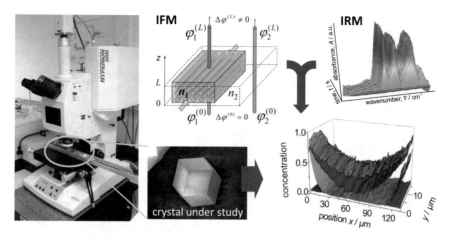

Fig. 10.3 Measuring principles of the techniques of microimaging: IR microscopy (IRM) and Interference microscopy (IFM). Having placed the crystal under study (bottom center) under the microscope (left) information about intracrystalline concentration is deduced in IFM (top center) by comparing optical path lengths through the crystal and the surroundings and in IRM (top right) from the intensity of the IR bands. As a result, one obtains a map of the concentration integrals in z direction in the (x, y) observation plane (bottom right)

ideally as small as 2.7 μm × 2.7 μm. Unavoidable disturbance by sample thickness reduce spatial resolution, in reality, often to values around 5–10 μm. Characteristic vibrational frequencies are employed for finger-printing the molecules under study so that the intensity of the respective band in the spectra (figure top right) may immediately be transferred into the corresponding concentrations (for more details see [11, 14]).

The determination of guest concentrations by IFM is based on their proportionality with the refractive index of the material. Changes in guest concentrations do, therefore, immediately appear in changes of the difference between the optical pathways through the crystal and the surroundings which are determined, as the primary quantity, by the images of interference microscopy. Note that, as a consequence of both measuring principles, it is in either technique the integral over the local concentration in the observation direction rather than the local concentration itself which is recorded. Both coincide, however, as soon as guest fluxes in observation direction are negligibly small in comparison with the perpendicular ones, since in such cases (implying the quite common case of parallel crystal faces on top and bottom) the concentration remains uniform in observation direction and the concentration integral is simply the product of the local concentrations with particle thickness. Absolute values of the concentrations need anyway comparison with the data of measurements of adsorption equilibria or theoretical predictions [3, 15]. Fluxes in observation direction may be excluded for nanoporous materials with channels arranged in only one or two direction(s) or, in 3d pore systems, by sealing top and bottom faces.

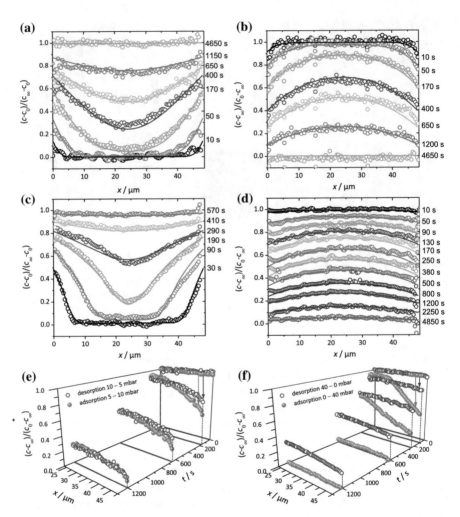

Fig. 10.4 Relative molecular uptake and release of methanol in ferrierite-type zeolites along the 8-ring channels: Comparison of simulated and experimental profiles for pressure steps **a** 5–10 mbar, **b** 10–5 mbar, **c** 0–40 mbar and **d** 40–0 mbar (after [16] and [17]). The points refer to experimental measurements, the lines are numerical solutions of Fick's 2nd law with concentration-dependent transport diffusivities and surface permeabilities. By plotting the concentrations from top to bottom for adsorption, in plots **e** and **f** profiles after selected times during ad- and desorption are shown in a unified representation. Here, for simplicity, only one half of the profiles (starting with $x = 23.8$ μm in the crystal centre) is shown. Reprinted with permission from Ref. [18], licensed under CC BY 3.0

For illustrating the wealth of information thus attainable, we refer to Fig. 10.4. It shows, with different examples, the evolution of the distribution of guest molecules within a nanoporous material. The situation in the chosen system is particularly simple since molecules get into or leave this material mainly via a set of parallel channels (referred to as 8-ring channels since their circumference is formed by 8 silicon (aluminum, phosphorous) atoms connected by oxygen bridges). Molecular spreading may thus be represented in the simple 1d representations of Fick's 1st and 2nd laws as given by Eqs. (2.6) and (2.9) or (2.14), respectively. The cases considered are molecular uptake (i.e. filling of the material by increasing gas pressure in the outer atmosphere: (a, c) and release (i.e. emptying by reducing the external pressure): (b, d). The pressure steps applied are between 5 and 10 mbar (a, b) and 0 and 40 mbar, respectively. The final concentrations of uptake are those established in dynamic equilibrium with the outside atmospheres at the respective pressure. Equilibrium concentrations in the given case are, at 40 mbar, notably higher than at 5 mbar. At the boundary between the nanoporous material and the surrounding atmosphere, in addition, we have to take account of the possible existence of an additional transport resistance. Such resistances prevent the guest concentration close to the boundary from instantaneously assuming the equilibrium value. It is quantified by the relation

$$j_x = \alpha\left(c_{eq} - c(x=0)\right) \tag{10.7}$$

for guest fluxes leaving or entering the material, with $c(x=0)$ and c_{eq} denoting, respectively, the actual concentration close to the boundary and the concentration in equilibrium with the external atmosphere. We note that infinitely high surface permeabilities ($\alpha = \infty$) automatically require boundary concentrations coinciding with the equilibrium values ($c(x=0) - c_{eq} = 0$). Only then fluxes through the surfaces (resulting, with Eq. (10.7), as $0 \times \infty$) remain finite which, for a physically reasonable quantity, has to be required.

In the representations shown in Fig. 10.4, the boundary concentrations are in fact seen to notably deviate from the equilibrium values. Since the equilibrium and boundary concentrations appear immediately from the profiles and since the flux entering the system results by dividing the area between two subsequent profiles by the respective time interval, surface permeabilities become immediately accessible via Eq. (10.7) by direct measurement. The rate of molecular uptake and release was in numerous nanoporous host-guest systems found to be affected by both the finite rate of permeation through the external surface and intracrystalline diffusion [19]. The existence of "surface barriers" did thus turn out to be the rule rather than the exception.

Transient concentration profiles as shown in Fig. 10.4 are well-known from standard diffusion textbooks [20] where they are used for visualizing the solutions of the diffusion equation, Fick's 2nd law (Eqs. (2.9) and (2.14)) (coinciding, in its mathematical form, with Fourier's law of heat conduction). Their direct measurement with guest molecules within nanoporous materials, however, has become possible during only the last few years, owing to the potentials of microimaging [1, 7, 13]. Depending on the uptake and release times of the host-guest systems under study, such sets of profiles may be recorded in the timespan of minutes to hours. This means, in comparison with, e.g., the serial sectioning techniques applied in solid-state diffusion studies [21], a dramatic enhancement in the speed of measurement and, thus, in also the wealth of information.

In the context of our book it might be allowed trying to transfer the phenomena displayed in Fig. 10.4 into a totally different field. Specialists may forgive us our audacity. Let us consider, in a thought experiment, large woodland crossed by two parallel rivers. The woodland is populated by a certain species (either animal or plant) with an essentially constant population density (measured in species per unit area) everywhere, except for the area between the two rivers (our "mesopotamia", which assumes the role of the zeolite crystallites in Fig. 10.4). The species are unable to cross these rivers. Moreover, the population density in the range beyond the two rivers (equivalent to the surrounding atmosphere in the uptake and release experiments) is ensured to remain—by whatever mechanism—constant. Starting with time 0, the species under consideration are able to cross the river via suitably constructed "bridges" and, thus, to penetrate into "mesopotamia" where they, so far, did not occur. Under such conditions the distribution of the new species may be expected to follow the patterns shown in Fig. 10.4a, c. The reverse phenomenon of molecular release might be simulated, in our hypothetical "habitat", by replacing the mechanism we had just implied for ensuring population constancy by another one ensuring complete extinguishing of the species outside of "mesopotamia". Now species distribution within "mesopotamia" should evolve similarly to the desorption patterns shown in Fig. 10.4b, d.

As a main message of this comparison, spreading phenomena outside physics and chemistry are immediately seen to be much more complicated. It is worthwhile emphasizing therefore that, after efforts of research over more than a century [22], it was only owing to the quite recent advent of the microscopic techniques of diffusion measurement that our modern view on molecular spreading in nanoporous materials was established.

The evidence of information thus accessible becomes particularly obvious in the 3d plots shown in Fig. 10.4 e, f. Here, opposite for Fig. 10.4a, c, the increase $(c - c_0)$ in concentration is plotted from top to bottom, resulting in plotting of $(c_\infty - c_0) - (c - c_0) \equiv c_\infty - c$. Thus, relative changes in concentration during adsorption, $(c_\infty - c)/(c_\infty - c_0)$ appear with exactly the same notation as applied for desorption in Fig. 10.4b, d. Plotted in this way, the curves with the small pressure steps (Fig. 10.4e) are seen to essentially coincide or, in other words, the amounts adsorbed and desorbed are similar. This is the situation well-known from tracer exchange experiments where one observes the exchange of labelled with unlabeled molecules (which might be different isotopes with, essentially, identical diffusion properties). Here, the total concentration (being the sum of the concentrations of both isotopes) remains constant anywhere in the sample. The observation may be correlated with the fact that Fick's second law in the form of Eq. (2.9) is a linear differential equation. For this type of equations, the sum of solutions is a solution again. Therefore, adding adsorption and desorption profiles initiated by reversed pressure steps must be expected to also lead to a solution, i.e. a physically reasonable scenario. In the given case, superposition reflects the simple case of invariance in surroundings and, hence, in guest concentration.

Reciprocity in uptake and release does not appear anymore with the large pressure step considered in Fig. 10.4c, d, f. Since the guest concentration, at equilibrium, must be expected to increase with increasing pressure in the surroundings, the range of concentrations considered in Fig. 10.4c, d, f does notably exceed that of Fig. 10.4a, b, e. For our further discussion we have to recollect that the diffusivity may, quite generally, be a function of the concentration of the diffusants. With the diffusion model considered in Sect. 2.1, Eq. (2.3), this concentration dependence might be brought about by the dependence of the mean time between successive steps and/or of the step length on the "density" of the diffusants. The effect of any concentration dependence on the diffusivity (and, similarly, on the permeability) shall clearly be the smaller, the smaller the covered concentration range. Thus, for the small pressure step between 5 and 10 mbar, assuming a constant diffusivity may in fact be expected being a good approximation. This is not the case anymore with the larger pressure step from 0 to 40 mbar, so that Fick's 2nd law must be applied in the form of Eq. (2.14). In comparison with the smaller pressure step from 5 to 10 mbar, molecular uptake is seen to be accelerated by about one order of magnitude while, most astonishingly, the desorption rate remains essentially unchanged. The remarkable impact of concentrations on diffusivity will concern us in more detail when, in the subsequent section, we are going to ask for the "driving forces" of diffusion.

10.4   The Driving Force of Diffusion

With Fick's 1st law, Eq. (2.6), diffusion fluxes are seen to be caused by a gradient in the concentration of the diffusing species under study. One might, therefore, consider gradients in concentration quite generally as the "driving forces" for diffusive fluxes. Though this is clearly true for homogeneous systems, just as a prerequisite for final equilibration of the diffusing species all over the system, it is not the case anymore in heterogeneous systems. As a most illustrative example, we may think of a liquid in equilibrium with its vapor phase. Here, dramatic differences in concentrations between the liquid and gaseous phases do, obviously, not give rise to diffusive fluxes in the direction of decreasing concentration. The situation is thus seen to be different from that considered by the Fourier's laws of heat conduction. Here, under equilibrium conditions, the temperature is known to be uniform all over the system so that gradients in temperature may in fact be considered as the "driving forces" for heat fluxes.

In our search for the driving force of diffusion we may adopt this reasoning by starting with a quantity which—similarly as the local temperature with respect to thermal equilibration—indicates equilibrium in composition. This quantity is known as the chemical potential μ. From elementary thermodynamics the chemical potential of a molecular species in an ideal gas phase at partial pressure p and temperature T is known to be given by the relation

$$\mu_g(p) = \mu_{g0} + RT \ln (p/p_0) \tag{10.8}$$

with μ_{g0} denoting a standard (coinciding with the chemical potential for pressure p_0). Being equal in all phases under equilibrium condition, the chemical potential $\mu(c)$ of the guest molecules at guest concentration c in the adsorbed phase, i.e. within the nanoporous host material, may thus—via Eq. (10.8)—be immediately noted as

$$\mu(c) = \mu_0 + RT \ln[p(c)/p_0]. \tag{10.9}$$

Here, $p(c)$ stands for the (partial) pressure of the guest molecules in the surrounding atmosphere in equilibrium with the actual guest concentration c. The correlations $p(c)$—or, vice versa, $c(p)$, referred to as the adsorption isotherm—are experimentally accessible, e.g. by gravimetric measurement. We may now follow the concept of diffusion as introduced by Maxwell [23], Stefan [24] and Einstein [25] and put, later on, into the formalism of irreversible thermodynamics [15, 26–28] and note the steady-state requirement for the flow velocity u_A of component A (see, e.g. Chap. 1 of Ref. [1])

$$fu_A = -\frac{\partial \mu_A}{\partial x} \tag{10.10}$$

as resulting by implying equality between the gradient of the chemical potential as the driving force and an opposing frictional force. μ_A and f stand for the chemical potential of component A and a friction coefficient. Inserting Eq. (10.9) into (10.10) and noting the flux as the product of concentration and velocity yields

$$j_{Ax} = u_A c_A = -\frac{RT}{f} \frac{d\ln p_A}{d\ln c_A} \frac{dc_A}{dx}. \tag{10.11}$$

The logarithmic derivative $\frac{d\ln p_A}{d\ln c_A} \equiv \frac{dp_A/dc_A}{p_A/c_A}$ appearing in this relation is referred to as the thermodynamic factor. Having noted the spatial derivative of the partial pressure p_A as the product of its derivative with respect to concentration and the concentration gradient, Eq. (10.11) is seen to be of the structure of Fick's first law, Eq. (2.6). Comparing Eqs. (2.6) and (10.11) yields, for the transport diffusivity,

$$D_T = \frac{RT}{f} \frac{d\ln p}{d\ln c}. \tag{10.12}$$

where the subscript A has been omitted since, in the given case, only a single component is considered.

For self-diffusion (i.e. for tracer exchange between microdynamically completely identical species A and A' under uniform overall concentration $c(x) = c_A(x) + c_{A'}(x)$) we have $c_A \propto p_A$ all over the sample, so that the relevant thermodynamic factor $\frac{dlnp_A}{dlnc_A}$ becomes 1. Now comparison with Fick's 1st law for self-diffusion, Eq. (2.7), yields

$$D = \frac{RT}{f}. \qquad (10.13)$$

Comparing Eqs. (10.12) and (10.13), the transport and self-diffusivities are seen to be correlated by the relation

$$D_T = D\frac{dlnp}{dlnc}. \qquad (10.14)$$

This relation has been introduced in Ref. [27] and was, subsequently, often referred to as the Darken relation [29], owing to its similarity with a relation used by Darken [30] in his study of interdiffusion in binary metal alloys. With Eq. (10.13), the self-diffusivity is explicitly seen to be nothing else than a measure of mobility, i.e. of the reciprocal value of the friction which the diffusants overcome on their trajectory. Most importantly, with Eq. (10.14) the transport diffusivity (also referred to, in other context, as collective, chemical or Fickian diffusivity) may be considered being subject to essentially two influences, namely the mobility of the diffusants (represented by D) and an extra driving force (above referred to as the thermodynamic factor) which, in the present context, is seen to emerge for a non-linear interdependence between (partial) pressure and guest concentration. In a more general context, one would consider the fugacity f_A rather than the partial pressure p_A (which only for an ideal gas coincides with the fugacity). With the activity coefficient γ_A introduced by the relation $f_A = \gamma_A c_A$, the thermodynamic factor would then assume the form $(1 + \frac{dln\gamma_A}{dlnc_A})$.

Coming back to Eq. (10.14), we have to mention that this simple relation is only correct if the "friction" between the individual molecules on their trajectories is negligibly small. Within the frame of irreversible thermodynamics, this requirement is equivalent with the absence of any cross correlations between the fluxes and the gradients of the chemical potential of different types of molecules (or of differently labelled ones).

Exactly such a situation is given with the host-guest system considered in Fig. 10.5. Here, molecular passages through the "windows" between adjacent cages are, as a consequence of their small size, the rate-controlling steps in the trajectories. Simultaneously, passages through the individual windows may be considered to be such rare events that the possibility of mutual molecular encounters within these windows may be neglected. Exactly this had to be required as a prerequisite for the validity of Eq. (10.14).

With the reciprocal values of the thermodynamic factor (broken line in Fig. 10.5a) as determined from the plots of the guest concentrations c as a function of the equilibrium pressure p in the surrounding gas phase (Fig. 10.5b), Eq. (10.14)

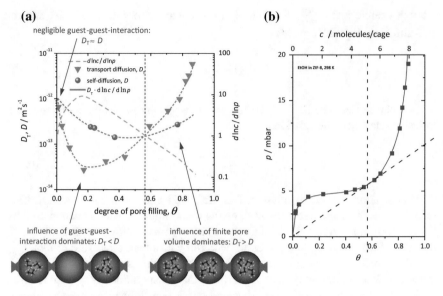

Fig. 10.5 Ethanol in a nanoporous host (metal-organic framework (MOF) of type ZIF-8):
a Experimental data of self-diffusivity (D) and transport diffusivity (D_T) and self-diffusivities
predicted via Eq. (10.14) from the transport diffusivities and the inverse d$\ln c$/d$\ln p$ of the
thermodynamic factor, plotted as a function of fractional loading $\Theta = \frac{c}{c_{max}}$ and **b** the adsorption
isotherm used for the determination of the thermodynamic factor. The cartoon illustrates the
mechanism of interaction by which, over the respective concentration range, molecular
propagation is dominated. The transition point between the two mechanisms appears in (**b**) as
the point of coincidence in the slope of the isotherm with that of the connecting line towards the
origin. After Refs. [6, 14], Fig. 10.5a translated and reproduced with permission from Ref. [6],
copyright (2013) Wiley-VCH Verlag GmbH & Co. KGaA

is in fact found to provide an excellent means for correlating self- and transport
diffusion.

In complete agreement with our expectation, self- and transport diffusivities are
seen to coincide at sufficiently low loadings, since any significant interaction
between the diffusants may in fact be neglected. With increasing loading, the
transport diffusivities are seen to drop even below the self-diffusivity. With the
schematics of transport and self-diffusion as provided by Fig. 2.2 in mind, this
finding appears, at a first glance, counter-intuitive: It is scarcely imaginable that
molecular fluxes (from left to right in Fig. 2.2a, b) are enhanced by a counter-flux
(self-diffusion, situation of Fig. 2.2b) rather than mitigated. We must have in mind,
however, that within our formalism the "friction" between the diffusants was
anyway assumed to be negligibly small—brought about by the dominating role of
window passages in molecular propagation. Molecular behavior is thus dominated

by the attractive forces between the diffusants (caused by their dipolar moment) by which they "prefer" sticking together (see left cartoon on bottom of Fig. 10.5) rather than exploring the less populated part of space. This situation is, obviously, changed at sufficiently high loadings where (within the frame of thermodynamics one would say by entropic reasons) the molecules, given the lack in free space, "prefer" passing into the region of lower guest concentration. In complete agreement with this view, at high concentrations the transport diffusivities are in fact found—both by experimental evidence and in the more rigorous theoretical prediction via Eq. (10.14)—to notably exceed the self-diffusivities.

For these larger concentrations, moreover, a significant increase in the transport diffusivities with increasing loadings may be noted. This behavior is a quite general feature of transport diffusivities since the factor $\frac{d\ln p}{d\ln c} \equiv \frac{dp/dc}{p/c}$ as appearing in Eq. (10.14) increases dramatically when, on approaching complete pore filling ($\Theta \rightarrow 1$), the slope dp/dc of $p(c)$ is progressively exceeding the slope of the connecting line towards the origin, p/c.

With this result in mind we may easily rationalize the pronounced difference in also the shape of transient concentration profiles during uptake and release as shown in Fig. 10.4: Diffusivities increasing with loading do automatically lead to steeper decays in the diffusion front during uptake, given the notably higher diffusivities on its top (namely at high concentration) in comparison with those of the very first molecules of the front, at lowest concentrations. Exactly the reverse is true on desorption where the efflux rate from the crystal center towards the boundary is continuously reduced by the decreasing diffusivities, tending to an equilibration. Correspondingly, in comparison with uptake and release with essentially constant diffusivities (Fig. 10.4a, b), the profiles with concentration-dependent diffusivities are found to be notably steeper for uptake (Fig. 10.4c) and shallower for release (Fig. 10.4d).

Though, owing to the emerging potentials of microimaging [13], these measurements had become possible quite recently, the theoretical framework applied for their analysis is well established for decades already, with the introduction of the formalism of irreversible thermodynamics [26, 31]. It is important to emphasize, however, that already such apparently simple objects like nanoporous host-guest systems are full of challenges from also the view point of their theoretical interpretation. This includes, in particular, the influence of the guest molecules on the structure of the host material if it cannot be assumed to be inert anymore. Such a situation may indeed be found with nanoporous host-guest systems [13, 32] and is the situation, quite in general, on considering the mutual diffusion of the various compounds in solids since here—apart from the observation of pure tracer exchange—diffusion is by its very nature always accompanied by a structural variation of the whole system.

10.5   Multicomponent Diffusion

In their main technological applications, notably including separation, purification and chemical conversion, nanoporous materials are accommodating mixtures rather than single components. Both microscopic measuring techniques presented in this chapter may be exploited for selective diffusion measurements with the individual components of such mixtures.

The option of selective diffusion measurement via PFG NMR is based on the resonance condition, Eq. (10.1). Distinction between different molecular species is particularly easy if the measurements may be based on different nuclei, besides protons (^{1}H) notably on deuterium (^{2}H) and the "NMR active" isotopes of carbon (^{13}C), nitrogen (^{15}N), fluorine (^{19}F), phosphorous (^{34}P) and xenon (^{129}Xe) [33]. Moreover, in recent PFG NMR studies with hydrated LSX zeolites (zeolites of type X of low silicon content, containing exchangeable lithium cations) by using both ^{1}H and ^{7}Li NMR [34] even cation diffusion has become accessible by direct measurement. Most interestingly, spreading of both the water molecules and the cations as appearing in the PFG NMR diffusion data (i.e., in the evolution of the respective propagators—see Fig. 10.2) was observed to be retarded by transport resistances existing in also the interior of the individual particles. This is illustrated in a cartoon-like manner by Fig. 10.6b. ^{1}H and ^{7}Li PFG NMR yielded similar spatial dimensions for the spacing between the internal barriers which, moreover, were of the order of the sizes of the individual crystallites which the zeolite particles (Fig. 10.6a) consist of.

The principle of mass separation of a binary mixture by permeation through nanoporous membranes may be easily rationalized on the basis of Fick's 1st law,

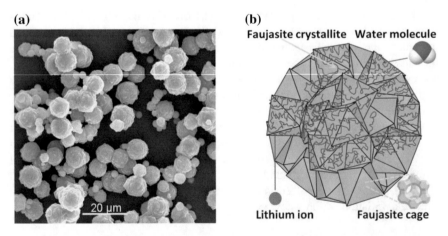

(a)

(b)

Faujasite crystallite   Water molecule

20 µm

Lithium ion   Faujasite cage

Fig. 10.6 a Scanning electron micrograph of a sample of zeolite LSX and **b** scheme of an individual zeolite particle with typical diffusion paths covered by lithium cations and the water molecules during comparable time intervals. Reprinted with permission from Ref. [35], copyright (2013) Elsevier (**a**) and from Ref. [34], copyright (2013) American Chemical Society (**b**)

Eq. (2.6). The partial pressures of the gas mixture at the "feed" side of the membrane are, respectively, p_A and p_B. The flow of a "carrier" gas ensures partial pressures of essentially zero on the "permeate" side. With the simplifying assumptions that, over the considered range of concentrations, the diffusivities D_A and D_B within the membrane are independent of concentration (and composition) and that there is a linear relation $c_{A(B)} = K_{A(B)}p_{A(B)}$ between the partial pressures and the concentration, Eq. (2.6) yields for the ratio of the fluxes of the two components through the membrane under steady-state conditions

$$\frac{j_A}{j_B} = \frac{D_A c_A}{D_B c_B} = \frac{D_A K_A p_A}{D_B K_B p_B}. \tag{10.15}$$

As a prerequisite of the validity of this relation, we have moreover implied the absence of any surface resistance on both the feed and permeate sides.

As a reasonable measure of the separation capability of the membrane with respect to components A and B, we may consider the enhancement in the ratio of the permeating fluxes (which is nothing else than the ratio of the concentrations of the two components "behind" the membrane, i.e. in the "permeate") in comparison with the concentration ratio (and, hence, the ratio of the partial pressures) on the feed side:

$$\frac{j_A/j_B}{p_A/p_B} = \frac{D_A}{D_B} \times \frac{K_A}{K_B}. \tag{10.16}$$

"Membrane separation selectivity" is thus found to be the product of "diffusion selectivity" $\frac{D_A}{D_B}$ and "adsorption selectivity" $\frac{K_A}{K_B}$ [1, 15]. We note in parentheses that the separation efficiency does, clearly, as well depend on the rate of separation, i.e. on the amount of "separated" gas passing the membrane. With reference to Fick's 1st law, Eq. (2.6), this amount is easily seen to be inversely proportional to the membrane thickness. Though not explicitly appearing in Eq. (10.16), membrane thickness is thus among the key parameter of membrane efficiency. Ensuring simultaneously small thicknesses and mechanical stability is the great challenge of membrane production [36, 37].

Both adsorption selectivity (by adsorption measurement or molecular modelling) and diffusion selectivity (by IRM or PFG NMR) on the one hand and membrane permeation selectivity, by flux measurement, on the other, have become experimentally accessible. The results of such a comparison presented in Fig. 10.7 show an order-of-magnitude agreement. Remaining differences may be easily referred to the influence of surface resistances and guest-guest interaction which both are neglected in Eq. (10.16).

While PFG NMR allows access to such interactions by selective two-component self-diffusion measurement under equilibrium conditions [38], microimaging is able to selectively follow molecular diffusion of the various mixture components under non-equilibrium conditions. These potentials are illustrated in Fig. 10.8 showing the results of two-component uptake measurements with zeolite DDR [39].

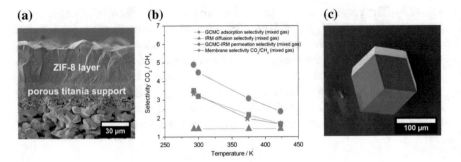

Fig. 10.7 **a** ZIF-8 membrane and **b** comparison of the membrane permeation selectivity for CO_2/CH_4 mixture (asterisks) with an estimate (squares) based on the adsorption (circles, data by molecular modelling) and diffusion (triangles) selectivities resulting from IRM measurements with (**c**) a "giant" ZIF-8 crystal. Reproduced with permission from Ref. [37], copyright (2010) Wiley-VCH Verlag GmbH & Co. KGaA

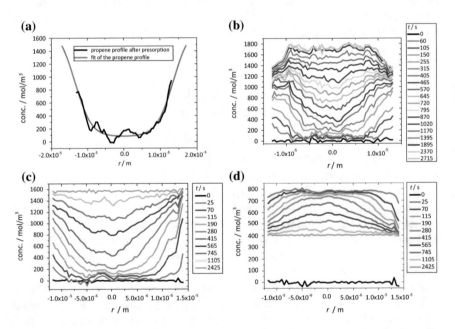

Fig. 10.8 Evolution of intracrystalline concentration profiles during guest overshooting in zeolite DDR: **b** Recording of buildup with ethane as the "driven" component initiated by a pressure step from 0 to 200 mbar in the surrounding atmosphere. **a** Before, the "driving" component propene was presorbed for 7 h at a pressure of 10 mbar. The propene profile (shown in the plot) remained essentially unaffected during ethane uptake. **c** Equilibration after overshooting with ethane as the "driving" and **d** CO_2 as the "driven" component during two-component adsorption with a step from 0 to 200 mbar in the partial pressures of either component in the surrounding atmosphere. Reprinted with permission from Ref. [39], licensed under CC BY 4.0

For rationalizing the outcome of these studies we have to return to Sect. 10.4 where the gradient of the chemical potential rather than the (mere) gradient of the concentration was identified as the "driving force of diffusion". Starting with Eq. (10.10), consideration of the spatial dependence of the chemical potential (known via Eq. (10.9) as a function of the spatial dependence of the concentration) leads—in the case of single-component adsorption—to Eq. (10.11). Now, however, on considering two-component diffusion, the chemical potential of a certain species (A) must be considered to be a function of the concentration of both components A and B! This new situation is immediately rationalized by having in mind that the chemical potential of A is related, via Eq. (10.9), to the pressure of A in the surrounding atmosphere which is necessary for keeping the concentration of A at the given value. Under the conditions of two-component adsorption, however, this pressure does, obviously, depend not only on the given concentration A. It does as well depend on the local concentration B. Quite intuitively, the partial gas pressure A in the surrounding atmosphere required to maintain the local guest concentration c_A will be expected to be the higher, the higher the local concentration c_B. This means, in other words, that the chemical potential of guest component A at a given concentration c_A tends to increase with increasing concentration c_B. In this reasoning we easily recognize a confirmation of also the validity of our assumption of simple additivity at sufficiently small concentrations (as exploited, as a first approximation, on considering membrane permeation) since there is, clearly, no interaction between the different guests so that the behaviour of A does in fact remain unaffected by the presence of B.

Following the procedure having led from Eqs. (10.9) and (10.10) to Eq. (10.11), for two-component adsorption we arrive at

$$j_{Ax} = u_A c_A = -\frac{RT}{f}\left(\frac{\partial ln p_A}{\partial ln c_A}\frac{dc_A}{dx} + \frac{\partial ln p_A}{\partial ln c_B}\frac{dc_B}{dx}\right) = -D_{AA}\frac{dc_A}{dx} - D_{AB}\frac{dc_B}{dx}. \quad (10.17)$$

where, with the last equation, Fick's 1st law is now presented in matrix notation. With this notation, molecular fluxes of a given component (A) are seen to be—at least potentially—driven by the gradients of both concentrations. The "driving efficiency" depends on the magnitudes of the elements D_{AA} and D_{AB} of the diffusion matrix which, with Eq. (10.17), are given by the logarithmic partial derivatives of the pressure of the component under consideration with respect to either guest concentration.

With Eq. (10.17) we do, in particular, note that diffusion fluxes of a certain species may even occur "uphill", i.e. into the direction of increasing concentration. This shall become possible as soon as the first term on the right of Eq. (10.17) (which, in such cases, is negative) is exceeded by the second one. Exactly such a situation may be recognized in Fig. 10.8. Figure 10.8b shows the evolution of the concentration of ethane in a crystal of zeolite DDR during adsorption within an external ethane atmosphere, following a pre-positioning of the crystal over 7 h in a propene atmosphere with the final propene distribution shown in Fig. 10.8a. As a consequence of the vast difference in the diffusivities of the two components in zeolite DDR, propene concentration may be assumed to remain invariant during the whole process of ethane uptake.

As an, at first sight, rather astonishing result we note that, after about 600 s (or 10 min), ethane concentration in the interior of the zeolite crystal continues to increase, irrespective of the fact that this increase necessitates an "uphill" diffusion flux, namely ethane fluxes into the direction of increasing ethane concentration. With Eq. (10.17), however, exactly such a behavior is predicted by the influence of the propene concentration gradient (Fig. 10.8a, second term on the right-hand side of Eq. (10.17)).

For an intuitive appreciation of this remarkable situation we refer to Fig. 10.9: Irrespective of an increase in ethane concentration, the equilibrium pressure and thus, via Eq. (10.9), the chemical potential is seen to decrease towards the crystal

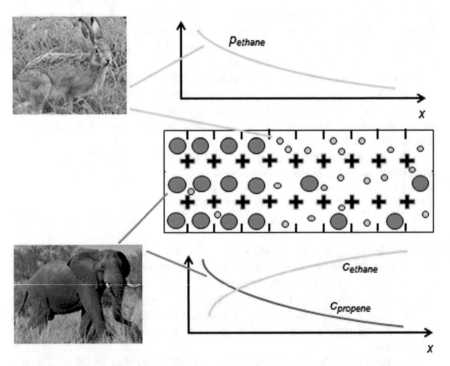

Fig. 10.9 Rationalizing "uphill diffusion" of ethane: Cartoon showing the distribution of propene (bulky and, hence, less mobile molecules represented by large bullets in grey) and ethane (small yellow bullets) close to the crystal boundary (center right), corresponding to the concentration profiles shown on bottom right. Top right shows the ethane pressure required for establishing, at equilibrium, the ethane concentrations shown on bottom right. Similarly as, by the influence of propene, the ethane molecules are driven to diffuse into the direction of increasing ethane concentration, the presence of the voracious big animals (bottom left) might be imagined to make the smaller (and more mobile) ones (top left) preferentially move into the direction of increasing populations of their conspecifics. Reprinted with permission from Ref. [40], copyright (2016) GDCh, Frankfurt am Main

interior, driving the ethane diffusion flux in "uphill" direction. As a consequence of the low propene diffusivities in the given type of zeolites, final equilibration (i.e. attainment of uniform concentration in propene and, thus, in also ethane) would require too large time spans. With the choice of the other "pair", namely with ethane now as the "slow" and CO_2 as the "fast" molecule, exactly this process of equilibration could be recorded. Thus, with their very first profile, the representations in Figs. 10.8c and d begin with the situation shown with the last one in Fig. 10.8b. With the subsequent profiles in Fig. 10.8c it is shown how the slower component continues to equilibrate over the crystal and how, correspondingly, the concentration of the faster component (CO_2) decreases. This decrease is the consequence of normal "downhill" fluxes into the direction of decreasing CO_2 concentration since, together with ethane equilibration, also the ethane concentration gradients are fading and, thus, also the second term on the right side of Eq. (10.17). The phenomenon of "overshooting", i.e. the observation of transient molecular uptake where individual components may, intermediately, exceed their equilibrium, has well-known predecessors with, e.g., methane-nitrogen mixtures in zeolite 4A [41], heptane/benzene mixtures in zeolite NaX [42] and, quite recently, n-alkane/ iso-alkane mixtures in MFI [43]. However only quite recently, with the advent of microimaging, could the relevant profiles be directly recorded.

10.6 Diffusion and Conversion

Since their use as highly selective, environmentally benign catalysts is among the most important technological applications of nanoporous materials, the investigation of both diffusion and conversion and the search for transport optimized materials is a challenging task of both fundamental and applied research. In fact, with the options of microimaging by IRM, catalysis research today disposes of a most powerful technique for the in situ investigation of diffusion and conversion [44].

Figure 10.10 introduces into the scheme of measurement, exemplified with the hydrogenation of benzene to cyclohexane, i.e. the transformation of an unsaturated hydrocarbon into the corresponding saturated one within a nanoporous host material (porous glass), serving as a support of metallic nickel (the catalyst). The process is initiated by bringing a mixture of benzene and hydrogen in contact with the initially empty host material. Diffusive spreading of benzene as the reactant molecule is accompanied by conversion to cyclohexane, following Eq. (2.19) with 1 and 2 referred to as benzene and cyclohexane, respectively. Hydrogen is offered in excess and may, owing to its high mobility, be assumed to be anywhere instantaneously present. The backward reaction rate is negligibly small, whence $k_{12} = 0$.

Figure 10.11 shows, together with the experimental data, also the solution of the diffusion-reaction equation as given by Eq. (2.19), with the simplifying assumption that the diffusivities of both components are independent of concentration and

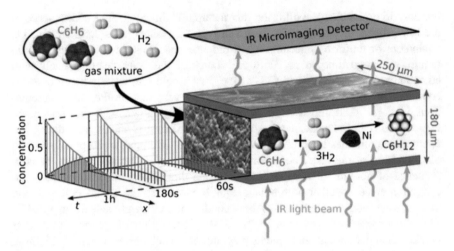

Fig. 10.10 Schematics of monitoring reactant and product profiles during the conversion of benzene (red) into cyclohexane (blue) in nanoporous materials by microimaging, with the arrows in green indicating the spatial extensions relevant for the experiments. Bottom and top faces are covered by an IR-transparent layer, impermeable for the reactant and product molecules. Reproduced with permission from Ref. [44], copyright (2015) Wiley-VCH Verlag GmbH & Co. KGaA

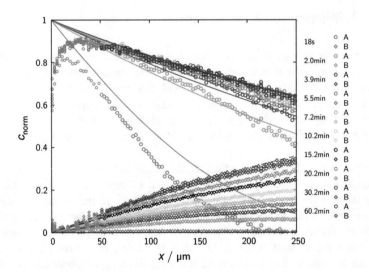

Fig. 10.11 Transient concentration profiles during hydrogenation of benzene to cyclohexane at 75 °C. The experiments are started by contacting an initially empty catalyst with a benzene-hydrogen atmosphere ($p_{benzene} = 27$ mbar; $p_{hydrogen} = 977$ mbar). Data points represent the experimental results obtained by IR microimaging (circles: benzene (A), diamonds: cyclohexane (B)), reflecting meaningful concentrations for $x \geq 50$ μm). The solid (benzene (A)) and dashed (cyclohexane (B)) lines are results of the analytical solution of Eq. (2.19) with the relevant initial and boundary conditions. Reproduced with permission from Ref. [44], copyright (2015) Wiley-VCH Verlag GmbH & Co. KGaA

composition and coincide. Initial and boundary conditions to be obeyed due to the chosen experimental conditions are, respectively, $c_1(x, t = 0) = c_2(x, t = 0) = 0$ and $c_1(x = \pm L, t) = 1$, $c_2(x = \pm L, t) = 0$, with $2L$ denoting the distance between the two faces (platelet edges opposing each other), which are in contact with the surrounding atmosphere. With the given boundary condition it is, further on, assumed that the total amount of benzene in the gas phase (and its diffusivity) is high enough so that, during the whole of the reaction, the concentration of cyclo-hexane—though emerging, during the reaction, within the nanoporous material—remains negligibly small in the gas phase.

Together with reasonable agreement between measurement and prediction, the data in Fig. 10.11 show the expected trend. Reactant benzene penetrates into the host material, following the direction of decreasing concentration. Simultaneously, a fraction is converted into cyclohexane. The emerging flux of cyclohexane, fol-lowing the direction of its decaying concentration, is directed towards the external surface. The cyclohexane molecules leaving the catalyst spread sufficiently fast in the surrounding atmosphere so that, also during the course of the experiment, their concentration in the gas phase may be assumed to remain negligibly small in comparison with that of benzene. The boundary conditions do thus coincide with those usually met in flow reactors as conventionally used in catalysis research. Steady state is attained when benzene influx is compensated by cyclohexane efflux, with the latter being identical with the total amount of cyclohexane (i.e. of product molecules) within the catalyst host particle "produced" per unit time. This amount is the key quantity of the process and, in technical applications, pursued to be as large as possible.

The highest overall reaction rate is obviously achieved if the reactant molecules attain highest concentrations all over the catalyst. This implies the absence of any diffusion resistance so that the product molecules can instantaneously disappear out of the catalyst, leaving space for new reactant molecules. The ratio between the actual overall reaction rate and the maximum possible one is referred to as the effectiveness factor. Though this factor is crucial for catalytic conversion, its direct measurement was so far impossible. This appears already in the given definition since it refers to conditions—namely the absence of any transport resistance—which cannot be fulfilled in reality. Rather than by comparison with reaction rates under total exclusion of transport resistances, effectiveness factors are conven-tionally determined by measurements with varied transport resistances. From the effect of such variations on the reaction rate, the effectiveness factor may in fact be estimated [1, 2]. As a prerequisite of such measurements, all other parameters must be kept constant. In addition to the need of performing several measurements rather than only a single one, data analysis is thus based on assumptions which are scarcely to be confirmed.

Both these constraints do not exist with the options of microimaging as demonstrated with Fig. 10.11. Here, the effectiveness factor of the chemical reac-tion results immediately as the area under the (normalized) concentration profile of the product molecule under steady-state conditions. It becomes, in this way, accessible by a one-shot experiment in a most direct way. It reminds of the expression "Poren-Nutzungsgrad" (degree of pore space exploitation) used as the German translation of "effectiveness factor". This term nicely recollects the way how microimaging is able to determine effectiveness factors, namely by recording exactly this part of the pore space which is occupied by the reactant molecules and which is, thus, exploited for the reaction.

The solution of the rather simple Eq. (2.19) nicely reflects the experimentally observed behaviour, irrespective of all inherent approaches. Slight deviations may easily be referred to a feature discussed already on analysing the shape of the concentration profiles shown in Fig. 10.4c: Diffusivities increasing with concentration lead to a sharpening of the diffusion front. And exactly such an (even if only slight) increase of the diffusivities with increasing loading may be also expected for benzene in porous glass. It is important to emphasize, however, that agreement between measurement and prediction is in no way a prerequisite for the determination of effectiveness factors by microimaging. It is rather the great advantage of IR microimaging that effectiveness factors result directly from experimental evidence, without need of any modelling!

10.7   Transport Enhancement in Pore Hierarchies

In the examples of technological application for matter upgrading by separation and conversion, process efficiency was found to be affected by the rate of exchange between intracrystalline pore space and the surroundings. Transport enhancement is thus among the key options of efficiency enhancement. Transport enhancement may be clearly achieved by operating with larger pores and, correspondingly, with reduced diffusional resistances. However, this option is in general not applicable since it is—in both separation and (selective) conversion—the intimate contact between the inner surface of the porous material and the molecules, on which performance is based. This contact would get lost with increasing pore radii.

Alternatively one might consider applying the nanoporous material as sufficiently small crystals/particles, with correspondingly small uptake and release times of the guest molecules. For spherical particles, e.g., the time constant of diffusion-controlled uptake and release is known to be [1]

$$\tau_{\text{uptake, release}} = \frac{R^2}{15D} \tag{10.18}$$

with R denoting the particle radius. Except for the factor $1/15$, the validity of Eq. (10.18) may immediately be rationalized as the only option for combining D and R to yield a quantity with the dimension of time. Equation (10.18) yields a good approach for even particles of arbitrary shape if R is understood as an "equivalent" radius

$$R_{eq} = \frac{3V}{A},$$ (10.19)

with A and V denoting the (external) surface area and the volume of the particle. With this relation, R_{eq} is immediately seen to be the radius of a sphere with the same surface-to-volume ratio. There are, however, narrow limits to transport enhancement through particle size reduction since too small particles enhance the risk of pipe plugging, with influx and efflux impediment.

Processes with hierarchically organized pore spaces have therefore, over the past few years, attained particular concern [45, 46]. In such arrangements, the space of micropores (with pore diameters in the range of the diameters of the molecules) is traversed by larger pores. Following the ingenious examples given by nature (see Ref. [47] and Chap. 11) these large pores serve as "highways", ensuring fast matter exchange between micropores and surroundings.

The complexity of such systems prohibits, in general, the treatment of mass transfer with analytical expressions as introduced in Sects. 2.1–2.3 and used, so far, also in this chapter. As an alternative one may adopt the option of simulating spread on cellular grids as introduced in Sects. 2.4 and 2.5. A scheme of such simulations is presented in Fig. 10.12. Here, the diffusivities D_{micro} and D_{meso} in the space of micro- and mesopores are quantitated by Eq. (2.3) (with 2 replaced by 6, corresponding with the 3 dimensions considered). The difference in the magnitudes ($D_{micro} \ll D_{meso}$) is taken account of by corresponding differences in the mean life times ($\tau_{micro} \gg \tau_{meso}$) between subsequent "jumps". Spacing ("jump length" $l_{micro} = l_{meso} = l$) is assumed to be uniform throughout the simulation grid. Differences in life times correspond to the differences in the energetic barriers between adjacent sites shown bottom right. The difference in the energy levels in micro- and mesopore space corresponds to the heat of adsorption which the guest molecules have to afford on leaving the intracrystalline space. Concentration in micropores is, correspondingly, much larger than in mesopores. This is ensured, in our model, by a corresponding choice of the exchange probabilities between the respective grid nodes ($P_{micro \rightarrow meso} \ll P_{meso \rightarrow micro}$) at the interface between micro- and mesopores.

Molecular spreading in such systems is, in general, a rather complex process. In addition to the pore architecture, it depends on the diffusivities (D_{micro} and D_{meso}) and the relative populations (p_{micro} and $p_{meso} \equiv 1 - p_{micro}$). As a first-order estimate, it is often possible to consider one of the two limiting cases of "fast exchange" and "slow exchange" between the two pore spaces [46, 48]. Estimates are in either case based on Eq. (10.18), however, with different meanings.

For exchange rates sufficiently fast in comparison with overall uptake and release, R ($=R_{cryst}$) is the crystal radius and D ($=p_{meso} D_{meso} + p_{micro} D_{micro}$) is the mean value of the diffusivities in the two pore spaces. With propane in zeolite NaCaA, e.g., transport enhancement in the limiting case of fast exchange was found to give rise to transport enhancement over more than 2 orders of magnitude in comparison with the purely microporous zeolite [49]. Further increase in the

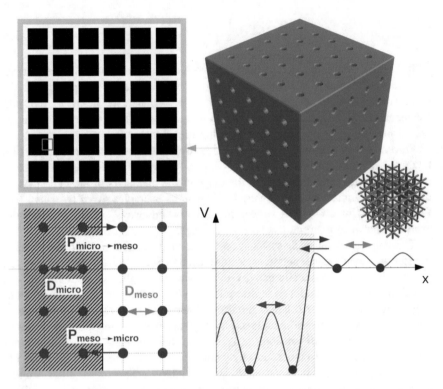

Fig. 10.12 Scheme for simulating molecular uptake by a hierarchically organized, regular pore network. A continuos microporous phase (top right) is penetrated by mesoprous channels (outermost right). Corresponding with the potential landscape on bottom right, the micropores (shaded area bottom left) are distinguished from the mesopores by a higher population density and reduced jump rates. For visual convenience, only 5 (rather than the actually considered 18) channels in parallel are considered. Reproduced with permission from Ref. [48], copyright (2015) Wiley-VCH Verlag GmbH & Co. KGaA

contribution $p_{meso} D_{meso}$ to overall diffusion (e.g. by temperature increase and corresponding increase in p_{meso}) is thus seen to lead to further transport enhancement.

During uptake in the opposite extreme, guest molecules are assumed to be essentially instantaneously spread over the whole of the internal surface of the mesopores, with uptake by the micropores occurring as a second step. This means, in terms of Eq. (10.18), a dramatic reduction in the extension (R) of the range over which, in the course of uptake, the guest molecules are now going to be distributed, whilst D maintains its meaning as the micropore diffusivity. Equation (10.19), with A and V now denoting the total area of the interface between the two pore spaces (=inner mesopore surface) and the crystal volume occupied by the space of mesopores (=crystal volume minus total mesopore volume), once again, yields a reasonable estimate [48]. Opposite to fast exchange, transport enhancement is a function of only the pore space geometry and remains unaffected by any further enhancement of mesopore mass transfer.

10.8   Anomalous Diffusion

With Sect. 2.4 we have introduced into the reasons which may lead to spreading mechanisms deviating from the simple logic of random walk and, thus, of normal diffusion, as introduced in Sect. 2.1. As an important feature leading to the occurrence of anomalous diffusion we did recognize the diffusant's "memory" [50]. Though mass transfer in nanoporous materials is in general subject to normal diffusion, in the broad context of this book a celebrated deviation, referred to as single-file diffusion [51, 52], should be mentioned. Mass transfer under single-file conditions is subject to the restriction that the diffusants keep their sequential arrangement, just like pearls on a neckless (or geese in "single file"). In nanoporous host-guest systems single-file diffusion occurs as soon as within channel pores adjacent molecules are unable to mutually exchange their positions [53]. Figure 10.13a illustrates such a situation. A rather unconventional way of simulating single-file diffusion is shown in Fig. 10.13b.

Let us start with considering an infinitely large single-file, with known particle positions at a certain instant of time. Since the particles are not allowed to change their order, displacement of an arbitrarily selected molecule into one direction requires a corresponding shift of all particles "in front of it". As a consequence, particle density in front of the particle (i.e. in the direction where it is shifted to) will, in general, be higher than "behind". Hence, for a particle shifted in one direction further shifts are more likely to occur in backward direction than into the direction to which it has been shifted already since the presumably higher particle densities in front of it will more likely impede further particle propagation into this direction than shifts backwards. This tendency increases with increasing shifts (corresponding with the increase in density "in front" of a particle with increasing shifts). Subsequent shifts are therefore not uncorrelated anymore. The starting assumption of random walk and, hence, of normal diffusion with mean square displacements increasing linearly with time, must thus be dropped. It rather turns out that the mean square displacement increases with only the square root of time. The distribution function remains a Gaussian (Eq. (2.10), now, however, with Dt replaced by $F\sqrt{t}$ and F referred to as the single-file mobility [51, 54]).

A totally different reasoning is necessary on considering mass transfer in single files of finite extension. This is the common situation with nanoporous particles of single-file structure. Though being still subject to the confinement just described in the channel interior, molecules at the file ends are allowed to leave and new molecules to enter. Such events occur essentially uncorrelated, with equal probabilities on either side. On considering the molecules initially in the file as labelled

(a)

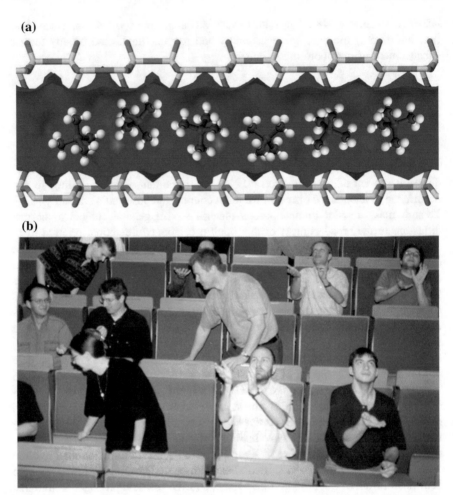

(b)

Fig. 10.13 "Particle" transport under the constraint of single-file diffusion: **a** As soon as guest molecules in channels are too bulky to pass each other, mass transfer is subject to the requirement of invariable order in "particle" arrangement. **b** Unconventional way of simulating single-file diffusion in the tiers of the physics lecture hall of Leipzig University: "Particle jump attempts", with the direction determined by throwing coins, are only successful if directed to a vacant seat. Reproduced with permission from Ref. [56], copyright (1998) Wiley-VCH Verlag GmbH & Co. KGaA

and all the other ones, entering later, as unlabelled, the process of molecular entering and leaving is easily seen to give rise to a random movement of the centre of mass of the labelled molecules. This movement is, moreover, seen to be subject to the same statistics as a random walker. For the corresponding (the "centre-of-mass") diffusivity one finds [55]

$$D_{cm} = D\frac{1-\Theta}{n} = D\frac{1-\Theta}{\Theta}\frac{\lambda}{L}.$$ (10.20)

D is the diffusivity of an isolated particle in a file of length L with site distance (determining the jump length) λ. Θ denotes the site occupation probability and n the number of particles in the file. $D(1-\Theta)$ results as the so-called mean-field approach of the diffusivity where, by the factor $1-\Theta$, it is taken into account that only jump attempts to vacant sites (occurring with just this probability) are successful. We note with Eq. (10.20) that, for taking account of the confinement under single-file conditions, this value must, in addition, be divided by the total number of particles in the file.

As a most remarkable result, the cross-over from mean-square displacements scaling with the square-root of time (typical of genuine single-file diffusion) to displacements following the normal diffusion of the centre of mass (Eq. (10.20)) may be estimated to be

$$\langle \Delta x^2 \rangle_{sf \leftrightarrow cmdiff} = \frac{2}{\pi}\frac{1-\Theta}{\Theta}\lambda L.$$ (10.21)

With Eq. (10.21), displacements over distances already much smaller than the crystal extensions L are seen to be controlled by normal diffusion, with the (effective) diffusivity given by Eq. (10.20). Exchange dynamics with nanoporous particles subject to single-file confinement is thus seen to follow normal diffusion conditions with, however, the "intrinsic" diffusivities replaced by the centre-of-mass diffusivity given by Eq. (10.20). Combining Eqs. (10.18) and (10.20) yields the important result that, under single-file constraint, molecular exchange rates (and, thus, processing efficiency in separation and conversion) are reduced with even the third power of the particle size [57] and not only their square as appearing from Eq. (10.18)! Since quite a number of zeolite catalysts do in fact operate under single-file conditions, efforts for reducing the extensions of the purely microporous ranges in nanoporous materials as motivated in the previous section are thus seen to be of even larger relevance!

10.9   Conclusions

Advent of the techniques of microscopic measurement has enabled insight into an impressive wealth of transport phenomena accompanying and controlling molecular spreading in nanoporous materials. The new options thus provided allow illustrating quite a number of mechanisms relevant for diffusive mass transfer quite in general. The presentations were mainly based on the possibility to record the evolution of transient guest profiles with microscopic resolution provided by the techniques of microimaging [13] and on the potentials of PFG NMR for exploring the propagation patterns of molecular spreading by monitoring the probability distribution of molecular displacements (the "mean propagator" [9]). In addition to these options, progress in our knowledge of diffusion in nanoporous materials is based on the fascinating developments in molecular modelling over the past few decades [1, 58, 59] as well as on the input by manifold variants and innovations of conventional measurements based on recording uptake and release [60] and, notably, by quasi-elastic neutron scattering (QENS [59]) and single-particle tracking [61]. Complementing the information of PFG NMR, QENS is able to trace spreading phenomena over nano- rather than micrometers, while single-particle observation, notably in combination with the novel options of microspectroscopy [62], is able to focus on individual molecules rather than on molecular ensembles.

Thus, by the simultaneous application of PFG NMR and single-particle tracking, a very fundamental hypothesis of statistical physics has become, in the field of diffusion, accessible by direct experimental investigation. As the very first equation of the book, Eq. (2.1) considered the mean value of the squared particle displacement. There exist, however, two ways for determining such values: one may take the average of the displacements of either *all particles* during *one and the same interval* of time, or of only *one particle* over *several, subsequent intervals* of time. It is required by the theorem of ergodicity [63] that, under equilibrium, both averages have to coincide. Applying both techniques to one and the same system, namely a suitably selected fluorescing molecule as guest and the porous glass considered in already Sect. 10.6 as the host, the agreement of both types of averages could in fact be confirmed [64].

Concerted applications of various measuring techniques are doubtlessly among the primary prerequisite of further progress in diffusion research with nanoporous materials in both theory and application. This is mainly due to challenges rising with the increasing complexity of the systems of interest and, hence, of the phenomena inherent to such systems. With the hierarchical pore spaces, a most prominent example has been introduced in Sect. 10.7, which shall be dealt with further on in Chap. 11.

The increase in pore diameters is accompanied by a phenomenon which we have completely left out in our presentation. It concerns the occurrence of phase transitions in pore spaces, both between the solid and liquid and the liquid and gaseous states and the preservation of such states outside of their thermodynamic equilibrium over essentially infinitely long time spans [65]. Under such conditions, equilibration turns out to necessitate the shift of molecular aggregates rather than of only individual molecules as so far generally considered. Investigating the spreading and diffusion of such aggregates [66] is among the great challenges of future research with nanoporous materials.

References

1. J. Kärger, D.M. Ruthven, D.N. Theodorou, *Diffusion in Nanoporous Materials* (Wiley-VCH, Weinheim, 2012)
2. G. Ertl, H. Knözinger, F. Schüth, J. Weitkamp (eds.), *Handbook of Heterogeneous Catalysis*, vol. 3, 2nd edn. (Wiley-VCH, Weinheim, 2008)
3. F. Schüth, K.S.W. Sing, J. Weitkamp (eds.), *Handbook of Porous Solids* (Wiley-VCH, Weinheim, 2002)
4. J. Weitkamp, Solid State Ionics **131**, 175 (2000)
5. W. Siegmund, *Anacortes Refinery (Tesoro)* (2008), https://upload.wikimedia.org/wikipedia/commons/5/51/Anacortes_Refinery_31911.JPG
6. J. Kärger, C. Chmelik, R. Valiullin, Phys. J. **12**, 39 (2013)
7. J. Kärger, ChemPhysChem **16**, 24 (2015)
8. P.T. Callaghan, *Translational Dynamics and Magnetic Resonance* (Oxford University Press, Oxford, 2011)
9. J. Kärger, W. Heink, J. Magn. Reson. **51**, 1 (1983)
10. R. Kimmich, *Principles of Soft-Matter Dynamics* (Springer, London, 2012)
11. C. Chmelik, J. Kärger, Chem. Soc. Rev. **39**, 4864 (2010)
12. T. Titze et al., Angew. Chem. Int. Ed. **54**, 14580 (2015)
13. J. Kärger et al., Nat. Mater. **13**, 333 (2014)
14. C. Chmelik et al., Phys. Rev. Lett. **104**, 85902 (2010)
15. R. Krishna, Micropor. Mesopor. Mater. **185**, 30 (2014)
16. P. Kortunov et al., J. Phys. Chem. B **110**, 23821 (2006)
17. C. Chmelik et al., ChemPhysChem **10**, 2623 (2009)
18. J. Kärger, D.M. Ruthven, New J. Chem. **40**, 4027 (2016)
19. (a) L. Heinke, P. Kortunov, D. Tzoulaki, J. Kärger, Phys. Rev. Lett. **99**, 228301 (2007); (b) L. Heinke, J. Kärger, Phys. Rev. Lett. **106**, 74501 (2011); (c) J. Cousin Saint Remi et al., Nat. Mater. **10** (2015); (d) J.C.S. Remi et al., Nat. Mater. **15**, 401 (2015)
20. (a) J. Crank, *The Mathematics of Diffusion* (Clarendon Press, Oxford, 1975); (b) H.S. Carslaw, J.C. Jaeger, *Conduction of Heat in Solids* (Oxford Science Publications, Oxford, 2004)
21. (a) J. Philibert, *Atom Movement —Diffusion and Mass Transfer in Solids* (Les Editions de Physique, Les Ulis, Cedex A, 1991); (b) H. Mehrer, *Diffusion in Solids* (Springer, Berlin, 2007)
22. J.M. van Bemmelen, Z. Anorg, Allg. Chem. **13**, 233 (1897)
23. J.C. Maxwell, Philos. Mag. **19**, 19 (1860)
24. J. Stefan, Wien Ber. **65**, 323 (1872)
25. A. Einstein, Ann. Phys. **19**, 371 (1906)
26. S.R. DeGroot, P. Mazur, *Non-Equilibrium Thermodynamics* (Elsevier, Amsterdam, 1962)
27. R.M. Barrer, W. Jost, Trans. Faraday Soc. **45**, 928 (1949)
28. J. Kärger, Surf. Sci. **57**, 749 (1976)
29. (a) R.M. Barrer, Adv. Chem. Ser. **102**, 1 (1971); (b) R.M. Barrer, *Zeolites and Clay Minerals as Sorbents and Molecular Sieves* (Academic Press, London, 1978)
30. L.S. Darken, Trans. Am. Inst. Mining Metall. Eng. **175**, 184 (1948)
31. I. Prigogine, *The End of Certainty* (The Free Press, New York, London, Toronto, Sydney, 1997)
32. F. Salles et al., Angew. Chem. Int. Ed. **121**, 8485 (2009)
33. (a) R.Q. Snurr, J. Kärger, J. Phys. Chem. B **101**, 6469 (1997); (b) F. Rittig, C.G. Coe, J.M. Zielinski, J. Phys. Chem. B **107**, 4560 (2003)
34. S. Beckert et al., J. Phys. Chem. C **117**, 24866 (2013)
35. D. Freude et al., Micropor. Mesopor. Mater. **172**, 174 (2013)
36. (a) J. Caro, Micropor. Mesopor. Mater. **125**, 79 (2009); (b) H. Bux et al., J. Membr. Sci. **369**, 284 (2011)

37. H. Bux et al., Adv. Mater. **22**, 4741 (2010)
38. (a) C. Chmelik, A. Mundstock, P.D. Dietzel, J. Caro, Micropor. Mesopor. Mater. **183**, 117 (2014); (b) U. Hong, J. Kärger, H. Pfeifer, J. Am. Chem. Soc. **113**, 4812 (1991); (c) C. Chmelik, D. Freude, H. Bux, J. Haase, Micropor. Mesopor. Mater. **147**, 135 (2012)
39. A. Lauerer et al., Nat. Commun. **6**, 7697 (2015)
40. J. Kärger, Nachr. Chem. **64**, 620 (2016)
41. H.W. Habgood, Can. J. Chem. **36**, 1384 (1958)
42. J. Kärger, M. Bülow, Chem. Eng. Sci. **30**, 893 (1975)
43. T. Titze et al., J. Phys. Chem. C **118**, 2660 (2014)
44. T. Titze et al., Angew. Chem. Int. Ed. **54**, 5060 (2015)
45. J. García-Martínez, K. Li (eds.), *Mesoporous Zeolites. Preparation, Characterization and Applications* (Wiley-VCH, Weinheim, 2015)
46. S. Mitchell et al., Nat. Commun. **6**, 8633 (2015)
47. M.O. Coppens, *Nature Inspired Chemical Engineering (Inaugural Lecture)* (Delft University Press, Delft, 2003)
48. D. Schneider et al., Chem. Ingen. Technol. **87**, 1794 (2015)
49. D. Mehlhorn et al., ChemPhysChem **13**, 1495 (2012)
50. J. Klafter, I.M. Sokolov, Phys. World August, 29 (2005)
51. P.S. Burada et al., ChemPhysChem **10**, 45 (2009)
52. N. Leibovich, E. Barkai, Phys. Rev. E **88**, 32107 (2013)
53. (a) K. Hahn, J. Kärger, V. Kukla, Phys. Rev. Lett. **76**, 2762 (1996); (b) V. Kukla et al., Science. **272**, 702 (1996)
54. (a) J. Kärger, Phys. Rev. E **47**, 1427 (1993); (b) M. Kollmann, Phys. Rev. Lett. **90**, 180602 (2003)
55. (a) K. Hahn, J. Kärger, J. Phys. Chem. **102**, 5766 (1998); (b) P.H. Nelson, S.M. Auerbach, J. Chem. Phys. **110**, 9235 (1999)
56. J. Kärger, K. Hahn, V. Kukla, C. Rödenbeck, Phys. Blätter **54**, 811 (1998)
57. S. Vasenkov, J. Kärger, Phys. Rev. E **66**, 52601 (2002)
58. F.J. Keil, R. Krishna, M.O. Coppens, Rev. Chem. Eng. **16**, 71 (2000)
59. H. Jobic, D. Theodorou, Micropor. Mesopor. Mater. **102**, 21 (2007)
60. (a) D.M. Ruthven, S. Brandani, M. Eic, in *Adsorption and Diffusion*, ed. by H.G. Karge, J. Weitkamp (Springer, Berlin, Heidelberg, 2008), p. 45; (b) J. van den Bergh, J. Gascon, F. Kapteijn, in *Zeolites and Catalysis: Synthesis, Reactions and Applications*, ed. by J. Cejka, A. Corma, S. Zones (Wiley-VCH, Weinheim, 2010), p. 361
61. C. Bräuchle, D.C. Lamb, J. Michaelis (eds.), *Single Particle Tracking and Single Molecule Energy Transfer* (Wiley-VCH, Weinheim, 2010)
62. E. Stavitski, B.M. Weckhuysen, Chem. Soc. Rev. **39**, 4615 (2010)
63. G.D. Birkhoff, Proc. Nat. Acad. Sci. **17**, 656 (1931)
64. F. Feil et al., Angew. Chem. Int. Ed. **51**, 1152 (2012)
65. (a) R. Valiullin et al., Nature **430**, 965 (2006); (b) S. Naumov, R. Valiullin, P. Monson, J. Kärger, Langmuir **24**, 6429 (2008); (c) D. Schneider, R. Valiullin, P.A. Monson, Langmuir **30**, 1290 (2014); (d) D. Kondrashova, R. Valiullin, J. Phys. Chem. C **119**, 4312 (2015)
66. (a) A. Lauerer et al., Micropor. Mesopor. Mater. **214**, 143 (2015); (b) D. Kondrashova et al., Sci. Rep. **7**, 40207 (2017)

Chapter 11
Nature-Inspired Optimization of Transport in Porous Media

Marc-Olivier Coppens and Guanghua Ye

11.1 Introduction

Transport of molecules across multiple length scales is of great practical importance, from food products and building materials to the recovery, production and distribution of chemicals and energy. Many relevant processes involve porous media; these include catalytic and separation processes, oil and gas recovery, and the delivery of pharmaceuticals. An effective transport system should be scalable, efficient and robust. These properties depend on the multiscale architecture of the transport system, that is, its morphology (shape) and topology (connectivity) at multiple length scales. An optimized transport system boosts production, saves time and cost, and reduces waste. This holds true for the infrastructure for transporting goods and information, as much as for the transport of molecules in porous media. To be optimal, the transport system needs to be suited to serve the other processes in the system, where production or consumption occurs. If these are not properly matched, transport limitations occur. This includes processes involving porous catalysts in chemical engineering, which we focus on in this chapter, although much of the discussion can be translated to other processes involving porous media as well.

The optimized transport system for such technical applications is not easy to obtain, but we can seek inspiration from biology. Indeed, through billions of years of evolution, plants and animals have acquired highly effective transport systems,

M.-O. Coppens (✉)
Department of Chemical Engineering, University College London,
London WC1E 7JE, UK
e-mail: m.coppens@ucl.ac.uk

G. Ye
State Key Laboratory of Chemical Engineering, East China University
of Science and Technology, Shanghai 200237, China
e-mail: guanghuaye@ecust.edu.cn

© Springer International Publishing AG 2018
A. Bunde et al. (eds.), *Diffusive Spreading in Nature, Technology and Society*,
https://doi.org/10.1007/978-3-319-67798-9_11

crucial to their survival. Although a chemical engineering application is different from a biological one in terms of materials (which could be inorganic instead of organic) and operating conditions (which could involve high temperatures and pressures instead of mild, ambient conditions), they share fundamental features. All rely on effectively connecting the action at microscopic scales (of cells, in the case of biology, or active sites, in the case of catalysts) with the overall system (the organism in biology or the reactor in a catalytic process). Robustness and scale independence are important in both instances. For this, they both rely on multiscale architectures, an example of which is illustrated in Fig. 11.1, while the dominant transport mechanism at each scale is governed by physics that are length scale dependent. Based on these common features, it is desirable to seek guidance from nature, to help improve the transport architectures of porous media for engineering applications. Another reason is the ability to design and optimize porous materials for these applications from scratch. This is possible in certain applications, like catalysis or fuel cells, as opposed to applications where transport networks would have to be adapted from existing plans, and are thus more difficult to change, as in city planning or in resource exploration in porous rocks.

Unlike transport architectures in nature, which were introduced in Part II of this book, transport pathways in artificial porous media are currently not the product of organic evolution. However, they can be optimized by mathematical modeling and computation, which could, for that matter, employ genetic algorithms inspired by evolution. Incredible progress in materials synthesis and manufacturing methods, with increasing control over structure extending down to ever-smaller scales, provides the opportunity to boost the performance of processes employing these materials. To do so effectively, requires guidance from theoretical insights and computational optimization.

In this chapter, some fundamental features of transport networks in porous media are introduced, and the structure-function relationships in these systems are briefly reviewed to introduce the available "handles" that can be used to manipulate molecular transport and improve the performance of processes that depend on it,

Cell (μm)   Leaf (cm)   Tree (m)

Fig. 11.1 The multiscale architecture of trees. Cells, 10–100 μm in size [1], are the basic building blocks (microscale) containing "active sites" for photosynthesis (nanoscale), converting CO_2 and water into carbohydrates; leaves, 1–100 cm in size, function as "porous photo-catalysts" with a veinal architecture for transport (mesoscale); the tree itself, with its water- and nutrient-transporting tree-crown, 1–100 m in size, functions as a living "reactor", providing mechanical strength and scalability during growth

like catalysis and molecular separations. These two sections are presented first, because a good understanding is essential to optimize transport phenomena in engineered systems. Subsequently, the nature-inspired chemical engineering (NICE) approach for transport optimization is introduced, and applications to heterogeneous catalysis and proton exchange membrane (PEM) fuel cells are given to illustrate this methodology.

11.2 Fundamental Features of Mass Transport Phenomena in Porous Media

Mass transport in porous media occurs primarily by two mechanisms, namely convective flow and diffusion. In wide pore channels, convective, pressure-driven flow is often the principal transport mechanism [2]. In narrower channels, diffusion is the dominant transport mechanism. Self-diffusion is a result of the thermal motion of molecules, while transport diffusion results from a chemical potential gradient; for non-interacting molecules, at sufficiently low pressure, self- and transport diffusivities are the same [3, 4]. For many processes in chemical engineering and beyond, involving porous catalysts, membranes, building materials and pharmaceutical tablets, for example, diffusion takes place in porous materials containing a hierarchical pore network, and diffusion can be subdivided into molecular diffusion, Knudsen diffusion, surface diffusion, and configurational diffusion, according to the interactions between the molecules and the pore walls [5–7]. Molecular diffusion dominates when the mean free path of a molecule is much smaller than the local pore size, so that the frequency of intermolecular collisions exceeds that of molecule-wall collisions. Knudsen diffusion becomes dominant when molecule-wall collisions are important. Surface diffusion describes the movement of adsorbed molecules along pore wall surfaces, and becomes important for very narrow pores and strongly adsorbed molecules [8]. Configurational diffusion dominates in zeolites and other microporous materials [7, 9], in which the effect of pore walls on the movement of molecules is so strong that diffusion is typically an activated process and, therefore, can be well described in terms of a succession of hops. Some state-of-the-art technologies, like interference microscopy (IFM) and IR microscopy (IRM), are now available to record such transport processes experimentally, even in single particles, which is introduced in Chap. 10. In addition, viscous flow plays an important role for transport in porous materials with wide pores, such as porous membranes for microfiltration and ultrafiltration [6].

Depending on the length scale, different transport mechanisms are involved. These different transport mechanisms often take place simultaneously, which complicates the optimization of the transport network. Let us take mass transport and reactions in a fixed-bed reactor packed with catalyst pellets as an example to illustrate this multiscale transport, shown in Fig. 11.2. Reactions take place on the so-called "active sites", which are of atomic or nanoscale dimensions, and dispersed

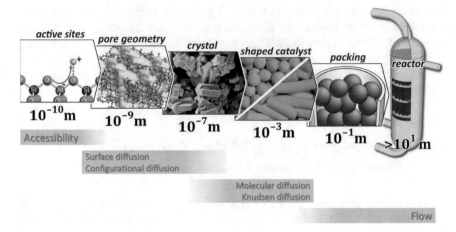

Fig. 11.2 Multiscale structure of a fixed bed reactor packed with zeolitic catalyst pellets and the dominant mass transfer mechanisms on different length scales. Reproduced from [10], with permission

on the internal surface of the porous pellets. The geometric and electronic properties of the active sites determine how some species are bound and converted on the catalyst surface. The local physicochemical conditions around these active sites, like the local species concentrations and temperature, affect the local reaction rates and, thus, the catalytic activity and selectivity. This local environment is influenced by the multiscale transport of reactant and product molecules toward and away from these sites, which frequently leads to spatially non-uniform distributions of reactants and products:

(1) Reactants are transported into catalyst pellets from the bulk phase by overcoming external film mass transfer resistance, and subsequently diffuse into the macropore (>50 nm diameter) and mesopore (2–50 nm) network, where molecular diffusion and Knudsen diffusion dominate. In, for example, the case of zeolites, molecules further diffuse into the micropore network (pores < 2 nm diameter), where surface diffusion and configurational diffusion become dominant. Simultaneously, molecules adsorb and react on active sites on the pore walls. Products desorbed from the active sites are transported out of the catalyst pellets in the opposite direction. Intrinsically fast reactions may lead to transport limitations, meaning that the resistance to molecular transport dominates the overall rate of the combined process.

(2) The flow of molecules in the fixed-bed reactor removes products that have been transported out of the catalyst pellets, and brings in reactants that enter the catalyst pellets. This leads to a decrease in reactant concentrations and an increase in product concentrations in the direction of the flow. Due to this, the boundary conditions at the interface between the catalyst pellets and the bulk flow in the reactor change from reactor inlet to outlet.

This multiscale transport of molecules is one of the most important, fundamental features for various engineering processes, beyond the example of fixed-bed reactors. In Sect. 11.6, the example of multiscale transport in PEM fuel cells is also depicted. Already, we can see a parallel with the tree shown in Fig. 11.1, something we will come back to in Sect. 11.4.

11.3 Basic Description of Transport in Porous Media

The effective transport properties (permeability for viscous flow and diffusivity for diffusion) depend on the structure of the porous medium, especially the pore size distribution. This provides abundant room for designing porous media with optimized transport properties. To do so, it is necessary to formulate relationships between material structure and transport properties. There is a huge literature on this subject, which will not be discussed here in detail. A brief introduction is given in order to aid the understanding of the following sections in this chapter. Readers can refer to a number of review articles for more details [5, 7, 11, 12].

11.3.1 Geometrical Description of Porous Media

Molecular transport networks can be ordered or disordered at different length scales. For example, at the macroscale, fixed bed reactors typically consist of random packings of catalyst particles, in between which the various species flow through a disordered void space. Other reactors employ structured packings, the most common type of which are monolithic structures with parallel channels; the catalytic converter to clean up car exhaust is an example of such a structured packing. In these monoliths, the walls of the channels are porous themselves, or are covered by a catalytic washcoat. At smaller scales, within porous catalysts and other porous materials, molecules diffuse through a network of macro-, meso- and/or micropores. In most amorphous catalyst supports and adsorbents, this pore network is disordered. However, pores can also form a regular network, such as in crystalline zeolites, metal-organic frameworks, and amorphous materials with ordered mesopores [13]. It is easier to model ordered systems and investigate the effects of the regular pore network properties on transport than to model disordered transport networks.

To describe transport in disordered porous media, two types of models are used: continuum models, which treat the porous medium as an effective continuum of reduced permeability or diffusivity, and discrete models, which explicitly account for the pores. Both have been extensively reviewed by Sahimi et al. [14] and Keil [5]. One of the earliest pore models is the parallel pore model proposed by Wheeler [15]. In his model, the pore space is represented by parallel pores with mean radius

$\bar{r}$ and length $\bar{L}$. The sum of the surface areas of all the parallel pores is set equal to the BET surface area of the particle, and the sum of the pore volumes is set equal to the experimentally determined pore volume of the particle:

$$\bar{r} = \frac{2V_g}{S_g}\sigma(1-\varepsilon) \qquad (11.1)$$

$$\bar{L} = \sqrt{2}V_p/S_x \qquad (11.2)$$

where: V_g and V_p are the specific pore volume and the total volume of the porous particle, respectively; S_g and S_x are the BET specific internal surface area and the external surface area of the porous particle, respectively; σ is the pore wall roughness factor; ε is the particle porosity. After that, numerous other pore models (including the cylindrical pore model [16], the tortuous pore model [17], the model of Wakao and Smith [18, 19], the model of Foster and Butt [20], the grain model [21], and the micro/macropore model [22]) have been proposed to account for more features of real porous materials, such as the tortuosity of the pores or a bidisperse pore size distribution.

Although these early models can describe certain morphological features and, in some cases, account for the pore size distribution, they do not account for the pore network connectivity (topology) and the spatial distribution of the pores. This becomes possible by using pore network models, in which equations for diffusion, adsorption and reaction are explicitly solved, kinetic Monte-Carlo simulations are employed, or various approximations based on statistical physics, like the effective medium approximation or renormalization group theory, are used [23–26].

However, even most pore network representations are still an abstraction of the real porous structure, based on macroscopic data, such as the measured pore size distribution, the porosity and the BET surface area. Recently, with the advent of powerful computers and more sophisticated experimental tools, it is becoming possible to digitally reconstruct a real porous structure with increasing accuracy. Some computational methods, including statistical methods (e.g., Monte Carlo method) and process-based methods (e.g., discrete element method) [27, 28], have been developed to digitally reconstruct porous materials with high accuracy, as shown in Fig. 11.3a. Cutting-edge experimental technologies, such as X-ray microtomography, directly provide us with three-dimensional (3D) images of porous materials, without even destroying the samples [29, 30], as shown in Fig. 11.3b. X-ray nanotomography and electron tomography allow to push the boundaries even further, to unprecedented resolution, although sample sizes are still limited, and care needs to be taken for samples that are anisotropic or macroscopically heterogeneous. In each case, the digitally reconstructed porous structure can be represented by the phase function $f(x)$, which takes the form of a 3D matrix, containing the information of the phase state in each voxel:

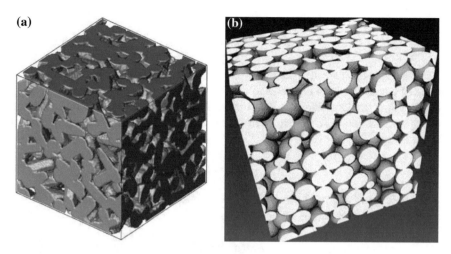

Fig. 11.3 Porous media generated by **a** virtual particle packing and **b** X-ray microtomography scans. From Refs. [30] and [29], with permission

$$f(x) = \begin{cases} 1 & \textit{if } x \textit{ belongs to pore space} \\ 0 & \textit{otherwise} \end{cases} \tag{11.3}$$

where x is the position vector of a voxel from an arbitrary origin. The digitally reconstructed pore structures can be successfully used to represent rocks [31], membranes [32], fuel cell electrodes [33], porous catalysts [30], fixed beds [29], and many other porous media. Furthermore, these digitally reconstructed pore structures can be reduced to pore networks, using network extraction algorithms, such as the thinning algorithm [34], the medial axis based algorithm [35, 36], and the maximal ball algorithm [37, 38]. Such digital reconstruction techniques have become a powerful tool for investigating various processes in porous media, especially mass transport.

11.3.2 Influence of the Structure of Porous Media on Transport Properties

Using the geometrical models briefly introduced in Sect. 11.3.1, we are able to describe transport in porous media, no matter whether they are ordered or disordered. As a prelude to the optimization studies discussed further on, it is important to understand how the structure of the pore network changes the transport properties. The effects of the geometry of a porous medium on viscous flow and diffusion are briefly recalled. Viscous flow of simple fluids through a single channel or, by extension, a porous medium can be described by Darcy's law, which has been derived from the Navier-Stokes equations via homogenization.

$$v = -\frac{k}{\eta}\nabla p \tag{11.4}$$

Here, v is the so-called Darcy velocity (the average velocity over a volume element containing both fluid and solid matrix), η is the viscosity of the fluid, ∇p is the pressure gradient, and k is the permeability of the (part of a) porous medium under consideration. The permeability for a cylindrical capillary can be calculated using Poiseuille's law [39]:

$$k = d^2/32 \tag{11.5}$$

where d is the diameter of the capillary. For a suspension of spheres with diameter d_0, the permeability can be obtained from the Richardson-Zaki correlation [40]:

$$k = (d_0^2/18)\varepsilon^{2.7} \tag{11.6}$$

For an aggregated bed of spheres with diameter d_0, the Carman-Kozeny relation can be used to calculate the permeability [6]:

$$k = (d_0^2/180)\left[\varepsilon^2/(1-\varepsilon)^2\right] \tag{11.7}$$

These are approximations; if more is known about the geometry of the porous medium, the permeability can be estimated more accurately. Structures in which viscous flow occurs, vary in morphology, topology, and randomness, resulting in different equations for the permeability. With advances in techniques to reconstruct the pore space, such as X-ray tomography, and to model flow in porous media, such as Lattice Boltzmann modelling, mesoscopic structural information can be employed to estimate macroscopic viscous flow, including that of complex fluids that can no longer be represented by Darcy's law [29].

Transport diffusion in porous materials, such as porous catalysts and adsorbents, is phenomenologically described by Fick's first law:

$$J = -D_e\nabla c \tag{11.8}$$

$$D_e = \varepsilon D_m/\tau \tag{11.9}$$

where: J is the diffusion flux; ∇c is the concentration gradient; D_e is the effective diffusivity in the porous medium; D_m is the bulk diffusivity; and τ is the tortuosity, lumping various geometrical (and, possibly, also non-geometrical) factors that affect diffusion in porous materials. Pore size affects the diffusivity through molecule-wall interactions. In micropores, this influence can be so significant that (11.9) is no longer valid and the diffusivity is typically 4–10 orders of magnitude smaller than the one in the bulk phase. The statistical and spatial distributions of pore size also affect the effective diffusivity and tortuosity. Diffusion of molecules tends to be slower when the pore size distribution is wider [41, 42]. Tortuosity values as high as 138 have been calculated for a pore network with a connectivity of 3 [43], when the wide and narrow pores of a bimodal pore-size distribution are spatially randomly distributed within the same network; however, this value would be much smaller if a connected network of wide pores surrounds particles with narrow pores, as is more typical in catalyst pellets [43]. The effective diffusivity decreases with decreasing connectivity, but is less dependent on the pore network topology when the connectivity is high enough [42]. The randomness of pore networks also affects the effective diffusivity, especially when the connectivity is low [43–45]. The effective diffusivity of a regular pore network is larger than the one of an irregular pore network, because the diffusion path in the irregular pore network is more tortuous [44].

Amorphous porous materials have a disordered framework, so that their pore walls are not smooth, as is assumed in common cylindrical and spherical pore models, but rough. For many amorphous materials used as catalyst supports and adsorbents, the surface roughness can be described by fractal geometry, similar to natural coastlines [46–52]. Fractals possess scale invariance, that is, they look similar at multiple length scales: magnifying certain parts reveals a structure similar to the whole.

Benoit Mandelbrot coined the word "fractal," when he discovered that there is a common mathematical language describing such rugged objects, which are infinitely fragmented (like the Cantor set), are lines that are almost nowhere differentiable (like the Koch curve) or are nets with an infinite power law distribution of holes (like the Sierpinski gasket or the Menger sponge) [46]. Each of these objects is strictly self-similar, whatever the magnification. Most importantly, however, what seemed esoteric examples by mathematicians are, in fact, prototypes for similar shapes in nature, like those shown in Fig. 11.5 further on; examples of natural fractals are as diverse as ore distributions, broccoli, clouds, trees, bread, turbulent flow, mountains, or natural coastlines. These are statistically self-similar or self-affine, within a finite range of magnification (self-affine meaning that the similarity under magnification is different along perpendicular directions). Mandelbrot introduced the concept of fractal dimension, D; without going into detail, this number conveys, for example, for fractal lines (like the Koch curve or a coastline) the property that such lines have a length that depends on the resolution following a power law, because magnification of parts reveals similar features to the whole. Thus, in the limit of infinite magnification, fractal lines in a plane tend to become infinitely long, yet they still fill less than the plane; thus, they have a dimension that is generally larger than 1 but less than 2: a fractal dimension is usually a broken number. Some fractal lines, like the Peano curve or Brownian motion, are so twisted that they ultimately fill the plane, and have a dimension $D = 2$. Fractal surfaces have a dimension larger than 2, but always lower than 3, the dimension of the space the surface is contained in.

Many amorphous porous materials have such a fractal, self-similar surface. Hence, the accessible surface area for a molecule depends on its molecular diameter, δ (effectively, the resolution of observation), following a power law, $\sim$, where D is the fractal dimension of the surface, a number between 2 (for a smooth surface—here the surface is seen to be independent of the size δ of the molecules used as a measure) and 3 (for a space-filling surface). Clearly, for $D > 2$, the surface area becomes larger for smaller probe molecules, indicating that smaller and smaller irregularities alongside the pore walls become accessible, like fjords upon fjords along the Norwegian coastline are accessible to a small boat. The fractal scaling range, within which self-similarity holds, is too narrow to significantly affect molecular diffusion, but it has a considerable influence on Knudsen diffusion, because molecule-wall interactions dominate the diffusion behavior. The effect of surface roughness on the Knudsen diffusivity, D_K, can be approximated by:

$$D_K = D_{K0}\delta^{D-2} \tag{11.10}$$

where D_{K0} is the Knudsen diffusivity when the pore wall is smooth; a more detailed expression is presented in [51].

11.4 Nature-Inspired Engineering Approach

Some of the challenges faced by biological organisms are similar to those we seek to solve for manmade systems. This includes the problem of maintaining efficient operation across length scales, and the related need to efficiently transport molecules across a wide range of length scales. Through billions of years of evolution, biological organisms have developed traits that are particularly effective, especially where these are related to functions essential for survival. Unraveling the fundamental mechanisms underpinning these traits not only helps us to better understand life, and, in medicine, to discover ways to combat disease, but it can also serve as a source of inspiration to solve parallel challenges in technology.

To do the latter in the most effective manner, it is essential to appreciate both the context and the constraints of the biological model and the engineering application. Properties like remarkable efficiency, adaptability, scalability and resilience in nature may give us pause, when compared to the same properties of manmade systems. Blind imitation of natural features will, however, be highly ineffective. One reason is that the environment of living organisms is often not the same as that of engineering applications, whether it be temperature, pressure or chemical environment. Natural systems are immensely complicated, but not all biological components are necessary in a technical application, because the boundary conditions (available resources, ways to grow or build the system) differ. Also, most solutions need to satisfy multiple objectives simultaneously, while, again, these frequently differ between a biological and a manmade construct. The sources of complexity differ, where constraints of manufacturability, desired time scales, chemical building blocks and scale of operation are often vastly different. Therefore, while the remarkable efficiency of a cell membrane, the agility of a bird or the incredible selectivity of an enzyme may hold valuable information on improving the performance of artificial membranes, aircrafts or catalysts, respectively, purely imitating shape or other all-to-obvious features will rarely lead to a workable, let alone better solution than existing ones.

It is this combination of learning lessons from nature, by seeking to understand the fundamental mechanisms behind desirable features, and applying these mechanisms within the context of a technical application, cognizant of differences in boundary conditions, that we call "nature-inspired engineering" or, for chemical engineering applications, nature-inspired chemical engineering (NICE). It differs from biomimicry in its narrow sense, eschewing direct translation of biological features, seeking a deeper understanding of mechanisms and applying these to build a workable technical solution that is acceptable within the constraints that the product or process demands (economics, safety, practical applicability, manufacturability, etc.). Thus, our NICE methodology is very much rooted in fundamental physics and chemistry, and combines a holistic approach looking at natural systems with the solution-oriented reductionism and pragmatism of engineering. Our NICE methodology is discussed in a few recent papers [12, 53–56], and aims to be a resource for innovation, guiding solutions to challenging problems related to energy, water, health and sustainability in human society.

The complexity of nature is daunting. Its diversity is a fascinating source of beauty, but can also be overwhelming to those seeking to build solutions inspired by nature. Biologists tend to embrace this complexity in all its forms, cataloguing and categorizing it with increasing detail, aiming to be comprehensive. There is value in seeking exceptional behavior that can help us understand evolution as well as reveal rare mechanisms, exceptions to the rule, pushing the boundaries of the biologically achievable—the miracle of the platypus or the bombardier beetle. Such outliers can also inspire out-of-the-box ideas for engineering solutions to technical problems. However, in our NICE approach, in first instance, we look for universal mechanisms that are highly common, and, while biological organisms and systems come in different forms and shapes, the abstraction of physics and mathematical modeling reveals striking similarities.

One of those most striking, universal features in biology is hierarchically structuring, which is also crucial in technology, yet nature is vastly superior in how hierarchical structures are organized, bridging scales from atoms and molecules to organs and organisms, in a way that is essential to their functioning. For example, bone has a hierarchical structure containing seven levels of organization with distinct chemical properties. This allows bone to have unique mechanical properties and transport properties to sustain physiologically important cells, while keeping the overall weight of the bone low [57, 58]. Fratzl and Weinkamer [59] illustrate the structure-function relation of biological tissues, such as bone, tendon, and wood, at various hierarchical levels, and the importance of this adaptation to fracture healing. Such hierarchical biological structures are a great source of inspiration to materials scientists, seeking to emulate similar properties.

Inspired by the hierarchical structure of the femur, Gustave Eiffel designed the eponymous tower with a minimum amount of iron, but strong enough to rise 324 m

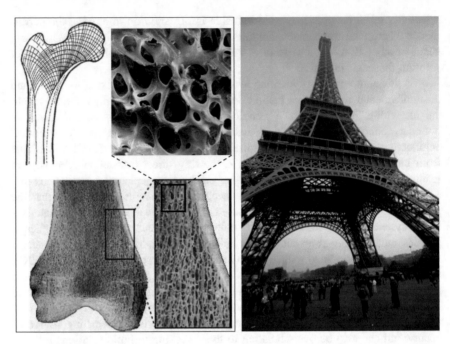

Fig. 11.4 The hierarchical structure of the femur (left [60]) and its inspiration to design of the Eiffel tower (right)

into the air. It is important to emphasize that Eiffel, quite obviously, did not copy the entire structure of the bone, but understood that it is the multi-scale balancing of forces in its trabecular structure that holds the secret to combining high strength, flexibility and low weight, as illustrated in Fig. 11.4. The size, shape, and materials used in the construction of the Eiffel tower are different from those of a bone, but it is the hierarchical design with balanced forces at each scale that lends the tower its unique mechanical properties. Scores of similar architectural examples could be cited that are nature-inspired in their design, in the engineering sense, from the work of Gaudí to Buckminster-Fuller and Calatrava. The most successful ones marry a nature-inspired design to other properties desired in their application, from functional in the technical sense, to esthetics.

Insights into hierarchical structures in biology provide us with a lot of ideas for the optimal design of hierarchically structured materials for processes that rely on efficient mass transport. A hierarchical network is widely adopted in biology to meet the challenge of transporting nutrients toward cells and products, including waste, away from cells through multiple length scales. At macroscopic scales, many of these networks have a fractal, self-similar branching structure, which interpolates between the scale of the organ or entire organism and a minimum length scale, the inner cutoff of the fractal scaling range. Examples are tree crowns (see Fig. 11.5a), the upper respiratory tract of the lungs (see Fig. 11.5b), and the vascular network

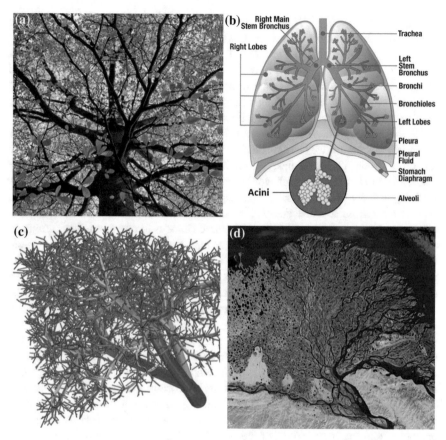

Fig. 11.5 Examples of fractal structures in the nature. **a** Tree crown; **b** lung [61]; **c** a vascular network of the human liver [62]; **d** Lena river delta [63]

(see Fig. 11.5c). Crucially, the lower bound or inner cutoff of the fractal scaling range also defines a cross-over in the dominant transport mechanism, from flow at large (macroscopic) scales, to diffusion at small (mesoscopic to microscopic) scales.

This is well illustrated by human and other mammalian lungs. The airway tree of a human adult lung repeatedly branches over approximately 23 generations. The upper airway tree is fractal; it consists of 14–16 levels of self-similar branching, counting from the trachea via the bronchi to the terminal bronchioles [64, 65]. The walls of these upper generations of bronchi are impermeable, and air through the bronchial tree is mainly transported via convective flow. As air flows through the bronchial tree, it gradually slows down from the trachea to the terminal bronchioles. This is because the radius of each branch only gradually decreases from generation to generation. More specifically, at each generation, $(r_p)^\Delta = m(r_d)^\Delta$, where r_p is the radius of the parent branch and r_d is the radius of one of the m daughters; in many cases, $m = 2$. The length of the branches decreases similarly from parent to daughter: $(l_p)^D = m(l_d)^D$. Thus, the upper airway tree is a space-filling, self-similar fractal with fractal dimension $D = 3$, which also has a diameter exponent $\Delta = 3$ [46, 66, 67]. If this diameter exponent, Δ, had been 2, as it is in most botanical trees (something da Vinci already showed), the flux and the transport velocity would remain constant, because the total cross-sectional area of all daughters remains constant, throughout all branching generations of the tree. However, for the lung, this cross-sectional area progresses with a factor $2^{4/3}$ from generation to generation, while the velocity decreases, correspondingly, by a factor $2^{-1/3}$. Ultimately, after about 14 generations, air has slowed down so much that diffusional transport, by the random motion of molecules, is as fast as convective transport; any further restriction in channel diameter would make diffusion more rapid than convection. At that point, the Péclet number, Pe, comparing convective with diffusive transport, crosses over from a value above 1 to one below 1. It is around this branching generation that the structure of the airway tree changes to one that is very compact, as shown in Fig. 11.5b: air enters the acinar airways, lined by alveoli, where exchange of oxygen and carbon dioxide with the bloodstream occurs. Throughout these lower 7–9 space-filling generations of acini, the channel diameter no longer changes much; there would be no advantage to such change, given that, unlike convective flow, diffusive transport by Brownian motion is not affected by the local channel diameter.

In summary, the airway tree acts as a fractal distributor and collector with a self-similar architecture between the macroscopic scale of the trachea to the mesoscopic scale of the bronchioles [68], while the channel size within the acini remains almost constant and the alveoli are uniformly distributed at mesoscopic length scales. A transition in dominant transport mechanism from convection to diffusion, corresponding to $Pe \sim 1$, occurs in parallel to this radical change in geometry, and the lower cutoff of the fractal scaling regime defines the cross-over between macroscopic and the mesoscopic length scales. This is a key insight that appears widely valid in biology, where characteristic length scales are tied to cross-overs in function, here exemplified by transport properties. Fractal interpolation between cross-over points bridging the mesoscopic and the macroscopic is common, because it enables preservation of function [56].

Trees show a similar cross-over in hierarchical structure to lungs (Fig. 11.1). The tree crown has a fractal, self-similar branching structure, which distributes water and nutrients, with leaves supported by its branch tips [46]. This self-similar structure is so advantageous in adaptability and scalability that it enables tree crowns to spread tall and wide, without change in structure at the micro- to mesoscale. The branches thicken and the number of branching generations advances with the age of the tree, while the size of the twigs and leaves does not change very much. In deciduous trees, the veinal architecture of leaves transitions from fractal to uniform, again corresponding to a change in dominant transport mechanism from flow to diffusion, where $Pe \sim 1$, similar to the case of lungs.

Thus, a key nature-inspired design principle emerges for artificial hierarchical transport networks in chemical reaction engineering applications and separation processes involving porous materials, namely to combine a fractal geometry at macroscopic scales, and a uniform one at mesoscopic scales, with the reaction, adsorption or exchange process occurring at microscopic scales. This particular hierarchical structure leads to inherent scalability, as the operation is scale independent, but, in addition, the system is also particularly efficient, if not optimal, as we will now discuss.

The ubiquity of transport networks that combine a fractal geometry at larger scales with uniformity at small scales, suggests the importance of understanding the physical reason behind a particular geometry before mimicking it to attempt optimization. Almost a century ago, it was already pointed out by Murray that there is, what he called a "physiological principle of minimum work" [69, 70]. He proposed to use the concept of "fitness" as a premise for physiological deductions, and hypothesized that physiological organization is such that the energetic cost of operation is minimized. More specifically, he showed that the hierarchical structure of the human vascular network is such that oxygen transport is most efficient. If the blood vessels are too narrow, too much work is needed for blood to flow through, due to high friction. If the vessels are too broad, however, the blood volume is similarly large, which is difficult to sustain as well. Efficiency is a compromise between the factors of work against friction, and the "cost" of upkeep of blood itself, which also requires metabolic energy. Minimizing the total amount of work (per unit of time and per unit of blood volume) as a function of the radius of the blood vessels led Murray to a similar value for the "cost" of blood (energy per unit time and per unit volume) for all arteries and capillaries. Although Murray does not use this term, it is very interesting to note that this implies equipartition of energy over the entire system of the vascular network, which is a thermodynamic principle.

We have shown a similar result for the architecture of the lung, and derived it in a different way, using irreversible thermodynamics, that is, second-law energy efficiency or minimization of entropy production [66]. In full agreement with physiological data for the respiratory network, the architecture of the lung is such that the pressure drop over each of the bronchi is the same, and the concentration drop over the acini is the same as well. This implies equipartition of thermodynamic forces over all constituting channels of the respiratory network. The space-filling architecture of the lung, $D = 3$, hence, $(l_p)^3 = 2(l_d)^3$, with also $\Delta = 3$, hence $(r_p)^3 = 2(r_d)^3$, throughout the bronchial tree, leads to minimum power dissipation, given a desired membrane surface area in the acini for exchange with the blood stream. This is a very important principle, which we will use in Sect. 11.6, when discussing nature-inspired fuel cells.

Underlying this analysis is the observation that we should be very cautious when learning from nature, and blind biomimetics should be avoided. Manmade designs that copy features of biological structures visually or intuitively to achieve similar properties are often referred to as *biomimetics* or *biomimicry*. The examples of the lung and the vascular network demonstrate that a physical analysis is necessary to understand the structural features leading to high efficiency and scalability. Straightforward biomimicry might, for example, assume that an infinitely self-similar fractal network is best, while our study showed a marked cutoff corresponding to $Pe \sim 1$. This adds to the different boundary conditions and context in technological applications, which must be accounted for when using the NICE approach to design and optimize artificial transport systems. We will now illustrate the NICE approach to optimizing transport in porous media, in the case of catalysts and fuel cells.

11.5 Nature-Inspired Optimization of Porous Catalysts

Desired properties of porous catalysts include high activity, selectivity, and stability. The geometric and electronic structure of the active sites determines the intrinsic kinetics (microscale), but the pore network structure significantly affects the apparent, effective kinetics (mesoscale), which, in turn, affects overall reactor yields and product distributions (macroscale), via the multiscale hierarchy illustrated in Fig. 11.2. Rational design at the mesoscale has not nearly received as much attention as the microscale, where spectroscopy, quantum chemistry and statistical mechanics have allowed for significant progress. Nevertheless, in a catalyst pellet, the concentrations of certain components might not be uniform, due to their long diffusion path, leading to considerable diffusion resistance. This, in turn, leads to a decreased volume-averaged reaction rate, compared to if the concentrations were uniform throughout the pellet and, therefore, the same to those at the outer surface. The *effectiveness factor* is defined to quantify the utilization of active sites in a catalyst pellet:

$$\eta = \frac{rate\ of\ reaction\ with\ diffusion\ limitation}{rate\ of\ reaction\ at\ outer\ surface\ conditions} \tag{11.11}$$

$$\eta = \frac{\int r(C)dV}{r(C_S)V_t} \tag{11.12}$$

where $r(C)$ is the reaction rate per unit volume at a (key) reactant concentration C at any position in the catalyst pellet, $r(C_S)$ is the reaction rate per unit volume at reactant concentration C_S at the external surface of the catalyst pellet, and V_t is the total volume of the catalyst pellet. A method to determine effectiveness factors by direct experimental inspection via IR imaging was given in Sect. 10.6.

Rational design at the microscale must be complemented by similar attention at the mesoscale. Indeed, an important objective is to maximize the effectiveness factor of a desired reaction, without changing the active sites themselves, thus preserving the intrinsic properties. A straightforward method is to shrink the size of the pellet. However, this method is rarely feasible in the chemical industry, because pellet size is typically dictated by reactor engineering requirements, such as pressure drop for fixed-bed reactors (increased for smaller pellets) and the minimum fluidization velocity in fluidized beds (controlled by particle size). Optimal design of the pore network, without affecting the pellet size, is, therefore, necessary to boost the effectiveness factor [71, 72].

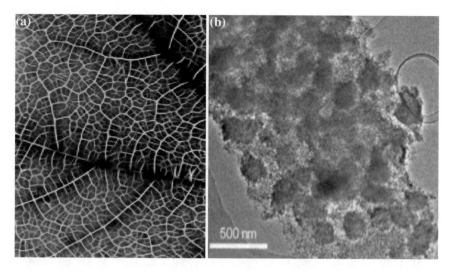

Fig. 11.6 Applying the NICE approach to optimal catalyst pellet design. **a** A leaf has a hierarchical network of veins to quickly transport reactants and products. **b** Inspired by the hierarchical transport network, a ZSM-5 zeolite catalyst was transformed into a hierarchically structured composite with microporous ZSM-5 nanocrystals embedded in a well-connected mesoporous matrix, thus facilitating diffusion. The zeolite composite was synthesized using the route reported in [80, 81]

Here, we can turn to nature for guidance. As illustrated in Fig. 11.2, a leaf bears similarities to a catalyst pellet, catalyzing carbon dioxide and water to sugar and oxygen, for which it is crucial to efficiently transport reactants and products in the leaf. To achieve fast transport, leaves have developed a hierarchical channel system, which we can use as a source of inspiration for the design of hierarchical pore networks in catalysts, as illustrated in Fig. 11.6. A hierarchical pore network in a nanoporous catalyst, like a zeolite, is generated by introducing macro- and meso-pores, which act as "highways" for fast transport (see also Sect. 10.7). However, important questions for the optimal design of these "highways" require an answer: Should they be distributed in a uniform or in a nonuniform way? Should they be of the same size or distributed in size according to an optimal distribution? What should the optimal macro- and mesoporosity be? How sensitive is the design to variations in these textural parameters? Should the optimal pore network be different if deactivation by fouling occurs at the same time? To address these questions, which guide the synthesis of improved catalysts, general features of the optimal pore network in porous catalysts were studied, using computational methods [73–79]. In a leaf, as in the lower airway of the lung, the transport network changes from fractal at large scale to uniform at small scales, where diffusion limits transport. The cells are strikingly uniformly distributed amongst the veins in a leaf. The theoretical and computational analysis that now follows does not prove that the leaf has an optimized structure, but we will see that similar features emerge from optimizing a hierarchical porous catalyst.

Gheorghiu and Coppens [73] used a two-dimensional model to computationally explore diffusion with first-order, isothermal reaction (A → B) in hierarchically structured catalysts, in which a wide-pore network is introduced into a nanoporous catalyst. They found that the catalyst with a fractal-like wide-pore network and broad pore size distribution operates very near optimality, in the sense that the effectiveness factor is maximized. However, the optimum is shallow, and, in these simulations, a constant number of large pores was assumed. This also does not guarantee that the total yield in the pellet is maximized.

Wang et al. [74] relaxed this constraint and compared monodisperse, bidisperse, and bimodal pore networks in a nanostructured catalyst for a first-order, isothermal reaction. For the bidisperse pore network, the large pores all have the same size; in the bimodal pore network, large pores vary in size throughout the pellet, as shown in Fig. 11.7. The computations showed that an optimized bidisperse catalyst could have a yield at least an order of magnitude higher than the one of the monodisperse catalyst (see Fig. 11.7), but also that local variations in pore diameter and porosity of the large pore network, as in general bimodal networks, do not appreciably increase the yield. Transport of molecules results from two diffusion processes, partly in series, partly parallel: (1) diffusion in the large pores penetrating the whole catalyst pellet, (2) local diffusion in the nanoporous "islands" surrounded by the large pores. In the optimal catalysts, the slowest, rate determining process is diffusion in the large pores, because the diffusion path in large pores is orders of magnitude longer than the one in narrow pores. Kärger and Vasenkov [82] reached a similar conclusion experimentally, based on PFG NMR, for catalysts used in fluidized bed catalytic cracking (FCC), namely that diffusion at the (high) reaction temperature in composite faujasite zeolite-containing particles is governed by diffusion in the large pores, rather than in the intracrystalline micropores, despite the intrinsically much smaller diffusivity in the latter. This is because the crystals are so small. Wang et al. [74] also found that the value of the total macro- and meso-porosity is essential, while the distribution of the wide (macro-/meso-)pore size is of secondary importance in determining the yield for the optimized hierarchical catalyst. In other words, a spatially uniform, wide pore distribution with uniform pore size (schematically represented by the bidisperse structure in Fig. 11.7) is preferred if the number of wide pores is large enough, while a fractal-like wide pore network may lead to higher yield and effectiveness factor if the number of wide pores is limited. This conclusion is also valid for the optimization of porous adsorbents [83].

Introducing macroporosity facilitates molecular transport, on the one hand, and reduces the amount of active catalytic material per unit volume, on the other hand. Hence, there is an optimal macroporosity when the objective is to maximize yield. Johannessen et al. [77] optimized the macroporosity analytically for a periodic bimodal porous catalyst (see Fig. 11.8) using optimal control theory and an effective one-dimensional model, with the assumptions of pure molecular diffusion in the large pore channels and first-order, isothermal reaction in the catalyst. For this model catalyst, the macroporosity (ε_{macro}) can be calculated by:

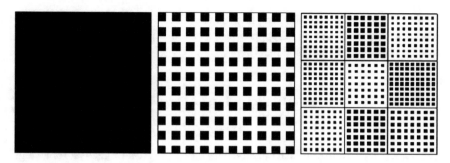

Fig. 11.7 Monodisperse (left), bidisperse (center), and bimodal (right) structures (nanoporous catalytic material: black; large diffusion channels: white). The monodisperse structure has a pore network with only narrow pores. The bidisperse structure has a hierarchical pore network, with narrow nanopores only in the black "islands" of the same size and wide pores of the same size surrounding these "islands". The bimodal structures are assemblies of $N \times N$ bidisperse substructures; in the illustration, N is 3. From [74], with permission

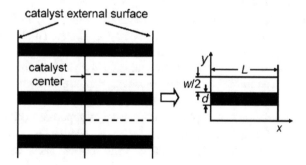

Fig. 11.8 Illustration of the bimodal catalyst (left) and one of its subunits (right). This bimodal catalyst is formed by repeating the subunit in the y direction. The white parts are nanoporous catalytic material; the black parts are large diffusion channels. L is half of the thickness of the catalyst, w is the thickness of the channel wall (i.e., the catalytic material), d is the diameter of the large channels. From [76], with permission

$$\varepsilon_{macro} = \frac{d}{d+w} \qquad (11.13)$$

where d is the diameter of large channels and w is the channel wall thickness, as shown in the right part of Fig. 11.8. The simulations show that the optimal macroporosity should always be less than 0.5. When channel diameter and channel wall thickness are optimized, concentration gradients are indistinguishable in the y (vertical) direction, which is consistent with the conclusion reached by Wang et al. [74]. Based on this result, a one-dimensional effective (continuum) model was developed; it was shown that this model is almost as accurate as the two-dimensional pore network model when optimizing the macroporosity of a bimodal catalyst.

The Thiele modulus method can be used to optimize the hierarchically structured porous catalysts. Wang and Coppens [76] defined a generalized distributor (i.e., macropore-based) Thiele modulus (Φ_0):

$$\Phi_0 = \frac{V}{S}\frac{r(C_0)}{\sqrt{2}}\left[\int_{C_c}^{C_0} D_m r(C)dC\right]^{-1/2} \qquad (11.14)$$

and related Φ_0 with the optimal effectiveness factor (η_{opt}) of the catalyst for a single reaction with general kinetics. In (11.14), V is the volume (3D) or area (2D) of the catalyst pellet; S is the external surface area (3D) or perimeter (2D) of the catalyst pellet; r is the reaction rate; C_0 is the concentration of a key reactant in the bulk phase; C_c is typically assumed to be zero for an irreversible reaction or the concentration in equilibrium for a reversible reaction; D_m is the diffusivity in macropores, rather than the effective diffusivity used in the conventional generalized Thiele modulus (Φ). They found that the $\eta_{opt} - \Phi_0$ relationship (see Fig. 11.9a) is analogous to the classical, universal $\eta - \Phi$ relationship (see Fig. 11.9b), that is, the effectiveness factor η is seen to decrease from 1 for a small Thiele modulus (corresponding to high diffusivities and low intrinsic reaction rates) to an inverse proportionality to Φ at high Thiele modulus. This yields a back-of-envelope approach to design a bimodal catalyst, because η_{opt} can be estimated solely from the value of Φ_0 without the need for case-by-case optimizations.

This $\eta_{opt} - \Phi_0$ relation was applied to optimize a mesoporous deNO$_x$ catalyst for the pseudo-first-order, isothermal reaction, $4NO + 4NH_3 + O_2 \rightarrow 4N_2 + 6H_2O$, which is used to reduce NO$_x$ pollutants from power plant emissions [76].

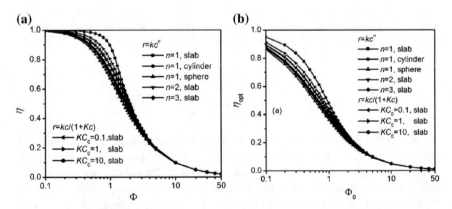

Fig. 11.9 **a** Effectiveness factor of a porous catalyst (η) as a function of the generalized Thiele modulus (Φ) for a single reaction with different reaction kinetics and in catalyst pellets of different shapes. **b** Optimal effectiveness factor of a porous catalyst (η_{opt}) as a function of the generalized distributor Thiele modulus (Φ_0) for a single reaction with different reaction kinetics and in catalyst pellets of different shapes. From Ref. [76], with permission

Fig. 11.10 Sensitivity of the effectiveness factor to the variations of channel diameter d $(0.5d_{opt}–1.5d_{opt})$ and channel wall thickness w $(0.5w_{opt}–1.5w_{opt})$, as labeled in Fig. 11.8. The colors indicate the loss in the effectiveness factor (i.e., a percentage of the optimal effectiveness factor). From Ref. [12], with permission

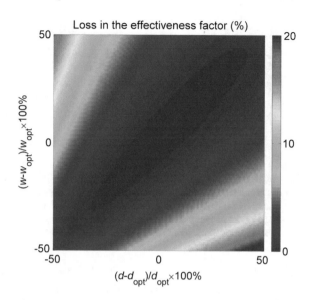

By introducing an optimal macropore network (occupying 20–40% of the total volume of the catalyst) into the washcoat consisting of the mesoporous deNO$_x$ catalyst, its overall activity can be increased by a factor of 1.8–2.8. Wang and Coppens [75] also optimized a commercial, mesoporous Ni/Al$_2$O$_3$ catalyst for the autothermal reforming of methane by introducing a macropore network. This process produces syngas (a mixture of, mostly, CO and H$_2$), which is the precursor to methanol, ammonia, artificial fuels and more, so it is one of the most important chemical processes. The computations show that the overall activity can be increased by a factor of 1.4–4 by only adjusting macroporosity and macropore size of the bimodal (or macro-mesoporous) catalyst. In addition, a larger macroporosity typically favors a lower CO/H$_2$ ratio (or a higher selectivity toward hydrogen), which indicates that the macroporosity can be used as a handle to control the CO/H$_2$ ratio.

When optimizing porous catalysts, the sensitivity of the catalyst performance to the structural parameters matters, as this shows how tightly the pore structure should be controlled during synthesis. Wang et al. [74] found that the optimal value of the macroporosity matters the most, while the distribution of large pore size around the optimal large pore size is less important than the size itself. Coppens and Wang [12] investigated how the effectiveness factor reacts to changes in channel diameter d and channel wall thickness w around the optimal values; Fig. 11.10 shows that the loss in effectiveness factor is less than about 5% within a rather broad region around the optimum. These results are important for the preparation of industrial catalysts, because it is much easier to precisely control the macroporosity, rather than the large pore size.

The performance of a catalyst often changes with time on stream, due to the deactivation of the catalyst by fouling, which covers active sites and blocks pore

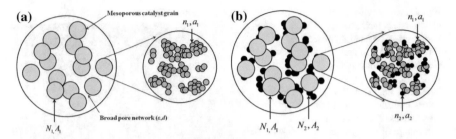

Fig. 11.11 Illustration of the pore structure of the catalyst before and after deactivation. **a** The hierarchically structured catalyst before deactivation is composed of overlapping mesoporous grains separated by a macropore network. Each grain consists of overlapping solid catalyst spheres separated by mesopores. **b** The catalyst after deactivation has a similar hierarchical structure as the one in Fig. 11.11a, but metal sulfide deposits (black spheres) cover the internal surface of the catalyst and can block pores. From [78], with permission

channels. Deactivation can be mitigated by optimizing the pore network of the catalyst, as suggested by Keil and his colleagues [84, 85]. Rao and Coppens [78, 79] computationally optimized a mesoporous hydrodemetalation catalyst by introducing an optimal hierarchical pore network, to maximize overall catalytic activity and robustness to deactivation over a given time on stream. This hierarchical pore network structure is illustrated in Fig. 11.11. A random sphere model was used to describe diffusion and reaction in the catalyst pellet. The results show that the lifetime of the hierarchically structured catalyst could be extended by 40%, while using 29% less catalyst than a non-optimized, purely mesoporous catalyst. Local variations in macroporosity and large pore size only negligibly change the overall yields, which is consistent with the optimization results of the porous catalysts without deactivation [74, 77]. Catalytic performance may also be affected by phase change, caused by capillary condensation in the pores. Ye et al. [86] proposed a pore network model to investigate diffusion, phase change, and reaction in a porous catalyst pellet. Hydrogenation of benzene to cyclohexane in the Pd/Al_2O_3 catalyst pellet was selected as a model reaction. Their results show that pore blocking by liquid can significantly affect the performance of the multiphase catalyst, indicating that pore blocking must be accounted for when modelling multiphase reactions. Ye et al. [87] also investigated the influence of pore network structure on the performance of the multiphase catalyst. These structural parameters include pore size distribution, connectivity, pellet size, spatial distribution of pores, and bimodal pore structure. The results show that the performance of the multiphase catalyst is very sensitive to these structural parameters, which indicates that the pore network structure should be well controlled to achieve a desired performance of the porous catalyst for multiphase reactions.

These studies demonstrate that an appropriate hierarchical catalyst pore network structure can substantially increase catalytic performance, whether it is in terms of activity, selectivity or stability. Within the context of NICE, our conclusions are in striking agreement with models in nature, such as leaves and the alveolar sacs of the lower airways of the lung: A uniform distribution and constant size of "cells" (translated to, e.g., zeolite crystal size) and wide pore channels (translated to macro-/mesopores) leads to maximum performance. The optimal porosity and pore channel size matters, as does the cell/crystal size, avoiding undesired further diffusion limitations within the crystals that would affect the (intrinsic) product distributions and prevent scalability, so important in nature and technology. The benefits are significant and should guide synthesis efforts. From a practical viewpoint, the optimum is shallow enough to allow for robust results, as some distribution around the optimum size distribution can be tolerated.

11.6 Nature-Inspired Optimization of PEM Fuel Cells

Proton exchange membrane (PEM) fuel cells are devices that convert chemical energy into electricity by electro-catalytic oxidation, at the anode, of hydrogen to protons, which diffuse through a membrane and electro-catalytically reduce, at the cathode, oxygen to water. Electrons produced at the anode move through an external circuit (where they are used to power a device) to the cathode, where they are consumed. Rather than direct combustion of hydrogen, the electro-catalytic route avoids Carnot's thermodynamic efficiency limit, thus, while more complicated, is potentially much more efficient, even at low temperatures. A PEM fuel cell consists of electrodes (anode and cathode), catalysts, proton exchange membrane, and gas diffusion layers for gas distribution on both sides of the electrodes. Since the average electric power from a single PEM fuel cell is limited to around 0.5 W/cm^2 [88], several cells must be stacked and bipolar plates are used to connect these cells, in order to achieve the desired power output in applications. During discharge, hydrogen (oxygen) are distributed over the anode (cathode) of the PEM fuel cell through the flow channels on bipolar plates, and then diffuse through the anode (cathode) gas diffusion layer and porous catalyst layer (often, Pt/carbon) before reaching the Pt active sites, where the reactions occur. At the same time, the product, water, is transported through the cathode catalyst and gas diffusion layer, to be collected and removed through the flow channels on the bipolar plates. Severe mass transfer limitations can cause rapid loss of voltage under high loads and significantly reduce power output [89]. Condensed water can clog the pores, but sufficient humidity of the membrane is necessary for the proton exchange to occur. Water management and alleviating, in particular, oxygen mass transfer limitations at the cathode is of great importance in PEM fuel cell design. Such problems have persisted over many decades. Can we turn to nature for inspiration in tackling them and redesign PEM fuel cells? We will discuss one aspect of this problem, taking the lung as a source of inspiration.

The required transport systems in PEM fuel cells and in lungs share some fundamental features: a hierarchy of transport channels is used, and dominant transport mechanisms include flow and diffusion. Hence, it is worthwhile to learn from lungs to guide the optimization of transport in PEM fuel cells, which is

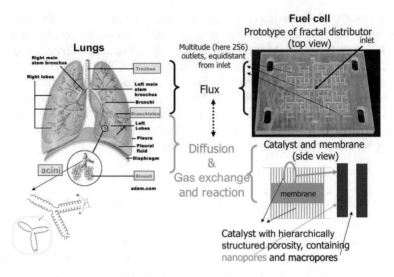

Fig. 11.12 Applying the NICE approach to the optimal design of PEM fuel cells. The hierarchical transport network of the lung, transitioning from fractal to uniform (left) inspired the design of a fractal distributor as bipolar plate (right top) and hierarchically structured nanoporous catalyst with uniformly distributed macropores (right bottom). From [55], with permission

illustrated in Fig. 11.12. As mentioned in Sect. 11.4, the upper respiratory tract (from trachea to bronchioles) has a self-similar, fractal architecture in which flow dominates. This fractal architecture connects the microscopic elements (i.e., the acini of the lung) to a single macroscopic element (i.e., the trachea of the lung) via equal hydraulic path lengths, leading to equal transport rates and minimized entropy production while breathing. Besides, this fractal architecture can be extended by simply adding a branching generation, without changing the microscopic building units (i.e., the acini). In the acini of the lung, transport of molecules is dominated by diffusion via the cell walls. Cell size is remarkably constant across mammals, in spite of considerable differences in size between organisms. As discussed, these fundamental properties of the hierarchical structure of the lung are tied to scalability and efficiency of the lung as a gas distributor and collector [66], and so can be utilized to design PEM fuel cells.

Inspired by the lung, a design was proposed to improve the energy efficiency and save the amount of expensive catalytic material in a PEM fuel cell [90]. In this design, the flow channels of a bipolar plate and the pore network architecture of a catalyst layer are optimized. The two parts can be decoupled and subsequently combined. To optimize the flow channel, criteria for minimum entropy production should be satisfied; to optimize the pore network structure, an optimized macroporosity should be introduced; both parts are ideally matched when $Pe \sim 1$ at the interface, as in the lung.

In the rest of this section, some examples of biomimetic and nature-inspired designs of flow channels of a bipolar channel are given and compared. An extension

(a)   **(b)**

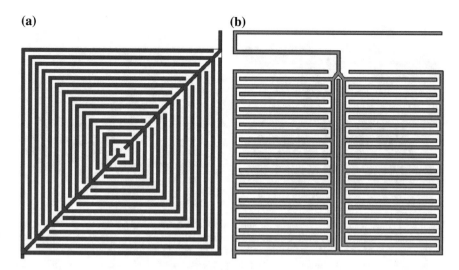

Fig. 11.13 **a** Leaf-inspired and **b** lung-inspired, biomimetic flow channel patterns. The inlet is at the top and the outlet is at the bottom. From [91], with permission

to electro-catalysis of the methodology discussed in Sect. 11.5 is used to optimize the design of the catalyst layer.

Some biomimetic designs of the flow channel pattern have been proposed to improve the flow of reactants and water in a PEM fuel cell [91–93]. Some designs combine the typically used serpentine (snake-like) and interdigitated patterns to form a "leaf-inspired" or "lung-inspired" channel pattern [91, 92], as shown in Fig. 11.13. The computations show that the leaf and lung flow channel patterns have a lower pressure drop and a more uniform pressure distribution, compared to the commercial serpentine and interdigitated designs. Experimental studies [91, 92] of these biomimetic designs show that the overall fuel cell performance can be increased by 30%. These biomimetic designs are important contributions to the improved design of PEM fuel cells, however, they only mimic certain natural features, without using the rigorous criteria behind the effectiveness of transport in leaves and lungs. Hence, they are essentially empirical, similarities with biology are superficial, and there is no reason for them to be optimal.

On the contrary, nature-inspired designs of the flow channels rely on funda-mental properties of pulmonary architecture and theories, such as Murray's law [69, 70]. A first step is to build flow channels into a fractal-like structure, just like the upper respiratory tract of the lung. Figure 11.14 shows a two-dimensional fractal distributor as bipolar plate, which can be built by rapid prototyping (Fig. 11.12). Reactants enter this distributor through a single inlet, flow through the branching channels, and eventually exit the distributor through a square array of outlets, which have the same hydraulic distance from the inlet. The diameter of the channels gradually changes, following a power law with exponent, Δ, as discussed in Sect. 11.4. In fractal distributor networks in nature, this exponent is different for

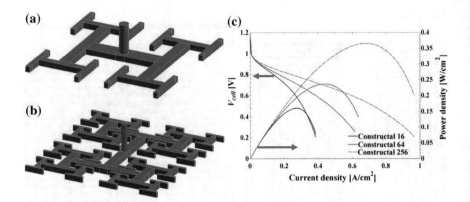

Fig. 11.14 Fractal flow distributors with single inlet and **a** 16 and **b** 64 outlets. **c** Polarization curves (that is, voltage as a function of current density) and power density (product of voltage and current density) of PEM fuel cells for three fractal (here, called constructal) distributors as bipolar plates. The legend "Constructal N" refers to a distributor with N outlets. From [97], with permission

botanical trees ($\Delta = 2$) [46, 94, 95], arteries ($\Delta = 2.7$) [46, 96] and lungs ($\Delta = 3$) [46, 96], because the function and transport mechanism in these natural distribution systems differ. Murray's law, where $\Delta = 3$, leads to the extraordinary efficiency of the lung. Ramos-Alvarado et al. [97] computationally compared the designs of fractal distributors with 16, 64, and 256 outlets. The fractal distributor with 256 outlets enhanced power generation by 200 and 50% over the ones with 16 and 64 outlets, respectively, because flow distribution was more uniform and the pressure drop was lower. Our own work [98] has used $\Delta = 3$ in a design that includes a number of branching generations guided by the boundary condition, $Pe \sim 1$, thus convective transport out of the last generation matches diffusion in the gas diffusion layer and the catalyst layer adjoining the bipolar plates—similar to the lung (Fig. 11.12).

The inefficient usage of expensive platinum catalyst caused by diffusion limitations not only adds to the total cost, but also decreases the power output. Marquis and Coppens [99] computationally optimized the microstructure by adjusting the platinum loading, platinum-to-carbon ratio, and catalyst layer void fraction. The results show that the optimization of catalyst microstructure can increase platinum utilization 30-fold over existing catalyst layer designs while maintaining power densities over 0.35 W/cm^2. An optimal large pore network should thus be introduced into the catalyst layer to further increase performance, similar to the results obtained in Sect. 11.5.

11.7 Conclusions

This chapter discussed a nature-inspired (chemical) engineering (NICE) approach to optimize mass transport, and illustrated it via a few examples relevant to chemical engineering, for catalytic systems employing porous media. In technology, as well as in nature, efficiently transporting molecules over multiple length scales, while maintaining scale-independent results, is of great importance. In each case, the performance of the transport systems is significantly affected by their structure over different length scales, which provides abundant room to optimize transport through manipulating the multiscale structure, such as transport channel size and distribution. Meanwhile, a fundamentally rooted methodology is still required to rationally design these transport systems for technological applications. Trees and mammalian lungs have evolved a hierarchical channel network for transport, which is efficient, robust, and scalable. At the macroscale, where flow dominates, the channel network is a self-similar fractal; at meso- to microscales, where diffusion dominates, the channel size becomes almost uniform. That these structural features are intertwined with functional optimality is a powerful basis for rational, nature-inspired design, beyond biomimicry by superficial imitation. We illustrated this for porous catalysts and PEM fuel cells.

Inspired by hierarchical diffusion networks in biology, an optimal large pore network can be introduced into nanoporous catalysts to maximize the usage of the catalyst, as well as overall yield. Computational and analytical studies indicate that an optimal hierarchically structured catalyst contains uniformly distributed wide pores in between nanoporous catalyst grains; the optimal macro/mesoporosity matters more than the optimal macro/mesopore size, and some distribution around the optimum is allowed, hence the result is robust. The same conclusions hold, irrespective of the reaction kinetics, and such a structure mitigates effects of deactivation by fouling. Learning from the fractal architecture of lungs and trees for fast transport across length scales where transport occurs by flow, bipolar plates with a fractal geometry and employing Murray's law were designed to improve the performance of PEM fuel cells, boosting their power output.

Rapid progress in synthesis and manufacturing technologies, from nanomaterials synthesis and microtemplating methods to additive manufacturing and micro-machining, increasingly allow to put theoretically optimized, three-dimensional, hierarchical architectures of porous materials and flow distribution networks into practice. Practical implementation of optimal transport networks, guided by the nature-inspired engineering, NICE, approach, is no longer a distant dream.

References

1. http://www.sciencephoto.com/media/801282/view (n.d)
2. T.L. Bergman, A.S. Lavine, F.P. Incropera, D.P. DeWitt, *Fundamentals of Heat and Mass Transfer*, 7th edn. (Wiley, New York, 2011)
3. D.D. Do, *Adsorption Analysis: Equilibria and Kinetics* (Imperial College Press, London, 1998)
4. J. Kärger, D.M. Ruthven, D.N. Theodorou, *Diffusion in Nanoporous Materials* (Wiley-VCH, Weinheim, 2012)
5. F. Keil, Catal. Today **53**, 245 (1999)
6. R. Krishna, J.A. Wesselingh, Chem. Eng. Sci. **52**, 861 (1997)
7. R. Krishna, Chem. Soc. Rev. **41**, 3099 (2012)
8. I. Medved, R. Černý, Microporous Mesoporous Mater. **142**, 405 (2011)
9. J. Kärger, ChemPhysChem **16**, 24 (2015)
10. W. Schwieger, A.G. Machoke, T. Weissenberger, A. Inayat, T. Selvam, M. Klumpp, A. Inayat, Chem. Soc. Rev. **45**, 3353 (2016)
11. F.J. Keil, Chem. Eng. Sci. **51**, 1543 (1996)
12. M.-O. Coppens, G. Wang, in *Design Heterogeneous Catalysis*, ed. U. Ozkan (Wiley, New York, 2009), pp. 25–58
13. M.-O. Coppens, in *Catalysis, Structure & Reactivity*, ed. A. Cybulski, J.A. Moulijn, 2nd edn. (CRC Press, Boca Raton, 2005), pp. 779–805
14. M. Sahimi, G.R. Gavalas, T.T. Tsotsis, Chem. Eng. Sci. **45**, 1443 (1990)
15. A. Wheeler, Adv. Catal. **3**, 249 (1951)
16. M.F. Johnson, W.E. Stewart, J. Catal. **4**, 248 (1965)
17. N. Epstein, Chem. Eng. Sci. **44**, 777 (1989)
18. N. Wakao, J.M. Smith, Ind. Eng. Chem. Fundam. **3**, 123 (1964)
19. N. Wakao, J.M. Smith, Chem. Eng. Sci. **17**, 825 (1962)
20. R.N. Foster, J.B. Butt, AIChE J. **12**, 180 (1966)
21. J. Szekely, J.W. Evans, Chem. Eng. Sci. **25**, 1091 (1970)
22. R. Mann, G. Thomson, Chem. Eng. Sci. **42**, 555 (1987)
23. V.N. Burganos, S.V. Sotirchos, AIChE J. **33**, 1678 (1987)
24. J. Wood, L.F. Gladden, Chem. Eng. Sci. **57**, 3047 (2002)
25. J. Wood, L.F. Gladden, Chem. Eng. Sci. **57**, 3033 (2002)
26. P. Rajniak, R.T. Yang, AIChE J. **42**, 319 (1996)
27. V. Novak, P. Koci, F. Štěpánek, M. Marek, Ind. Eng. Chem. Res. **50**, 12904 (2011)
28. F. Dorai, C. Moura Teixeira, M. Rolland, E. Climent, M. Marcoux, A. Wachs, Chem. Eng. Sci. **129**, 180 (2015)
29. F. Larachi, R. Hannaoui, P. Horgue, F. Augier, Y. Haroun, E. Youssef, E. Rosenberg, M. Prat, M. Quintard, Chem. Eng. J. **240**, 290 (2014)
30. V. Novak, F. Stepanek, P. Koci, M. Marek, M. Kubicek, Chem. Eng. Sci. **65**, 2352 (2010)
31. M.J. Blunt, M.D. Jackson, M. Piri, P.H. Valvatne, Adv. Water Resour. **25**, 1069 (2002)
32. G.T. Vladisavljević, I. Kobayashi, M. Nakajima, R.A. Williams, M. Shimizu, T. Nakashima, J. Memb. Sci. **302**, 243 (2007)
33. H. Sinha, C.-Y. Wang, Electrochem. Solid-State Lett. **9**, A344 (2006)
34. C.A. Baldwin, A.J. Sederman, M.D. Mantle, P. Alexander, L.F. Gladden, J. Colloid Interface Sci. **181**, 79 (1996)
35. A.R. Riyadh, T. Karsten, S.W. Clinton, Soil Sci. Soc. Am. J. **67**, 1687 (2003)
36. J.-Y. Arns, V. Robins, A.P. Sheppard, R.M. Sok, W.V. Pinczewski, M.A. Knackstedt, Transp. Porous Media **55**, 21 (2004)
37. H. Dong, M.J. Blunt, Phys. Rev. E **80**, 1 (2009)
38. D. Silin, T. Patzek, Phys. A **371**, 336 (2006)
39. F.A.L. Dullien, *Fluid Transport and Pore Structure*, 2nd edn. (Academic Press, San Diego, 1992)

40. J.F. Richardson, W.N. Zaki, Trans. Inst. Chem. Eng. **32**, 35 (1954)
41. P.N. Sharratt, R. Mann, Chem. Eng. Sci. **42**, 1565 (1987)
42. G.S. Armatas, Chem. Eng. Sci. **61**, 4662 (2006)
43. M.P. Hollewand, L.F. Gladden, Chem. Eng. Sci. **47**, 2757 (1992)
44. M.M. Mezedur, M. Kaviany, W. Moore, AIChE J. **48**, 15 (2002)
45. M.P. Hollewand, L.F. Gladden, Chem. Eng. Sci. **47**, 1761 (1992)
46. B.B. Mandelbrot, *The Fractal Geometry of Nature*, 2nd edn. (Freeman, San Francisco, 1983)
47. D. Avnir, *The Fractal Approach to Heterogeneous Chemistry* (Wiley, Chichester, 1989)
48. S. Havlin, D. Ben-Avraham, Adv. Phys. **51**, 187 (2002)
49. M.-O. Coppens, G.F. Froment, Chem. Eng. Sci. **50**, 1013 (1995)
50. M.-O. Coppens, G.F. Froment, Chem. Eng. Sci. **50**, 1027 (1995)
51. M.-O. Coppens, Catal. Today **53**, 225 (1999)
52. M.-O. Coppens, G.F. Froment, Chem. Eng. Sci. **49**, 4897 (1994)
53. P. Trogadas, V. Ramani, P. Strasser, T.F. Fuller, M.-O. Coppens, Angew. Chemie - Int. Ed. **55**, 122 (2016)
54. P. Trogadas, M.M. Nigra, M.-O. Coppens, New J. Chem. **40**, 4016 (2016)
55. M.-O. Coppens, Curr. Opin. Chem. Eng. **1**, 281 (2012)
56. M.-O. Coppens, in *Multiscale Methods Multiscale Methods Bridging the Scales in Science and Engineering*, ed. by J. Fish (Oxford University Press, New York, 2010), pp. 536–559
57. S. Weiner, H.D. Wagner, Annu. Rev. Mater. Sci. **28**, 271 (1998)
58. J.Y. Rho, L. Kuhn-Spearing, P. Zioupos, Med. Eng. Phys. **20**, 92 (1998)
59. P. Fratzl, R. Weinkamer, Prog. Mater Sci. **52**, 1263 (2007)
60. http://www.gla.ac.uk/ibls/US/fab/tutorial/generic/bone2.html (n.d)
61. http://juanribon.com/design/lung-Cancer-Diagram.php (n.d)
62. https://3dprint.com/7729/3d-Print-Organs-Vascular/ (n.d)
63. http://earthobservatory.nasa.gov/IOTD/view.php?id=2704 (n.d)
64. E.R. Weibel, *Morphometry of the Human Lung* (Springer, Berlin, 1963)
65. E.R. Weibel, *The Pathway for Oxygen* (Harvard University Press, Cambridge, MA, 1984)
66. S. Gheorghiu, S. Kjelstrup, P. Pfeifer, M.-O. Coppens, in *Fractals in Biology and Medicine*, ed. by T.F. Nonnenmacher, G.A. Losa, E.R. Weibel (Springer, Birkhäuser, 2005), pp. 31–42
67. C. Hou, S. Gheorghiu, M.-O. Coppens, V.H. Huxley, P. Pfeifer, in *Fractals in Biology and Medicine*, ed. by T.F. Nonnenmacher, G.A. Losa, E.R. Weibel (Springer, Birkhäuser, 2005), pp. 17–30
68. E.R. Weibel, Am. J. Physiol. **261**, L361 (1991)
69. C.D. Murray, Proc. Natl. Acad. Sci. U. S. A. **12**, 207 (1926)
70. C.D. Murray, Proc. Natl. Acad. Sci. U. S. A. **12**, 299 (1926)
71. F.J. Keil, C. Rieckmann, Chem. Eng. Sci. **54**, 3485 (1994)
72. S. van Donk, A.H. Janssen, J.H. Bitter, K.P. de Jong, Catal. Rev. Eng. **45**, 297 (2003)
73. S. Gheorghiu, M.-O. Coppens, AIChE J. **50**, 812 (2004)
74. G. Wang, E. Johannessen, C.R. Kleijn, S.W. de Leeuw, M.-O. Coppens, Chem. Eng. Sci. **62**, 5110 (2007)
75. G. Wang, M.-O. Coppens, Chem. Eng. Sci. **65**, 2344 (2010)
76. G. Wang, M.-O. Coppens, Ind. Eng. Chem. Res. **47**, 3847 (2008)
77. E. Johannessen, G. Wang, M.-O. Coppens, Ind. Eng. Chem. Res. **46**, 4245 (2007)
78. S.M. Rao, M.-O. Coppens, Chem. Eng. Sci. **83**, 66 (2012)
79. S.M. Rao, M.-O. Coppens, Ind. Eng. Chem. Res. **49**, 11087 (2010)
80. J. Wang, J.C. Groen, W. Yue, W. Zhou, M.-O. Coppens, J. Mater. Chem. **18**, 468 (2008)
81. J. Wang, W. Yue, W. Zhou, M.-O. Coppens, Microporous Mesoporous Mater. **120**, 19 (2009)
82. J. Kärger, S. Vasenkov, Microporous Mesoporous Mater. **85**, 195 (2005)
83. G. Ye, X. Duan, K. Zhu, X. Zhou, M.-O. Coppens, W. Yuan, Chem. Eng. Sci. **132**, 108 (2015)
84. F.J. Keil, C. Rieckmann, Hung. J. Ind. Chem. **21**, 277 (1993)
85. C. Rieckmann, T. Duren, F.J. Keil, Hung. J. Ind. Chem. **25**, 137 (1997)
86. G. Ye, X. Zhou, M.-O. Coppens, W. Yuan, AIChE J. **62**, 451 (2016)

87. G. Ye, X. Zhou, M.-O. Coppens, J. Zhou, W. Yuan, AIChE J. **63**, 78 (2017)
88. W.H.J. Hogarth, J.B. Benziger, J. Power Sour. **159**, 968 (2006)
89. J. Larminie, A. Dicks, *Fuel Cell Systems Explained*, 2nd edn. (Wiley, Chichester, 2003)
90. S. Kjelstrup, M.-O. Coppens, J.G. Pharoah, P. Pfeifer, Energy Fuels **24**, 5097 (2010)
91. J.P. Kloess, X. Wang, J. Liu, Z. Shi, L. Guessous, J. Power Sour. **188**, 132 (2009)
92. R. Roshandel, F. Arbabi, G.K. Moghaddam, Renew. Energy **41**, 86 (2012)
93. A. Arvay, J. French, J.C. Wang, X.H. Peng, A.M. Kannan, Int. J. Hydrogen Energy **38**, 3717 (2013)
94. K.A. McCulloh, J.S. Sperry, R.A. Frederick, Nature **421**, 939 (2003)
95. P. Domachuk, K. Tsioris, F.G. Omenetto, D.L. Kaplan, Adv. Mater. **22**, 249 (2010)
96. T.F. Sherman, J. Gen. Physiol. **78**, 431 (1981)
97. B. Ramos-Alvarado, A. Hernandez-Guerrero, F. Elizalde-Blancas, M.W. Ellis, Int. J. Hydrogen Energy **36**, 12965 (2011)
98. P. Trogadas, J.I.S. Cho, T.P. Neville, J. Marquis, B. Wu, D.J.L. Brett, M.-O. Coppens, Energy Environ. Sci. doi: 10.1039/c7ee02161e (2017)
99. J. Marquis, M.-O. Coppens, Chem. Eng. Sci. **102**, 151 (2013)

Chapter 12
NMR Versatility

Scott A. Willis, Tim Stait-Gardner, Allan M. Torres, Gang Zheng
and William S. Price

12.1 Introduction

Whether molecules (or ions) are in free solution or in some sort of composed porous system, translational diffusion (i.e., random thermal or Brownian motion) can be viewed as the most fundamental form of motion—Nature and Technology depend on it. In the presence or absence of other forms of motion (e.g., flow, convection, mutual diffusion ...) diffusion is always occurring. It is the background level of dispersion at the molecular level. Consequently, translational motion is fundamentally involved in most chemical reactions from the gas phase to those in condensed matter including those involved in metabolism. In general, diffusion brings reactants together and/or to the active site in the case of a catalyst or enzyme, and similarly moves the reactants away. Indeed, in very rapidly reacting systems it may even be that the kinetics are limited by the diffusion of the reactants to or the products from the point of reaction (see, e.g., the discussions in Sects. 10.6 and 11.5.) Thus, to understand reaction kinetics it is necessary to be able to probe the diffusive properties of the various species.

S. A. Willis · T. Stait-Gardner · A. M. Torres · G. Zheng · W. S. Price (✉)
Nanoscale Organisation and Dynamics Group, Western Sydney University,
Penrith, NSW 2751, Australia
e-mail: W.Price@westernsydney.edu.au

S. A. Willis
e-mail: Scott.Willis@westernsydney.edu.au

T. Stait-Gardner
e-mail: T.Stait-Gardner@westernsydney.edu.au

A. M. Torres
e-mail: A.Torres@westernsydney.edu.au

G. Zheng
e-mail: G.Zheng@westernsydney.edu.au

© Springer International Publishing AG 2018   233
A. Bunde et al. (eds.), *Diffusive Spreading in Nature, Technology and Society*,
https://doi.org/10.1007/978-3-319-67798-9_12

In addition to the direct role of diffusion in chemically based processes there are other very important reasons for wishing to be able to probe diffusion and much of the following considerations assumes knowledge of Chap. 2 Spreading Fundamentals. In particular, it is possible to obtain enormous information on the interactions of the diffusing species with each other (e.g., self-association, hydrogen bonding,...) and other species (binding, hydrogen bonding,...), and their interactions with the boundaries that such species may be diffusing within (e.g., cell walls). Naturally, the sort of information that can be obtained depends on the nature of the system, for example, whether it is a pure substance or diffusing within some kind of porous material.

Conceptual diagrams of diffusion in three systems of increasing complexity are shown in Fig. 12.1. In a pure liquid the measured diffusion coefficient D^0 corresponds to the so-called bulk diffusion coefficient. Diffusion coefficients range from $\sim 10^{-6}$ m^2 s^{-1} in the gas phase to $<10^{-15}$ m^2 s^{-1} in large polymers. Diffusion in solids (see Chap. 13) is thus very slow although even in crystals, albeit over enormous timescales, it is not completely zero [1]. More generally real systems are mixtures of different species (e.g., biological milieu and polymer systems) and the diffusion coefficient of a species will be reduced due to interparticle interactions, including binding, and obstruction. We can think of systems of interest in increasing order of complexity going from a pure substance (i.e., gas, liquid or solid), an essentially homogeneous mixture (e.g., ethanol and water), a structurally heterogeneous mixture (e.g., an emulsion) to ultimately some sort of composed

(a)   **(b)**   **(c)**

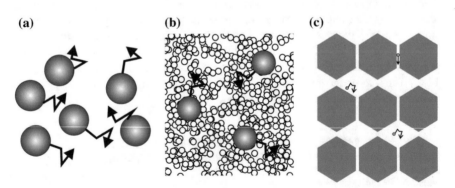

Fig. 12.1 Conceptual diagrams of three different diffusing systems. **a** The **free solution** system is representative of a pure liquid or a gas, except that in the case of a gas the distance between the species and their diffusion coefficients will be far greater. **b** A **solute dissolved in a solvent**. The diffusive behavior of the solute is more complicated as it depends on both its interaction with the solvent (i.e., solvation and charge) and the relative sizes and shapes of the solute to the solvent species. Further, depending on the concentration of the solute and the size and diffusion coefficient, it is possible that the diffusive paths of the solute molecules will be self-obstructed. **c** Small species diffusing in a **porous medium**. When the mean squared displacement is of similar order to the characteristic distances of the porous media (pores and throats) the interpretation of the diffusion coefficient can be strongly observation time dependent. Further, if the throats are narrow 'single file' diffusion can occur (see, e.g., Sect. 10.8 and Fig. 10.13)

system containing both diffusing and non-diffusing species (e.g., a porous medium such as biological tissue or zeolite).

To be able to extract information from the diffusion of a species requires both a means of measuring diffusion and a pertinent model to the specific environment so as to extract the information from the diffusion data. A detailed consideration on the information obtainable from diffusion measurements can be initiated from a few simple equations as largely introduced in Chap. 2. Firstly, the mean squared displacement (MSD) of a species diffusing isotropically with a diffusion coefficient D in three dimensions over a time t, is given by

$$\langle R^2 \rangle = 6Dt. \tag{12.1}$$

In essence, this equation tells us the volume a diffusing species samples over t. Thus, for the same t the MSD of a gas is vastly greater than that of a liquid by virtue of its diffusion coefficient being orders of magnitude larger. Importantly, as shown by Eq. (12.1), the MSD scales linearly with t. For Eq. (12.1) to hold there are some underlying assumptions: (i) any inter-particle interactions are completely averaged at a timescale much shorter than t and (ii) the diffusing particles are diffusing in an infinite medium (i.e., no interactions with boundaries). In the absence of these two assumptions there will be a non-linear dependence of the MSD on t. Obstruction of the diffusive path of a small molecule by a large molecule will lead to a decrease in the MSD over a time t, and in turn the measured diffusion coefficient ($D(t)$). The decrease in the measured diffusion coefficient can be characterized by an obstruction factor,

$$O_D = D(t)/D^0. \tag{12.2}$$

Obstruction can be thought of as diffusion in a complex time-dependent geometry. Relatedly if diffusion is occurring in a porous medium it is possible to observe a time-dependence of the measured diffusion coefficient of the diffusing species which depends in a complicated way on the geometry and connectivity of the voids (Fig. 12.2).

Another means of interpreting diffusion coefficients is through the Stokes-Einstein-Sutherland equation [2–5] (see also Sect. 8.2) which provides a link between molecular size and diffusion, viz.

$$D = \frac{kT}{6\pi\eta r} \tag{12.3}$$

where k is the Boltzmann constant and T is temperature (i.e., the numerator is the thermal energy). The denominator of Eq. (12.3) is the friction coefficient of a sphere where r is the Stokes (or effective hydrodynamic) radius of the diffusing species and η is the solvent viscosity. Although widely used to interpret diffusion of species at finite concentrations, Eq. (12.3) strictly only holds at infinite dilution, hence η refers to the solvent viscosity and not the solution viscosity. Further, it is

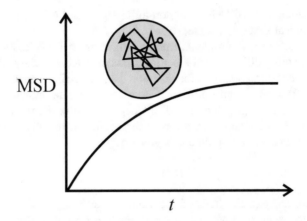

Fig. 12.2 Apparent time-dependence of the mean square displacement (MSD) for a species undergoing diffusion in a pore (i.e., a restricted geometry—in this case a sphere). The MSD can then be correlated to the diffusion coefficient through Eq. (12.1). At very short timescales the diffusion is unaffected by the pore boundaries and so analysis of the MSD will give the bulk diffusion coefficient. When collisions with the boundary become significant the MSD increases less than linearly with t resulting in a time-dependent diffusion coefficient. At long time the MSD is described solely by the pore dimensions

assumed that a solute is sufficiently large that the solvent can be seen as a continuum.

Equations (12.1)–(12.3) provide the conceptual basis of what can be probed with diffusion. For a particle diffusing in a finite medium (e.g., a 'pore', like water diffusing in a biological cell) then as t increases more and more particles would have the possibility of colliding with the boundaries (this occurs when the root mean square (RMS) displacement approaches the characteristic length scale of the bounding geometry). As a consequence the MSD would no longer scale linearly with t as specified by Eq. (12.1). As will be discussed below it is often experimentally difficult to probe D at sufficiently short t and thus RMS displacement to probe inter-particle interactions. Equation (12.3) states that in the case of a pure substance (gas, liquid, solid) diffusion reports on molecular size (and thus binding and exchange) and the solvent viscosity.

As can be understood from the above discussion, providing there is an applicable, accurate, non-invasive method for measuring diffusion then structural data may be extracted. However, there must be a defined measurement timescale (i.e., t) such that the RMS displacement probed is known. Diffusion measurements can provide a wealth of information including:

- The size of the diffusing species
- Activation energies of diffusion
- The environment that the species is diffusing in
- Ordering of the environment
- Geometry that a molecule is moving within

- Information on porous systems (e.g., characteristic distances, tortuosity, volume fraction and obstruction…)
- Macromolecular crowding and confinement

In the next section we will discuss methods of measuring diffusion and the NMR method in particular. Then, in order of increasing complexity, we explore the versatility and richness of information that can be obtained from NMR diffusion measurements (also known as NMR diffusometry and pulsed gradient spin-echo (PGSE) NMR or pulsed field gradient (PFG) NMR).

Diffusion measurement is far from straightforward and presents particular challenges. Due to the generally small size of diffusing species and low energies involved only methods that do not perturb the generally delicate thermodynamics of the system being studied are useful. Many methods exist for measuring diffusion including light scattering, centrifuge and capillary methods (see Table 4.1 in Ref. [6]). And it should be noted that there is some confusion in the literature between those that measure translational (or self-) diffusion and mutual (or concentration) diffusion in which the driving force is a chemical potential gradient. But most of these methods have limited applicability and this is even more pronounced when applied to biological systems and clinical studies. In essence, we require a technique that is non-invasive and chemically selective. For example, many systems are not easily amenable to be radio-labelled or others are only applicable over a limited concentration range or are capable of being measured over a certain timescale or within a limited range of diffusion coefficients.

As will be seen below NMR provides a powerful technique for rapidly measuring diffusion which obviates most of the problems associated with other techniques and can, in general, be applied directly to samples (i.e., without labelling) and under a range of conditions (e.g., temperature and pressure). Further, the necessary equipment is widely available as almost any recent NMR spectrometer includes the requisite hardware for performing basic NMR diffusion experiments. It is this combination of properties which makes NMR diffusion measurements so versatile.

12.2   NMR Diffusion Measurements

12.2.1   Basics of Diffusion NMR Measurements

NMR provides a particularly elegant and convenient approach to measuring translation diffusion [7]. Detailed accounts of the technique are given elsewhere including [8–10]. Traditional (i.e., non-diffusion) NMR experiments are conducted in a homogeneous static magnetic field ($\mathbf{B}_0$)—thence all spins of the same type resonate at the same frequency (i.e., the Larmor frequency ω_0) irrespective of their position ($\mathbf{r}$) in the sample. In an NMR diffusion measurement, in addition to the homogeneous static field, a magnetic field gradient is applied in pulses of duration δ

during the pulse sequence. In the presence of the pulsed field gradients the Larmor precession frequency of a nuclear spin is given by

$$\omega(\mathbf{r}) = \omega_0 + \gamma \mathbf{g} \cdot \mathbf{r} \tag{12.4}$$

where γ is the gyromagnetic ratio and $\mathbf{g}$ is the applied magnetic field gradient. In future discussion the ω_0 is ignored as it is common to all spins. What is important is the second term in Eq. (12.4). It indicates that the precession frequency changes both with respect to the position of the spin and the direction of the applied gradient (NB a vector quantity). This term ultimately not only gives NMR the ability to measure diffusion, but it also provides it the ability to measure diffusion in a certain direction. The basic concept of the diffusion measurement is easily visualized in the short gradient pulse approximation, where it is assumed that the gradient pulses are so short that diffusive motion during the pulses can be ignored. Although the direction of the gradient pulse can, hardware permitting (e.g., a triple axes gradient or imaging probe), arbitrarily be set to any direction, for the purpose of our present discussion it will be assumed that $\mathbf{g} = g_z\mathbf{k}$ (where $\mathbf{k}$ is the unit vector along z). Thus, initially coherent transverse magnetization is wound into a helix by the first gradient pulse and at a time $t = \Delta$ later a gradient pulse of equal magnitude but opposite direction is applied. If there has been no diffusive motion along the direction of the gradient pulse then the winding effect of the first gradient pulse is exactly counteracted by the second gradient pulse. Thus, the initial coherent transverse magnetization less any loss due to the spin-spin relaxation—as detected—constitutes a maximum echo signal, S_0. Diffusive motion during the diffusion time Δ causes irreversible attenuation of the magnetization helix resulting in an attenuated echo signal, S. The simplest PGSE NMR pulse sequence is depicted in Fig. 12.3. Analysis of the NMR data is facilitated by using the spin-echo attenuation E ($=S/S_0$) since the attenuation due to spin relaxation is normalized out leaving only the attenuation due to diffusion. In the case of free diffusion it is possible to arrive at an analytical result, even including the effects of finite length gradient pulses, connecting the experimental parameters, D and the signal attenuation,

$$E = \frac{S}{S_0} = \exp\left(-\gamma^2 g^2 D \delta^2 (\Delta - \delta/3)\right). \tag{12.5}$$

The dephasing effect of the gradient pulse, essentially the area of the gradient pulse scaled by γ, is sometimes referred to as q, viz.

$$\mathbf{q} = \frac{\gamma \delta \mathbf{g}}{2\pi} \quad (\mathrm{m}^{-1}). \tag{12.6}$$

Further analysis reveals that in the short gradient pulse limit the PGSE NMR experiment is sensitive to the MSD averaged over all of the measured species $\langle Z^2(\Delta) \rangle$,

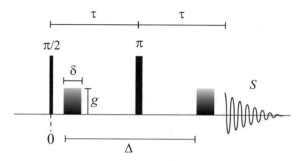

Fig. 12.3 The simplest PGSE NMR pulse sequence—the modified Hahn spin-echo. This sequence is often referred to as the Stejskal and Tanner sequence [7]. The $\pi/2$ radio frequency pulse transforms the initial magnetization into coherent transverse magnetization which is then wound into a helix by the first gradient pulse of duration δ and amplitude g. The π rf (radio frequency) pulse negates the spin phase acquired up until $t = \tau$, as a consequence the second gradient pulse is effectively of opposite sign to the first. Thus, in the absence of diffusion the helix created by the first gradient pulse is perfectly unwound by the second gradient pulse to give a spin-echo signal S_0 of maximum intensity which is acquired at $t = 2\tau$. In the presence of diffusion the magnitude of the acquired echo signal is reduced (i.e., $S < S_0$). The mechanism of the diffusion measurement is most easily visualized and modelled in the short-gradient approximation where the gradient pulses are infinitely short (i.e., $\delta \to 0$) but with finite area (i.e., δg is finite) since then diffusion during the gradient pulses can be ignored. The delay Δ defines the timescale of the diffusion measurement

$$E \approx \exp\left(-(2\pi q)^2 \frac{\langle Z^2(\Delta)\rangle}{2}\right). \tag{12.7}$$

Note, Eq. (12.7) is most accurate in the limit of small q values. Thus, if the diffusion is measured in a restricted system and the timescale of the measurement, Δ, is such that boundary effects become important then the attenuation will no longer be given by a simple single exponential. In such a case the shape and characteristic distance(s) of the restricting geometry will impart particular signatures to the echo attenuation profile. Further, if the data is (naively) analyzed using the equation for free diffusion (cf. Fig. 12.2) the measured diffusion coefficient will be in reality an apparent diffusion coefficient (ADC) and its value will be a function of Δ.

Consideration of Eq. (12.5) in conjunction with the spectroscopic properties of NMR reveals why NMR has become such a dominant force in diffusion measurements. To begin with, most species naturally contain an NMR sensitive isotope and thus do not require labelling that may not only be difficult to do but also may change the system being measured. Thus, sample preparation may be as simple as loading the sample into the NMR spectrometer. As the various isotopes resonate within distinct frequency ranges there is no ambiguity as to which nucleus is being detected. The NMR isotopes suited for use in NMR diffusion measurements include many of biological and industrial significance (e.g., ^{1}H, ^{2}H, ^{7}Li, ^{13}C, ^{17}O, ^{19}F, ^{23}Na, ^{31}P). Further only very small energies are used in NMR measurements that are unlikely to have any significant effects on the delicate thermodynamics of the system. Modern NMR spectrometers are capable of accurate temperature control. NMR has also two other extraordinary abilities that set it apart from all competing techniques: (i) in general full chemical shift information is acquired and so it is possible to probe the diffusive motion of more than one and sometimes all species present and (ii) it is possible to combine diffusion measurements with imaging. This is particularly relevant to spatially heterogeneous samples. And thus, it is possible to obtain information from specific areas. There are some caveats with respect to concentration and relaxation behavior—that is on a sample by sample basis. NMR measurements are ultimately limited by sensitivity and the relaxation properties of the measured species. However, the advances in NMR technology (esp. static magnetic field strength, hyperpolarization, and applied magnetic field gradient performance) means that more nuclei or measurements in less favorable samples become more practicable for almost solid samples to those in the gas phase.

The power of NMR for measuring diffusion and probing various composed samples is illustrated in the following sections.

12.2.2 Advancing Capability

Before around 1990 NMR diffusion measurements were only possible through the addition of home-made accessories (i.e., gradient coils, current amplifier and controlling hardware) to an NMR spectrometer. An example of a home-made PGSE system is shown in Fig. 12.4.

The last two and a half decades have seen enormous progress in the hardware for conducting diffusion measurements especially with respect to gradient coil design. Now virtually all modern NMR spectrometers come as standard with some ability to perform NMR diffusion measurements. A typical high resolution probe will come equipped with a gradient coil capable of generating a gradient of around 50 G cm^{-1} with excellent settling times after a gradient pulse. Such a system will be capable of measuring diffusion of molecules up to something like 50 kDa. Special NMR probe/amplifier combinations capable of producing gradients up to about 3000 G cm^{-1} are now commercially available and afford the possibility of measuring extremely slowly diffusing systems and/or rapidly relaxing systems or systems where the observed nucleus has a very low γ.

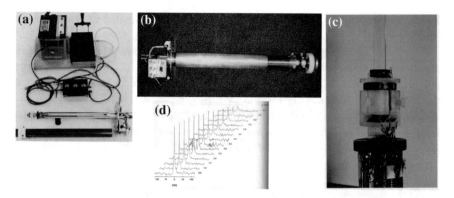

Fig. 12.4 Home-made additions to allow the performance of PGSE NMR experiments on a Varian XL400 spectrometer ca 1988. **a** A first attempt at a gradient generator—a car battery connected to a gradient coil via field effect transistors to provide current switching. **b** A modified heteronuclear NMR probe. The top part of the probe casing was replaced with Perspex in an attempt to remove conducting surfaces from the vicinity of the gradient coil and, thereby, reduce the chance of eddy currents being generated by the gradient pulses. **c** The top of the NMR probe with the casing removed. The gradient coil is positioned around the rf (radio frequency) coil and sample—note the NMR tube inserted into the top of the probe. **d** A series of ^{31}P PGSE NMR spectra showing attenuation of the echo signal with increasing gradient pulse duration. The spectrum acquired with $\delta = 3.5$ ms is particularly badly affected by RF interference to which this modified probe was especially susceptible. From W. S. Price Ph.D. thesis Univ. Sydney 1990

12.3 Applications to Gases, Liquids and Gels

12.3.1 Diffusion in Pure Substances, Mixtures and Solutions

12.3.1.1 Gases, Pure Liquids and Simple Mixtures

Diffusion in pure gases has been widely studied using NMR diffusion measurements. For instance, the diffusion of neat CO_2 has been measured at pressures up to 200 MPa between 223 and 450 K [11]. The diffusion of noble gases [12] in water and as well as diffusion in supercritical mixtures [13] have also been measured. A particular problem with gas diffusion measurements is the low spin-density and thus low signal-to-noise ratio. The rapidly emerging field of hyperpolarized gas NMR significantly obviates this problem and opens the possibility for measurements performed under clinical conditions (e.g., [14]).

Diffusion in pure liquids is still far from understood in many cases. A case in point is the diffusion of liquid water due to its complicated transient hydrogen bonds. Diffusion measurements provide an incisive means of studying the changing dynamics in liquids with temperature. Due to NMR's non-invasive nature it presents one of the rare means of studying diffusion in metastable states. ^{1}H and ^{2}H have been used to measure the diffusion of ^{1}H$_2$O and ^{2}H$_2$O down to 238 K and

244 K, respectively [15, 16]. Later work looked at the diffusion of water in brine [17].

Equation (12.3) can be extended to other shapes and thus a diffusion measurement can report on both the size and shape of a diffusing species. In fact this was put to good effect in a pioneering work by Moll in 1968 in which the helix to random coil transition of poly-L-glutamic acid was studied [18]. The literature now contains many studies of using diffusion to investigate protein folding and denaturation (e.g., [19, 20]), protein association and crowding (e.g., [21, 22]) and self-stacking and nanorod formation of platinum(II) intercalators [23].

12.3.1.2 Solution Structuring and NMR Diffusography

Diffusion is a powerful probe of solution structure. It might be naively assumed that something like a water-alcohol mixture is homogeneous at the molecular level. Yet, NMR diffusion measurements, with their ability to probe multiple components simultaneously, reveal that in such a hydrogen bonded environment the solution is far from homogeneous with concentration-dependent clustering of species. Further, the differences in hydrogen bonding ability of isomers in such systems can in turn be used to spectroscopically separate the isomers on the basis of diffusion coefficients as shown in Fig. 12.5 (i.e., to partition the spectrum into groups of equivalent diffusion coefficients).

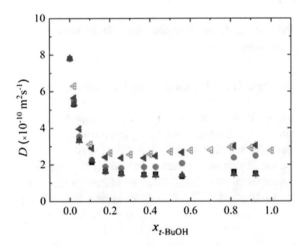

Fig. 12.5 'NMR Diffusography'—separation of hydroxybenzene isomers on the basis of diffusion. A plot of the diffusion coefficients of resorcinol (■), catechol (red ●), and hydroxyquinone (blue ▲) at ~ 15 mM and t-BuOH (purple left-facing ▲) in samples containing various tert-butanol mole fractions ($x_{t\text{-BuOH}}$) in D_2O measured at 298 K. The data for t-BuOH (green left-facing △) from the H_2O−alcohol system taken from Ref. [24] are shown for comparison. Reprinted with permission from Codling et al. [25] copyright (2013) American Chemical Society

12.3.2 Gels and Macroscopically Aligned Lyotropic Liquid Crystals

12.3.2.1 Liquid Crystals and Alignment

Molecules in liquid crystals are free to move but are ordered about an axis (the director), in contrast to in a liquid where there is neither positional nor orientational order. Liquid crystals are typically classified as thermotropic or lyotropic liquid crystals (LLC), the former has phase structures that depend on temperature (but also pressure or applied forces/fields) while the later contains solvent/second component as well as the liquid crystal molecules, and phases are additionally dependent on concentration. The molecule responsible for order/structure in LLCs is typically a surfactant (i.e., amphiphile) molecule where one end of the molecule is solvophillic (solvent-loving) and the other solvophobic (solvent-hating). Note an additional distinction between the thermotropic and lyotropic cases lies in the molecules and phase structures, where the molecules of thermotropic liquid crystals achieve order/display phases through molecular shape while in LLCs the phases are based on aggregate structures formed from the surfactant molecules. Nevertheless, whether it is diffusion of the ordered thermotropic liquid crystal molecules, the surfactant molecules in the LLC aggregates, a probe molecule diffusing among the ordered thermotropic liquid crystal molecules or a probe molecule diffusing around or in the LLC aggregates, the diffusion will now be direction dependent (i.e., anisotropic; see Sects. 12.3.2.2, 12.4.1 and 12.5.2).

For LLCs, different phase structures form as the concentration of surfactant is increased. For example, at low concentrations the surfactant molecules exist as monomers, but form micelles, in the simplest case these are spherical aggregates of surfactant molecules (whether the solvophillic or solvophobic faces outwards depends on the conditions), as the concentration is increased with the transition occurring at the critical micelle concentration. The most common phase structures are the lamellar and hexagonal phases. The lamellar phase is anisotropic and consists of stacked bilayers with water in-between. The surfactant bilayers cover large distances (>µm) and ideally the bilayers are planar and parallel but realistically there are defects present. The hexagonal phase is also anisotropic and consists ideally of infinitely long circular rods of the surfactant molecules in a hexagonal packing. Macroscopically aligned lamellar or hexagonal LLC samples can be made by stacking glass plates with films of the LLCs in between them. Samples may be aligned by a magnetic field (e.g., Ref. [26]) too (Fig. 12.6).

12.3.2.2 Diffusion NMR as a Tool to Monitor the Alignment and Transport in LLCs

Deuterium NMR spectra have long been used to study the phase transitions in LLCs [27–29] and diffusion NMR can be used to determine the critical micelle

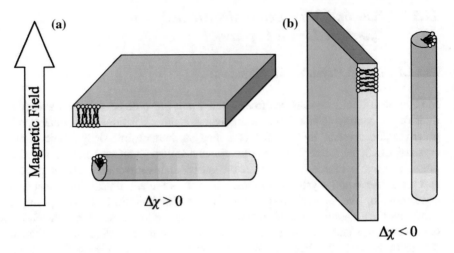

Fig. 12.6 The alignment of the LLC aggregate in a magnetic field depends on the sign of the anisotropic magnetic susceptibility, $\Delta\chi = \chi_\parallel - \chi_\perp$ of the surfactant molecules making up the aggregate (i.e., the response of the molecule in a magnetic field—attracted or repelled—may be different depending on the orientation of the molecule in a magnetic field). Normally the thermal energy of the molecules outweighs the magnetic energy and so they are free to tumble but as the size of an aggregate increases its overall magnetic energy results in a preferential alignment of the aggregates (and hence constituent molecules) by the magnetic field. **a** If $\Delta\chi > 0$ then the aggregate tends to align so that the long axis of the surfactant molecules aligns parallel to the magnetic field, in this case the lamellar phase would be the most aligned as the direction of the long axis of the cylinder in the hexagonal phase (i.e., the director) is undetermined in the plane perpendicular to the magnetic field while normal to the bilayer of the lamellar phase (i.e., the director) is parallel to the magnetic field. **b** If $\Delta\chi < 0$ then the aggregate tends to align so that the long axis of the surfactant molecules aligns perpendicular to the magnetic field, in this case the hexagonal phase would be the most aligned as the director of the lamellar phase is undetermined in the plane perpendicular to the magnetic field while the director of the hexagonal phase is parallel to the magnetic field

concentration [30]. If the spectrometer comes equipped with triple axis gradients, such as with an imaging probe, then it is possible to measure diffusion in arbitrary directions. This capability allows the determination of diffusion tensors and direction (and time) dependent diffusion coefficients. These can then be analyzed by comparison with predictions with models that account for any structural/geometric obstruction and obstruction due to the matrix surrounding the obstacles and knowledge of D^0. Note the term geometric obstruction here refers to an obstruction due to the shape of boundaries/obstacles that the molecules diffuse around (e.g., suspended solid spheres or cylinders) and the matrix obstruction refers to that arising from, for example, a gel network surrounding solid spheres where there is obstruction due to the polymer chains in addition to the geometric obstruction.

Consider the example of a macroscopically aligned lyotropic hexagonal phase in the presence of a secondary polymer gel network (e.g., [26]), where the obstruction from the aligned cylindrical aggregates of the hexagonal phase is $O_{D,\,Hex}$ (e.g., [31]) and the obstruction from the chains in the polymer gel network is $O_{D, \text{Matrix}}$ (e.g., [32–34]) and the probe molecule diffuses in the spaces around the cylinders. The diffusion coefficients expected for an off-axis direction in the principal frame (here the principal axis frame and the gradient/laboratory frame are identical), can be calculated using the corresponding principal diffusivities (i.e., D_x, D_y and D_z; NB diffusion perpendicular to the cylinder axes in a macroscopically aligned hexagonal phase would be $D_\perp = D_x = D_y$ and diffusivity parallel to the cylinder axes would be $D_{\text{II}} = D_z$) and direction cosines [8, 35] converted to spherical coordinates (i.e., the diffusion 'peanut' [36–38]),

$$D_{\theta, \phi} = D_x \cos^2 \phi \sin^2 \theta + D_y \sin^2 \phi \sin^2 \theta + D_z \cos^2 \theta, \qquad (12.8)$$

where θ and ϕ are the polar angle ($0°$ along the principal frame z-axis), and azimuthal angle ($0°$ along the principal frame x-axis), respectively. The angular dependency between any two principal axes n and m, i.e., x, y, or z, is [35, 36]

$$D_\theta = D_n \sin^2 \theta + D_m \cos^2 \theta, \qquad (12.9)$$

where θ is defined as the angle from m. Note an equation similar to this can be obtained for the MSD [39]. The appearance of the two-dimensional polar plot, or similarly a three-dimensional plot, is that of a 'peanut' shape depending on the anisotropy of the diffusion (e.g., see [36–38]). Compare this to the diffusion ellipsoid (see Sect. 12.5.2) which shows the expected displacement for a given time in three dimensions—for this reason the 'diffusion' ellipsoid is actually the root mean-squared displacement (RMSD) ellipsoid. Hence, the equation to calculate the diffusion at any angle (i.e., from Eq. (12.9)) using D^0, $O_{D, \text{Matrix}}$ and $O_{D, \text{Hex}}$ may be written as

$$\begin{aligned} D_\theta \left(D^0, O_{D, \text{Matrix}}, O_{D, \text{Hex}}, \theta \right) &= D_\perp \sin^2 \theta + D_{\parallel} \cos^2 \theta \\ &= D_{\parallel} O_{D, \text{Hex}} \sin^2 \theta + D_{\parallel} \cos^2 \theta \qquad (12.10) \\ &= D^0 O_{D, \text{Matrix}} \left(O_{D, \text{Hex}} \sin^2 \theta + \cos^2 \theta \right) \end{aligned}$$

where θ is defined as the angle from the cylinder axes as usual (but this can be applied to any LLC system provided the structural obstruction and obstruction/binding from the matrix is modelled). Note, $O_{D, \text{Hex}}$ is the ratio of $D_\perp / D_{\text{II}}$ since D_{II} is the 'free' diffusion (the diffusion in the absence of geometric obstruction) in this instance, and $O_{D, \text{Matrix}}$ is the ratio of D / D^0 where D is equivalent D_{II} and is the diffusion coefficient reduced from D^0 due to the matrix obstruction—with the assumption that the gel network obstructs the diffusion of the probe molecule in all directions equally. The obstruction from the matrix may also be modelled to include binding to the matrix. Among several assumptions for Eq. (12.10) is that the effects of defects in the LLC structure and averaging at the microdomain boundaries are not significant for the timescale the diffusion is measured over. Further, the obstruction factors are likely to be solute dependent [40].

12.3.3 Binding and Exchange

The measured diffusion coefficient of a species is sensitive to a change in the physical environment of a species such that it binds to another or exchanges into a region subject to different restriction of its motion as depicted in Fig. 12.7.

In the case of two-site exchange, the signals from the free, S_f, and the bound, S_f, sites can be modelled using the coupled differential equations [8, 41, 42],

$$\frac{dS_f}{dt} = -(2\pi q)^2 D_f S_f - \frac{S_f}{\tau_f} - \frac{S_f}{T_{2f}} + \frac{S_b}{\tau_b} \tag{12.11}$$

and

$$\frac{dS_b}{dt} = -(2\pi q)^2 D_b S_b - \frac{S_b}{\tau_b} - \frac{S_b}{T_{2b}} + \frac{S_f}{\tau_f} \tag{12.12}$$

where D_f and D_b are the (true) diffusivities in the two domains. Similarly T_{2f} and T_{2b} denote the spin-spin relaxation times in the two domains. The initial conditions are given by $S_f|_{t=0} = P_f = (1 - P_b)$ and $S_b|_{t=0} = P_b$ where P_f and P_b are the free and bound populations, respectively. Ignoring relaxation differences between the two domains, the solutions to these equations are well-known albeit complex multi-exponential functions. Often the signals from the two sites (domains) cannot be separated and thus the solution is of the form

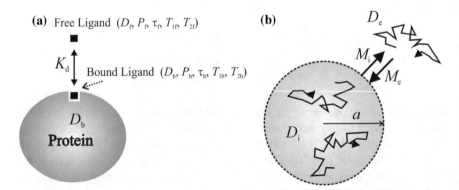

(a) Free Ligand $(D_f, P_f, \tau_f, T_{1f}, T_{2f})$

K_d Bound Ligand $(D_b, P_b, \tau_b, T_{1b}, T_{2b})$

D_b

Protein

(b) D_e

M_i

M_e

D_i a

Fig. 12.7 Binding and exchange. Diffusion of a species between **a** a free (f) and a bound (b) state as in ligand binding or **b** between the exterior (e) and interior (i) of a pore, can result in differences in the observed diffusion coefficient. In the former case the ligand will have a different diffusion coefficient (D), population (P), reorientational correlation time (τ) and spin-lattice (T_1) and spin-spin relaxation (T_2) times depending on whether it is in the free or bound state. Similarly a particle in the interior of a pore will likely have a different diffusion coefficient inside the pore—for example the viscosity inside a biological cell might be higher than in the exterior. Further, the pore boundaries can contribute to a lower observed diffusion coefficient

$$E(q, \Delta) = S_b + S_f = P_1 \exp\left(- (2\pi q)^2 D_1 \Delta\right) + P_2 \exp\left(- (2\pi q)^2 D_2 \Delta\right) \quad (12.13)$$

with well-defined mathematical expressions correlating the population fractions (relative signal intensities) P_1 and P_2 and the "apparent" self-diffusion coefficients D_1 and D_2 with the "true" populations (P_f and P_b) and diffusivities (D_f and D_b). Under fast exchange conditions (i.e., τ_f, $\tau_b \rightarrow 0$ and thus $D_2 \rightarrow \infty$) Eq. (12.13) reduces to

$$E(q, \Delta) = S_b + S_f = \exp\left(- (2\pi q)^2 \langle D \rangle_P \Delta\right), \quad (12.14)$$

where

$$\langle D \rangle_P = (1 - P_b) D_f + P_b D_b \quad (12.15)$$

is the population-weighted average diffusion coefficient.

The emergence of this formalism, today generally referred to as the Kärger equations was as early as the late 1960s [43] and these equations provide the means for studying an extraordinary range of binding and exchanging systems including estimation of molecular exchange in beds of nanoporous particles between the particles and the surrounding atmosphere [41, 44] (see also Chap. 10), binding of salicylate to albumin [45], dextran in polyelectrolyte capsule dispersions [46], water in brain tissue [47] and with a simple modification to account for restricted diffusion in one domain, transport through biological cells [48].

12.3.4 Electrophoretic NMR and Flow

The possibility of electrophoretic NMR (ENMR) in which ionic velocities are measured in the presence of electric fields using NMR was recognized by Packer in 1969 [49] and experimentally realized in 1982 [50, 51]. It opens up the possibility of resolving NMR spectra according to the individual electrophoretic mobilities (or drift velocities; μ) of the ionic species in the sample and has been widely reviewed [52, 53].

An ENMR experiment is in effect a PGSE sequence modified to include a pulsed electric field (E_{dc}). μ is related to the measured velocity (i.e., v) by

$$\mu_\pm = v_\pm / E_{dc} \quad (12.16)$$

where the subscripts '+' and '−' related to the cationic and anionic species, respectively. A simple ENMR pulse sequence is depicted in Fig. 12.8.

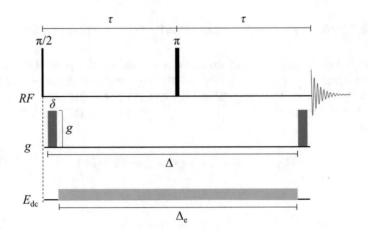

Fig. 12.8 A basic ENMR pulse sequence. The spins are first encoded spatially by the first magnetic gradient pulse, then the encoded spins experience diffusion and more importantly induced flow due to the electric field pulses. The dephasing effect due to diffusion, coherent phase shift due to flow, and phase modulation due to thermal convection are all accumulated after the application of the second pulsed gradient

The complex attenuation (i.e., attenuation and phase shift) of the echo signal for a particular species is given by [54]

$$E(E_{dc}) = \overbrace{\exp(-\gamma^2\delta^2 g^2 D(\Delta - \delta/3))}^{Attenuation}\overbrace{\exp(i\gamma\delta g E_{dc}\Delta_e\mu)}^{Phase-shift}. \tag{12.17}$$

Thus, diffusion results in echo signal attenuation as before and the electrophoretic mobility (and direction) is then determined from the complex phase modulation of the echo signal (Fig. 12.9).

Given that only charged species are affected by the pulsed electric field, it is possible to use ENMR as a mobility filter such that in a complex NMR spectrum of a multicomponent liquid mixture the resonances of the electrically charged species can be selectively filtered [55, 56]. Similar to PGSE measurements being displayed as DOSY plots (i.e., diffusion ordered spectroscopy plots in which one dimension is the usual chemical shift and the second dimension is the diffusion coefficient), it is possible to process the ENMR data to give a 2D plot (i.e., one dimension is the usual chemical shift and the second dimension is intensity versus Δ or E_{dc}) [57, 58]. Recently, ENMR was used to study ion association in aqueous and non-aqueous solutions [59]. Slice selection ENMR was used to investigate transport processes (e.g., electro-osmotic drag) in a fuel cell, consisting of several layers of Nafion [60]. Ion migration in bulk ionic liquid was measured by ENMR under the influence of small electric fields similar to those used in electric devices [61].

Fig. 12.9 ^{1}H (left) and ^{19}F (right) ENMR spectra showing, as expected, phase shifts of opposite signs for the cation and anion peaks in 10 mM aqueous solution of tetramethylammonium hexafluorophosphate. Reprinted from Bielejewski et al. [56] copyright (2014) with permission from Elsevier

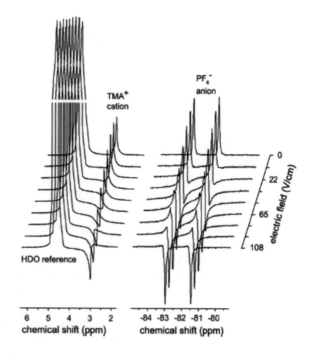

12.4 Porous Systems

12.4.1 Anisotropic Geometries

Diffusion in restricted environments is expected to be slower than that in unrestricted environments as the translational movements of diffusing molecules are hindered by the walls of the confining geometry or obstructions. The difference in diffusion between two types of environment is more pronounced when the diffusion time is set to values that are sufficiently long to allow the diffusing molecules to traverse the length of the confining geometry. In NMR, restricted diffusion can be observed for cavity dimensions in the order of tens to hundreds of micrometers. The shape and dimension of the confining geometry or pore and the type of surface wall all affect diffusion so that the equation describing the signal attenuation in the restricted geometry is much more complex than in free diffusion. For example, in restricted diffusion between two planes separated by length $2a$ and with surface relaxivity n, the diffusion propagator is given by [9, 62],

$$P(z_0, z_1, t) = \frac{1}{2a} + \frac{1}{a} \sum_{n=1}^{\infty} \cos\left(\frac{n\pi z_0}{2a}\right) \cos\left(\frac{n\pi z_1}{2a}\right) \exp\left(-\frac{n^2 \pi^2 D t}{(2a)^2}\right). \quad (12.18)$$

This propagator represents an eigenfunction expansion and is composed of a time-independent first term (from the zero eigenvalue) equal to the inverse of the characteristic distance of the pore whereas all of the latter terms are time dependent. The echo signal attenuation based on this propagator is

$$E(q, \Delta) = \frac{2[1 - \cos(2\pi q(2a))]}{(2\pi q(2a))^2} + 4(2\pi q(2a))^2 \sum_{n=1}^{\infty} \exp\left(-\frac{n^2\pi^2 D\Delta}{(2a)^2}\right) \frac{1 - (-1)^n \cos(2\pi q(2a))}{\left[(2\pi q(2a))^2 - (n\pi)^2\right]^2}.$$

$$(12.19)$$

The first term results from the time-independent part of the propagator (Eq. (12.18)) whereas the contribution of the second term, by virtue of containing a negative exponential, eventually disappears leaving only the profile from the first term in which one may, most impressively, recognize the pattern of the signals observed with radiation diffraction. This is in distinct contrast to the single exponential attenuation observed for free diffusion (Eq. (12.5))—and an immediate consequence of the similarity of the expressions resulting from the respective mathematical analyses. Such diffraction patterns, for various idealized geometries such spheres, cylinders, and annular geometries under a variety of boundary conditions (e.g., reflecting or absorbing), have been obtained either by direct derivation (analytically or numerically) or by simulation (e.g., [9, 10, 63–65]).

An experimentally derived 'diffusive diffraction' echo attenuation for diffusion between planes is shown in Fig. 12.10 for the model system of water in a Shigemi NMR tube. This tube consists of an outer tube and inner piston-like insert which was positioned to give a separation of ~140 μm. To be able to observe interesting features in the NMR signal attenuation profile, a long diffusion time was required so that the echo attenuation profile would be sensitive to the planar boundaries. The gradient was directed perpendicular to the planes.

The observed echo attenuation profile is interesting as it shows repetitive maxima and minima whose position is related to the interplanar separation and is well-described by the theoretical prediction. This NMR diffusion phenomenon is analogous to optical single slit diffraction patterns and is predicted by the cosine modulated terms in Eq. (12.19). It is easy to see that this NMR "diffusive-diffraction" pattern can provide useful information about the nature and dimensions of the restricting geometry. In fact such repetitive NMR diffusion patterns have already been observed experimentally in liquids containing samples of polystyrene beads, red blood cells, and capillaries [67–69]. This is possible in these particular cases because in that they incorporate a large number of restricting geometries which are regular or homogenous in size, shape and orientation. For non-ideal systems such as cells in tissues which are heterogeneous (e.g. they contain cells with various sizes and shapes), the diffusion diffraction pattern may not be observed. For such systems the 'apparent' diffusion coefficient (ADC) can be obtained; this is also very useful as it provides information about the mean dimensions of the cells in a given sample.

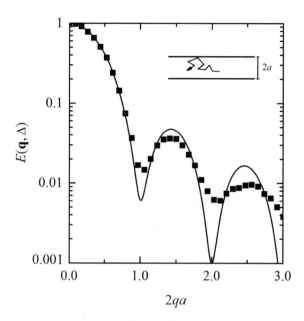

Fig. 12.10 Experimental verification of 'diffusive diffraction' spin-echo attenuation (■) for water diffusing between parallel planes separated by distance $2a$ plotted as a function of applied magnetic field gradient q. The sample consists of water in a Shigemi NMR tube. The data were acquired with $\Delta = 2$ s and $\delta = 2$ ms, which is very close to short gradient pulse conditions. The data were simulated (solid line) using Eq. (12.19) using $D = 3.69 \times 10^{-9}$ m^2 s^{-1} and a separation of $2a$ of 128.4 µm. As the diffractive effects are only evident at high attenuations (i.e., $E < 0.1$) the ordinate is plotted logarithmically. Modified from Price et al. [66] copyright (2003) with permission from Elsevier

12.5 Diffusion and Magnetic Resonance Imaging (MRI)

If the sample being investigated is relatively small and homogenous then the regular NMR diffusion method can be used for characterization. However in general most matter, including interesting and relevant samples ranging from Biology, Medicine to Chemical Engineering, are of a porous composition and are often heterogeneous systems so that it is desirable to be able to obtain diffusion information for a localized volume (or voxel) of a given sample and at the same time use diffusion as a form of contrast in imaging. Diffusion (esp. mutual diffusion and kinetics of adsorption) can be gauged by measuring concentration profiles and this can be done directly using NMR as an imaging method, referred to as MRI (magnetic resonance imaging, e.g., see [70–73] and Fig. 12.3 in [74] as early examples of its application in chemical engineering) although it is also noted that other methods can be used to complement MRI measurements to obtain information such as concentration profiles especially at shorter length and timescales such as IR Microscopy and Interference Microscopy (see Fig. 10.3). Fortunately, the NMR diffusion method can be readily incorporated into the MRI pulse methods as these two methods are

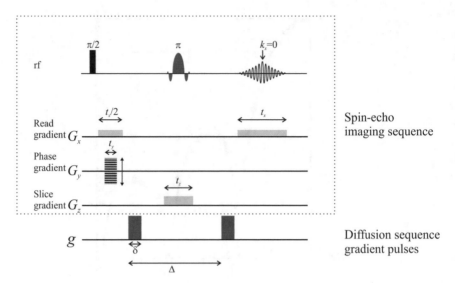

Fig. 12.11 Incorporation of the diffusion sequence into the MRI sequence. The diffusion sequence gradient pulses are combined into the spin echo imaging sequence

somewhat related in that both utilize magnetic field gradients, radio frequency pulses and delays (see Fig. 12.11).

To date two important diffusion-based MRI methods are now commonly used in many facilities, namely Diffusion Weighted Imaging (DWI) [75, 76] and Diffusion Tensor Imaging (DTI) [77, 78].

12.5.1   DWI and Isotropic Diffusion

Together with T_2 and T_1-weighted imaging, DWI has been used as an MRI contrast method to differentiate various regions in a given sample. The DWI method has been useful by itself and a complementary method in differentiating various tissue samples by providing ADC (apparent diffusion coefficient) maps. It is known as the isotropic diffusion map as the measured diffusion values are assumed equal in all directions within the sample.

12.5.2   DTI and Anisotropic Diffusion

DTI, on the other hand, is found to be a very versatile tool used for Fiber Tracking Mapping or Tractography in MRI. This is possible in fibrous samples because of the anisotropy of the ADC meaning that the measured diffusion is dependent on the

Fig. 12.12 Restricted
diffusion in a cylinder. The
apparent diffusion coefficient
(ADC) depends on the
orientation. Diffusion
measured across the cylinder
will appear to be slower

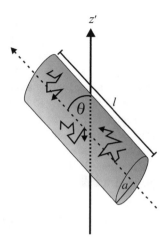

orientation of the fiber. The diffusion anisotropy is a result of the restricted elongated geometry where the ADC in the long section is greater than that in the short section as exemplified by diffusion in a cylinder, presented in Fig. 12.12.

For taking account of diffusion anisotropy we have to realize that now the probability of molecular propagation (see Eq. (2.10) for unidirectional diffusion and/or diffusion in an isotropic system) has become a function of the direction. As a most important consequence of this situation, deviating from Fick's 1st law in the simple notation of Eqs. (2.6) and (2.7), diffusive fluxes are not anymore necessarily directed in parallel with the concentration gradient. The position of D in Fick's law must now rather be assumed to be occupied by a diffusion tensor $\mathbf{D}$ yielding,

$$\mathbf{j} = -\mathbf{D}\,\nabla c \qquad (12.20)$$

or, in explicit notation with the individual components in the x, y and z directions and the diffusion tensor appearing in matrix notation,

$$\begin{pmatrix} j_x \\ j_y \\ j_z \end{pmatrix} = - \begin{pmatrix} D_{xx} & D_{xy} & D_{xz} \\ D_{yx} & D_{yy} & D_{yz} \\ D_{zx} & D_{zy} & D_{zz} \end{pmatrix} \begin{pmatrix} \frac{\partial c}{\partial x} \\ \frac{\partial c}{\partial y} \\ \frac{\partial c}{\partial z} \end{pmatrix}. \qquad (12.21)$$

Note that in Eq. (12.21) the coordinate system is for an arbitrary frame of reference. However, it is always possible to find a coordinate system in which the diffusion matrix assumes diagonal form, i.e., with all off-diagonal elements equal to zero (i.e., a principal axis system). Concentration gradients in the x, y, and z directions of the principal frame/coordinate system are easily seen to give rise to diffusive fluxes in exactly these directions as required by Fick's 1st law in the notations of Eqs. (2.6) and (2.7), now, however, with different parameters, i.e., principal diffusivities, D_x, D_y, D_z, in general. They are referred to as the main or principal tensor elements, attributed to the principal tensor axes x, y and z (for more on diffusion tensors see also [37, 38, 79, 80]). For Fig. 12.12, the z' axis represents an example laboratory frame of reference while the cylinder axis would be the principal axis for the major element of the principal diffusion tensor. In this case, the diffusion tensor is of rotational symmetry, i.e., with coinciding principal tensor elements in the radial direction which, moreover, are subject to restriction.

We are now going to determine the probability density of molecular shifts over a distance $\mathbf{r} = \{x; y; z\}$ (with reference to a coordinate system given by the principal axes of the diffusion tensor). This (combined) probability is simply given by the product of the respective probabilities of displacements in the principal x, y and z directions. With Eq. (2.10) and the principal tensor elements D_x, D_y and D_z in place of the diffusivity D we thus obtain,

$$P(\mathbf{r}, t) = \left(4\pi t \sqrt[3]{D_x D_y D_z}\right)^{-3/2} \exp\left\{ -\frac{1}{4t}\left[\frac{x^2}{D_x} + \frac{y^2}{D_y} + \frac{z^2}{D_z}\right]\right\}. \tag{12.22}$$

By adopting the procedure practiced already with Eq. (2.11), with Eq. (12.22) the mean square displacement in the directions of each of the principal axes of the diffusion tensor is easily found to follow the Einstein relation, now with the main elements of the diffusion tensor in place of the diffusion coefficient. This gives rise to a representation known as the 'diffusion ellipsoid'—actually the RMSD ellipsoid—and in the principal reference frame is represented by [37, 38],

$$\left(\frac{x}{\sqrt{2D_x t}}\right)^2 + \left(\frac{y}{\sqrt{2D_y t}}\right)^2 + \left(\frac{z}{\sqrt{2D_z t}}\right)^2 = 1, \tag{12.23}$$

and an example diffusion ellipsoid is shown in Fig. 12.13. Note that in Fig. 12.13 the laboratory frame of reference does not coincide with the principal frame of reference. For the special case where the time for diffusion, t, equals 0.5 s it can be seen that the major axes of the ellipsoid are given by the square root of the corresponding main elements of the diffusion tensor. If $t = 0.5$ s, the effective diffusivity along an arbitrary direction (i.e., in particular in the direction of the pulsed field gradient) can be determined from the square of the value of the ellipsoid along that direction. The 'diffusion' ellipsoid may also be constructed using the diffusion tensor obtained in an arbitrary reference frame (i.e., where the laboratory or gradient frame of reference does not coincide with the principal frame of reference) but in this case all tensor elements, D_{xx}, D_{yy}, D_{zz}, D_{xy}, D_{xz} and D_{yz}, are required to plot the ellipsoid so that orientation information is preserved/shown [78, 81]. The "diffusion ellipsoid" (i.e., the RMSD ellipsoid) can be plotted for a given MRI voxel size to easily visualize the magnitude and direction of anisotropic diffusion. Recall that the effective diffusivity along an arbitrary direction may also be determined via the diffusion 'peanut' (e.g., Eqs. (12.8) and (12.9)) (NB: derivations of the PGSE attenuations for anisotropic diffusion, i.e., with Eq. (12.22), can be found in [8])

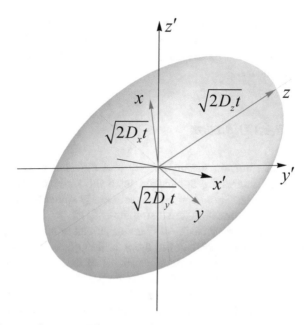

Fig. 12.13 The 'diffusion' or RMSD ellipsoid deduced from a diffusion tensor. The major axes of the ellipsoid are proportional to the square root of the corresponding main elements of the principal diffusion tensor, D_x D_y and D_z, and length of time diffusion occurs, t. Here the principal axes are x, y and z, and the laboratory axes are x', y' or z'. The effective RMSD in any arbitrary direction can easily be seen from the ellipsoid

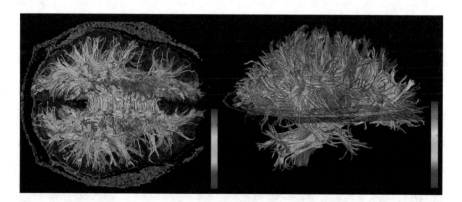

Fig. 12.14 One of the many applications of DTI is fiber-tracking in the brain. The fiber tracks are calculated from the principle eigenvector orientation (which is obtained from the diffusion tensor) in each voxel. Reprinted from Kubicki et al. [82] copyright (2007) with permission from Elsevier

Fiber Tracking maps of the brain can show its long range white matter connectivity as the maps created from data acquired in the DTI experiment shown in Fig. 12.14 illustrate.

The application of diffusion in the form of DWI and DTI (or Fiber Track Mapping) has gained popularity in MRI in Medicine in the last decade or so as it provides useful information on the nature and structure of tissues which cannot be obtained by other means.

12.5.3 Localized Diffusion

The applications of diffusion tensor imaging and diffusion MRI extend beyond medicine. One such interesting application is to the study of grape berry morphology and pathologies. Grape berries are particularly suited to MRI due to their high water content and intricate internal structure.

Two studies of grape berries using DTI are described briefly here [83, 84]. The first examined developmental changes in the Semillon grape tissue structure using DTI. Twenty-one grape berries at different stages of development were scanned with a standard PGSE echo-planar DTI sequence (see Fig. 12.15 where the diffusion preference along the radial direction is quite evident). Also evident is the high degree of anisotropy within the grape seed. The anisotropy patterns correlate with the known microstructure of the grapes at the various stages of development including an increase in diffusion vector coherence at 28–41 days after flowering when the mesocarp cells transition to a radially elongated state. The other study used MRI and DWI to track the progression of berry splitting.

Fruit split or berry split is a particularly important grape pathology due to its economic costs. During fruit split an excess uptake of water causes an increase in

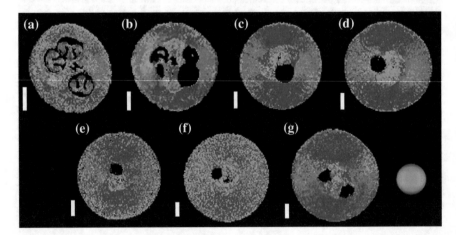

Fig. 12.15 Diffusion tensor images of Semillon grape berries at seven different stages of berry development (pre-véraison—prior to ripening—in (**a**) to post-harvest in (**g**). Scale bar: 3 mm. The color in the axial slices indicates the preferred diffusion direction where the 'color sphere' beside grape (**g**) correlates direction with color (from Dean et al. [83])

turgor pressure within the grape berry eventually causing the skin to split. The berry is then susceptible to pathogen infections and, in dryer conditions, to desiccation. Splitting adversely affects berry quality and yield and is a significant cost to the viticulture industry. Climate change is expected to exacerbate the conditions under which fruit splitting occurs in Australia.

Part of this study analyzed the development of a single split in the skin of a ripe table grape (Thompson Seedless) and the effect of this split on the mesocarp tissue using a sequence of ADC maps spaced over a period of 8 h. The ADC increased near the split and this splitting was linked to cell death.

Diffusion MRI can also be, and has been, applied to many other botanical species (for example olives, maize stems, barley seeds, carrot roots, celery and asparagus stems) [85–88].

12.6 Conclusions

The growth and range of application of NMR-based diffusion studies has been phenomenal with applications spanning a vast range of sciences, engineering and clinical medicine. The advancements in both hardware and method development have indeed propelled NMR/MRI as a popular analytical or clinical tool. No other diffusion measuring technique can come close to NMR in its versatility and practicality. Thus, the versatility of NMR means that its range of applications encompasses much of nature and technology.

Not only is NMR a powerful method for chemical identification, it is able to study binding events and kinetics that are important for biochemical systems. Recent advances in combining NMR/MRI with electrophoretic measurements allows visualization of dynamics processes which underpin battery and electrolyte development. NMR diffusion measurements not only provide information necessary to characterize binding but also provide structural information of the environment the molecules are diffusing within. NMR could therefore play a crucial role in advancing battery technology. This could have many benefits especially on the way society uses renewable energy leading to a better future for mankind and the environment.

For most of the history of NMR the push has been to higher static field strengths as this naturally gives greater sensitivity and affords greater resolution from an imaging perspective. However, the last decade has also seen increasing interest in low field equipment and the realization that in some cases not all of the information that is obtainable from a high field machine is required. Naturally modern low field equipment has far greater sensitivity and function than low field equipment from the past and comes with advanced permanent magnets. This resurgence in low field equipment has important consequences for the future of NMR. In particular, low field equipment is in general cheaper and more portable making it easier to adapt and use directly in agricultural and industrial practices and advanced models now have diffusion measuring capability. Thus portable "benchtop" NMR will become widespread not only in education but in industry as well.

References

1. E.B. Watson, Geochim. Cosmochim. Acta **68**, 1473 (2004)
2. W. Sutherland, Philos. Mag. **S6**(9), 781 (1905)
3. A. Einstein, *Investigations on the Theory of Brownian Movement* (Dover, New York, 1956)
4. J. Renn, Ann. Phys. **14**, 23 (2005)
5. G.A. Truskey, F. Yuan, D.F. Katz, *Transport Phenomena in Biological Systems* (Prentice Hall, New York, 2003)
6. S.A. Willis, T. Stait-Gardner, A.S. Virk, R. Masuda, M. Zubkov, G. Zheng, W.S. Price, *Chapter 4—Diffusion: definition, description and measurement*, in *Modern NMR Techniques for Synthetic Chemistry*, ed. by J. Fisher (CRC Press, Taylor & Francis, 2014)
7. E.O. Stejskal, J.E. Tanner, J. Chem. Phys. **42**, 288 (1965)
8. J. Kärger, H. Pfeifer, W. Heink, Adv. Magn. Reson. **12**, 1 (1988)
9. W.S. Price, *NMR Studies of Translational Motion: Principles and Applications* (Cambridge University Press, Cambridge, 2009)
10. P.T. Callaghan, *Translational Dynamics & Magnetic Resonance* (Oxford University Press, Oxford, 2011)
11. T. Groß, J. Buchhauser, H.-D. Lüdemann, J. Chem. Phys. **109**, 4518 (1998)
12. M. Holz, R. Haselmeier, R.K. Mazitov, H. Weingärtner, J. Am. Chem. Soc. **116**, 801 (1994)
13. P. Etesse, W.G. Chapman, R. Kobayashi, Mol. Phys. **80**, 1145 (1993)
14. T. Meersmann, E. Brunner (eds.), *Hyperpolarized Xenon-129 Magnetic Resonance* (Royal Society of Chemistry, Cambridge, 2015)
15. W.S. Price, H. Ide, Y. Arata, J. Phys. Chem. A **103**, 448 (1999)
16. W.S. Price, H. Ide, Y. Arata, O. Söderman, J. Phys. Chem. B **104**, 5874 (2000)
17. P. Garbacz, W.S. Price, J. Phys. Chem. A **118**, 3307 (2014)
18. R.E. Moll, J. Am. Chem. Soc. **90**, 4739 (1968)
19. W.S. Price, F. Tsuchiya, C. Suzuki, Y. Arata, J. Biomol. NMR **13**, 113 (1999)
20. J. Balbach, J. Am. Chem. Soc. **122**, 5887 (2000)
21. W.S. Price, Curr. Opin. Colloid Interface Sci. **11**, 19 (2006)
22. S. Barhoum, S. Palit, A. Yethiraj, Prog. NMR Spectrosc. **94–95**, 1 (2016)
23. A.M. Krause-Heuer, N.J. Wheate, W.S. Price, J.R. Aldrich-Wright, Chem. Commun. 1210 (2009)
24. W.S. Price, H. Ide, Y. Arata, J. Phys. Chem. A **107**, 4784 (2003)
25. D.J. Codling, G. Zheng, T. Stait-Gardner, S. Yang, M. Nilsson, W.S. Price, J. Phys. Chem. B **117**, 2734 (2013)
26. S.A. Willis, G.R. Dennis, G. Zheng, W.S. Price, React. Funct. Polym. **73**, 911 (2013)
27. J. Seelig, Q. Rev, Biophysics **10**, 353 (1977)
28. P.T. Callaghan, M.A. Le Gros, D.N. Pinder, J. Chem. Phys. **79**, 6372 (1983)
29. S. Müller, C. Börschig, W. Gronski, C. Schmidt, D. Roux, Langmuir **15**, 7558 (1999)
30. O. Söderman, P. Stilbs, W.S. Price, Concepts Magn. Reson. A **23**, 121 (2004)
31. H. Jóhannesson, B. Halle, J. Chem. Phys. **104**, 6807 (1996)
32. B. Amsden, Macromolecules **32**, 874 (1999)
33. M.I.M. Darwish, J.R.C. van der Maarel, P.L.J. Zitha, Macromolecules **37**, 2307 (2004)
34. D. Sandrin, D. Wagner, C.E. Sitta, R. Thoma, S. Felekyan, H.E. Hermes, C. Janiak, N. de Sousa Amadeu, R. Kuhnemuth, H. Lowen, S.U. Egelhaaf, C.A.M. Seidel, Phys. Chem. Chem. Phys. **18**, 12860 (2016)

35. H. Mehrer, Continuum theory of diffusion, in *Diffusion in Solids, Part I* (Springer, Berlin, Heidelberg, 2007)
36. D.K. Jones, P.J. Basser, Magn. Reson. Med. **52**, 979 (2004)
37. L. Minati, W.P. Węglarz, Concepts Magn. Reson. A **30**, 278 (2007)
38. P.B. Kingsley, Concepts Magn. Reson. A **28**, 101 (2006)
39. P.T. Callaghan, K.W. Jolley, J. Lelievre, Biophys. J. **28**, 133 (1979)
40. S.A. Willis, G.R. Dennis, T. Stait-Gardner, G. Zheng, W.S. Price, J. Mol. Liq. **236**, 107 (2017)
41. J. Kärger, Adv. Colloid Interface Sci. **23**, 129 (1985)
42. A.R. Waldeck, P.W. Kuchel, A.J. Lennon, B.E. Chapman, Prog. NMR Spectrosc. **30**, 39 (1997)
43. J. Kärger, Ann. Phys. **24**, 1 (1969)
44. J. Kärger, Z. Phys. Chem. **248**, 27 (1971)
45. W.S. Price, F. Elwinger, C. Vigouroux, P. Stilbs, Magn. Reson. Chem. **40**, 391 (2002)
46. T. Adalsteinsson, W.-F. Dong, M. Schönhoff, J. Phys. Chem. B **108**, 20056 (2004)
47. R.V. Mulkern, H.P. Zengingonul, R.L. Robertson, P. Bogner, K.H. Zou, H. Gudbjartsson, C. R.G. Guttmann, D. Holtzman, W. Kyriakos, F.A. Jolesz, S.E. Maier, Magn. Reson. Med. **44**, 292 (2000)
48. W.S. Price, A.V. Barzykin, K. Hayamizu, M. Tachiya, Biophys. J. **74**, 2259 (1998)
49. K.J. Packer, Mol. Phys. **17**, 355 (1969)
50. M. Holz, C. Müller, Ber. Bunsenges. Phys. Chem. **86**, 141 (1982)
51. M. Holz, J. Magn. Reson. **58**, 294 (1984)
52. M. Holz, Field-assisted diffusion studied by electrophoretic NMR, in *Diffusion in Condensed Matter*, ed. by J. Kärger, P. Heitjans (Springer, Berlin, 2005)
53. P.C. Griffiths, A. Paul, N. Hirst, Chem. Soc. Rev. **35**, 134 (2006)
54. Q. He, Z. Wei, J. Magn. Reson. **150**, 126 (2001)
55. S.R. Heil, M. Holz, Angew. Chem. (Int. Ed. Engl.) **35**, 1717 (1996)
56. M. Bielejewski, M. Giesecke, I. Furó, J. Magn. Reson. **243**, 17 (2014)
57. C.S. Johnson, Jr., Electrophoretic NMR, in *Encyclopedia of Nuclear Magnetic Resonance*, ed. by D.M. Grant, R.K. Harris (Wiley, New York, 1996)
58. E. Li, Q. He, J. Magn. Reson. **156**, 181 (2002)
59. M. Giesecke, G. Meriguet, F. Hallberg, Y. Fang, P. Stilbs, I. Furó, Phys. Chem. Chem. Phys. **17**, 3402 (2015)
60. F. Hallberg, T. Vernersson, E.T. Pettersson, S.V. Dvinskikh, G. Lindbergh, I. Furó, Electrochim. Acta **55**, 3542 (2010)
61. K. Hayamizu, Y. Aihara, J. Phys. Chem. Lett. **1**, 2055 (2010)
62. J.E. Tanner, E.O. Stejskal, J. Chem. Phys. **49**, 1768 (1968)
63. D.S. Grebenkov, Rev. Mod. Phys. **79**, 1077 (2007)
64. B.F. Moroney, T. Stait-Gardner, B. Ghadirian, N.N. Yadav, W.S. Price, J. Magn. Reson. **234**, 165 (2014)
65. B. Ghadirian, A.M. Torres, N.N. Yadav, W.S. Price, J. Chem. Phys. **138**, 094202 (2013)
66. W.S. Price, P. Stilbs, O. Söderman, J. Magn. Reson. **160**, 139 (2003)
67. A.M. Torres, R.J. Michniewicz, B.E. Chapman, G.A.R. Young, P.W. Kuchel, Magn. Reson. Imaging **16**, 423 (1998)
68. L. Avram, Y. Assaf, Y. Cohen, J. Magn. Reson. **169**, 30 (2004)
69. A.M. Torres, B. Ghadirian, W.S. Price, RSC Adv. **2**, 3352 (2012)
70. W. Heink, J. Kärger, H. Pfeifer, Chem. Eng. Sci. **33**, 1019 (1978)
71. R. Kimmich, *NMR: Tomography, Diffusometry, Relaxometry* (Springer, Berlin, 1997)
72. S. Stapf, S.-I. Han (eds.), *NMR Imaging in Chemical Engineering* (Wiley, New York, 2006)
73. P.T. Callaghan, *Principles of Nuclear Magnetic Resonance Microscopy* (Oxford University Press, Oxford, 1991)
74. J. Kärger, D.M. Ruthven, D. Theodorou, *Diffusion in Nanoporous Materials* (Wiley, New York, 2012)
75. R. Bammer, Eur. J. Radiol. **45**, 169 (2003)

76. S. Mori, P.B. Barker, Anat. Rec. **257**, 102 (1999)
77. P.J. Basser, J. Mattiello, D. Le Bihan, J. Magn. Reson. B **103**, 247 (1994)
78. P.J. Basser, J. Mattiello, D. Le Bihan, Biophys. J. **66**, 259 (1994)
79. P.B. Kingsley, Concepts Magn. Reson. A **28**, 123 (2006)
80. P.B. Kingsley, Concepts Magn. Reson. A **28**, 155 (2006)
81. P.J. Basser, NMR Biomed. **8**, 333 (1995)
82. M. Kubicki, R. McCarley, C.-F. Westin, H.-J. Park, S. Maier, R. Kikinis, F.A. Jolesz, M.E. Shenton, J. Psychiat. Res. **41**, 15 (2007)
83. R.J. Dean, T. Stait-Gardner, S.J. Clarke, S.Y. Rogiers, G. Bobek, W.S. Price, Plant Methods **10**, 35 (2014)
84. R.J. Dean, G. Bobek, T. Stait-Gardner, S.J. Clarke, S.Y. Rogiers, W.S. Price, Aust. J. Grape Wine Res. **22**, 240 (2016)
85. N. Ishida, H. Ogawa, H. Kano, Magn. Reson. Imaging **13**, 745 (1995)
86. C.H. Sotak, Neurochem. Int. **45**, 569 (2004)
87. S. Boujraf, R. Luypaert, H. Eisendrath, M. Osteaux, Magn. Reson. Mater. Phys. Bio. Med. **13**, 82 (2001)
88. J. Lätt, M. Nilsson, A. Rydhög, R. Wirestam, F. Ståhlberg, S. Brockstedt, Magn. Reson. Mater. Phys. Bio. Med. **20**, 213 (2007)

Chapter 13
Diffusion in Materials Science and Technology

Boris S. Bokstein and Boris B. Straumal

13.1 Introduction

Diffusion in materials (i.e. the random movement of molecules or atoms activated by thermal fluctuations) is an extremely important process. Macroscopically it appears in changes in the concentration profiles. Diffusion processes can be observed in gases and liquids, just as also in amorphous or crystalline metals, ceramics, polymers, semiconductors etc. [1, 2]. The concentration profiles caused by the diffusion contain important information about the atomic structure of materials as well as about defects within them. It is of particular relevance that diffusion can control the kinetics of the synthesis of materials and their modification, just as the processes by which these materials may fail.

The driving force of thermal diffusion in simple systems emerges from the random distribution of its components, which by physicists is referred to as the "entropy of mixing". Diffusion takes place due to the thermal motion of atoms and molecules. Therefore, its rate increases with increasing temperature. The mechanisms of thermal motion in gases and liquids do, notably, lead to a mixing of the constituent components. We refer here to the random collisions of atoms and

B. S. Bokstein · B. B. Straumal (✉)
National University of Science and Technology "MISIS",
Leninski prosp. 4, 119049 Moscow, Russia
e-mail: straumal@issp.ac.ru

B. S. Bokstein
e-mail: bokstein@mail.ru

B. B. Straumal
Institute of Solid State Physics, Russian Academy of Sciences,
Ac. Ossipyan str. 2, 142432 Chernogolovka, Russia

B. B. Straumal
Karlsruhe Institute of Technology (KIT), Institute of Nanotechnology,
Hermann-von-Helmholtz-Platz 1, 76344 Eggenstein-Leopoldshafen, Germany

© Springer International Publishing AG 2018
A. Bunde et al. (eds.), *Diffusive Spreading in Nature, Technology and Society*,
https://doi.org/10.1007/978-3-319-67798-9_13

molecules in gases (see, e.g., Chap. 10) or to the Brownian motion in liquids (see, e.g., Chaps. 8 and 12). To the contrary, mixing mechanisms of atoms in a solid are much more complicated and by far not as obvious [2]. Thermal motion in solids is driven by the vibrations of atoms around their equilibrium positions in a crystalline lattice. However, the amplitude of such vibrations is usually very small, in comparison with the nearest-neighbor distances. One would expect, therefore, that such thermal motions cannot lead to interatomic mixing. Answering the question "how can atoms migrate in solids" is thus by far not as simple as it might appear on first sight. It is the aim of this chapter to introduce into the main approaches suggested and exploited for describing atomic motion under such conditions.

13.2   Mathematical Description of Diffusion

The equations describing diffusion (see also Sects. 2.1 and 2.2 in Chap. 2) were proposed by Adolf Fick in 1855 [3]. They originate from the equations of heat transfer as suggested by Joseph Fourier in 1824 [4]. Fick replaced, in the Fourier equations, (i) the amount of heat by the number of atoms, (ii) the thermal conductivity by the diffusivity and (iii) the temperature by concentration.

First Fick's law predicts the relationship between molecular or atomic fluxes and concentration gradients. One can understand this relation by using the analogy between diffusion and thermal or electrical conduction. Recollect that the heat flux (in the simple one-dimensional case) is proportional to the temperature difference in the same area, just like the electric flux is proportional to the difference in electric potentials. Since fluxes are vectors, they are more generally noted as being (in the given case) proportional to the gradients of temperature and electric potential. As a main effect of diffusive motion and mixing, concentrations within a system become homogeneous.

In solids one generally distinguishes between three types of diffusivity. They correspond to three possible, physically different situations (Fig. 13.1). In the first case, we follow the diffusion of "labeled" atoms. It is true that, generally, atoms are undistinguishable. However, we can use a certain isotope (stable or radioactive) and follow its spreading in a material, which is composed of a natural isotope mixture. Typically, in such an experiment one deposits a film of radioactive isotopes on the surface of a sample composed of the natural isotope mixture. Then one follows the penetration of the radioactive isotope into the depth of a sample by measuring its radiation. The nuclei of stable and radioactive isotopes have almost the same mass. Therefore, they possess nearly the same physical and chemical properties and we can call this case self-diffusion. The respective coefficient in Fick's law is, correspondingly, referred to as the self-diffusion coefficient. The second case shown in Fig. 13.1 is the diffusion of one species in another, for example the diffusion of copper in nickel. For following this kind of diffusion one may coat a large nickel sample with a copper film, followed by monitoring the succession of the

Fig. 13.1 The three situations under which diffusion in solids is generally considered

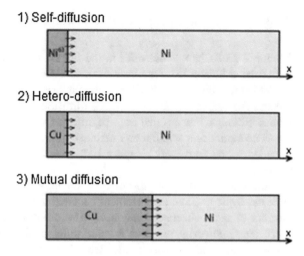

1) Self-diffusion

2) Hetero-diffusion

3) Mutual diffusion

concentration profiles of copper in the nickel sample. This process is called hetero-diffusion, it can be described by a hetero-diffusion coefficient. The third type of diffusion considers the mutual diffusion of two different species into one another. For monitoring diffusion in this case, one has to join, e.g., two large pieces of nickel and copper and to monitor, subsequently, the evolution of the concentration profile on either side of the interface. Such processes are commonly referred to as mutual diffusion and they are quantitated by introducing a coefficient of mutual diffusion.

The presentations of Fig. 13.1 may be easily correlated with the scheme provided by Fig. 2.2 in Chap. 2 for a quite general introduction of diffusivities. While Fig. 2.2b is seen to reproduce the situation of self-diffusion as illustrated by Fig. 13.1(1), Fig. 2.2c refers to a second possibility to determine self-diffusivities, namely by measuring the mean square displacement of the individual diffusants. The situation shown in Fig. 2.2a corresponds, in some way, with Fig. 13.1(2), since in both cases diffusive mass transfer is seen to emerge from a non-uniform particle distribution. However, while diffusion in solids implies the existence of another atomic species in, e.g., nanoporous materials (see Chap. 10) diffusion experiments driven by a concentration gradient may be performed with only a single molecular species (with the "second species" being nothing else than the "holes" left by the guest molecules in the pore space). As a matter of course, diffusive motion in pore spaces can also be observed under the conditions of counter diffusion, yielding the exact counterpart of the situation shown in Fig. 13.1(3).

It has been demonstrated in Sect. 2.2 of Chap. 2 that, by combination with the continuity equation, i.e. with the law of matter conservation, Fick's 1st law is converted into Fick's 2nd law (Eqs. (2.9) or (2.14)). While Fick's 1st law correlates diffusive fluxes with existing concentration gradients, Fick's 2nd law goes a step further and allows predicting the evolution of concentration profiles based on knowledge of their present stage.

13.3 Diffusion as a Random Walk Process

Section 2.2 of Chap. 2 did refer to the model developed by Einstein in 1905 for molecular diffusion [5]. One year later, in 1906, Smoluchowski [6] extended this model in such a way that it allowed the inclusion of atomic diffusion in crystalline metals. According to Smoluchowski, diffusion is the result of random hops of atoms. "Random" means that hops of different atoms as well as subsequent hops of the same atom occur without any mutual correlation. A series of such random hops is called a "random walk". Figure 13.2 shows three examples of such "random walks".

Atoms do, clearly, also hop in complete absence of concentration gradients. In such cases the average displacement becomes zero (see Fig. 13.3) and there is no net flux of mass observable anymore. The mean value of the squares of the displacements, however, assumes a value different from zero. As a most important result of the random walk model (see Sect. 2.1 of Chap. 2), the mean square displacement is found to increase linearly with time, following the Einstein relation in one dimension (Eq. (2.4))

$$\langle x^2(t) \rangle = 2Dt \tag{13.1}$$

The diffusion length, i.e. the square root of the mean square displacement, is thus seen to increase with only the square root of time. The "rate of diffusion", defined as the derivative of the "diffusion length" with respect to the "diffusion time" does, correspondingly, decreases with time!

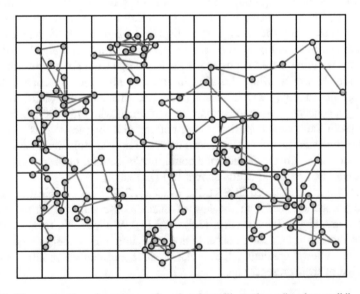

Fig. 13.2 Three examples of sequences of random hops illustrating a "random walk"

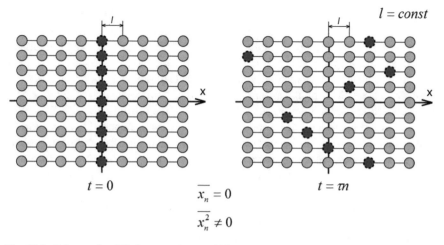

Fig. 13.3 Scheme of a diffusion experiment within a lattice of spacing l showing an example of particle distribution after n time steps of duration τ (right) when these particles are known to be initially (at $t = 0$) positioned at the origin ($x = 0$, left). The location (x) of such particles does, obviously, coincide with their displacement

The random walk model cannot, as a matter of course, specify the mechanism by which the atoms hop. From detailed experiments, however, one has learned that self-diffusion and hetero-diffusion in substitutional solutions are commonly promoted by the existence and formation of "vacancies", i.e. by holes in the scaffold constituting the solid. The "vacancy mechanism" has thus become key for understanding diffusion in solids quite in general and, thus, for the exploration of such important processes as diffusion creep, sintering, pore formation and annihilation, grain boundary migration, grain growth, phase transitions and precipitation.

On considering one-dimensional mass transfer with steps of length l and time intervals τ between subsequent steps, the diffusivity is easily shown (see Sect. 2.1) to be given by the relation

$$D = l^2/2\tau. \tag{13.2}$$

The temperature dependence of the diffusivity follows, in most cases, the Arrhenius law [7]

$$\frac{dlnD}{dT} = E/RT^2 \quad \text{or}$$
$$D = D_0 exp(-E/RT). \tag{13.3}$$

D_0 and E denote, respectively, a pre-exponential factor and the activation enthalpy and do not vary, as a rule, with varying temperature. The diffusivity is known to increases with increasing temperature so that E (just like D_0) has to be positive, quite in general. For the majority of solid metals the self-diffusion coefficient close to the melting temperature is approximately $D = 10^{-12}$ m^2/s. It is interesting to compare this value with the diffusivities in liquid metals ($D_L = 10^{-9}$ m^2/s) and in gases (approximately $D_g = 10^{-5}$ m^2/s under "normal" conditions) which, significantly deviating from the diffusivities in solids, are almost temperature independent. The activation enthalpy of self-diffusion in solid metals is, as another remarkable common feature, roughly (i.e. with deviations of about ±20%) proportional to the melting temperature, following the relation $E = 18RT_{melt}$. With the melting temperatures of aluminum ($T_{melt} = 933$ K), copper ($T_{melt} = 1356$ K) and nickel ($T_{melt} = 1728$ K) one thus obtains, as an estimate of the respective activation energies, values of, respectively, 140, 203 and 259 kJ/mole, in amazingly good agreement with the experimental data of 142, 197, and 275 kJ/mole.

For estimating diffusivities we may use, for most solid metals, a value of $D_0 = 10^{-5}$ m^2/s. From this and with the above given estimate of the activation enthalpy we obtain, at temperatures $T = 0.4T_{melt}$ and $0.7T_{melt}$, self-diffusion coefficients of, respectively, $D = 10^{-23}$ m^2/s and $D = 10^{-16}$ m^2/s.

The diffusion length, i.e. the square root of the mean square displacement as a reasonable measure of the mean diffusion path length of the atoms under study, is, with Eq. (13.1), seen to be given by the relation $(2Dt)^{0.5}$. Using the above estimates of the diffusivities at $T = 0.4T_{melt}$, $0.7T_{melt}$, and T_{melt} the diffusion lengths attained in an experiment over 100 (!) hours amount to, respectively, 10 nm (which is just about 30 interatomic distances!), 10 μm, and 1 mm. We have to conclude that diffusion in solids is, indeed, very slow, even at high temperatures!

13.4 Diffusion Mechanisms in Metals

We remember that at non-zero temperature atoms usually vibrate around their equilibrium position with quite small amplitudes. They are much smaller than the distance between nearest neighbors in a lattice. Thus, vibrations alone cannot give rise to the diffusional motion of the atoms. There are several mechanisms which have been proposed to ensure the elementary steps of diffusion in solids. The most important ones are shown in Fig. 13.4.

Experiments show that self-diffusion and hetero-diffusion in substitutional solutions commonly occur via the vacancy mechanism. The main hetero-diffusion mechanism in interstitial solutions is, obviously, the interstitial mechanism. Note that, in a binary substitutional solution, sites are occupied by atoms of both types. The vacancy mechanism was proposed by Frenkel [8]. He was the first to recognize that, in a solid under equilibrium conditions, there have to exist vacancies or, in other words, that the free energy assumes its minimum with a finite number of vacancies incorporated in the system.

In the vacancy mechanism of diffusion, obviously, an atom can only jump if one of the nearest-neighbor lattice sites is empty. The hopping frequency of an atom is thus given by the relation

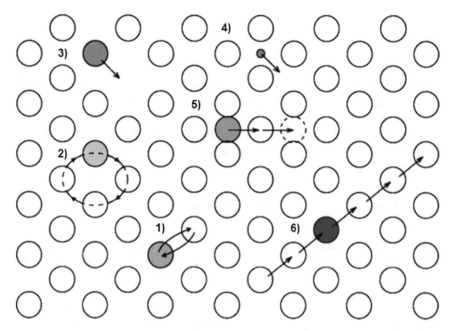

Fig. 13.4 Main atomic mechanisms of diffusion in a crystal, including (1) simple exchange (a couple of neighboring atoms exchange their positions), (2) cyclic exchange (several atoms change their positions one after another in a four- or six-member ring), (3) exchange with a vacancy (vacancy is an empty lattice site), (4) interstitial jump (an atom in an off-lattice site between the lattice atoms jumps to a neighboring, empty off-lattice site), (5) concerted movement by which an atom on an interstitial site assumes the position of an atom sitting on a regular site which, in turn, is shifted into an adjacent interstice, (6) atoms in a row are brought in closer contact due to an inserted ("crowded in") extra atom (to appear at the beginning of the row of arrows)

$$\Gamma = Z\omega X_v, \tag{13.4}$$

where Z is the coordination number, ω is the frequency with which an atom may jump into an adjacent vacancy (coinciding with the "jump of a vacancy") and X_v is the vacancy concentration. The jump rate Γ appearing in Eq. (13.4) is easily seen to be much less than the frequency ω of vacancy jumps, since $X_v \ll 1$ (otherwise a crystal could not exist).

Experiments show that in most metals at the melting point there is, among 1000 occupied sites, about one vacancy, i.e. $X_v = 10^{-3}$. This value decreases rapidly with decreasing temperature. For copper at $T = 300$ K, e.g., one has $X_v = 10^{-19}$. With Eqs. (13.4) and (13.2)), the diffusion activation enthalpy in the case of the vacancy mechanism is the sum of the enthalpy of vacancy formation and the activation enthalpy of "vacancy migration" (i.e. for the jump of an atom into an adjacent vacancy).

Let us now discuss the mechanism of interstitial diffusion. An interstitial atom is shown in Fig. 13.4 under number (4). For self-diffusion, we may follow the reasoning applied already with the vacancy mechanism, with the concentration and jump rate of vacancies now replaced by the equilibrium concentration of interstitials and their migration rate. The physical picture for the diffusion of solved atoms in interstitial solid solutions, however, is quite different. Such atoms are small and they are already in interstitial sites (number 4 in Fig. 13.4). For the diffusion of such solved atoms there is no need for creating additional interstitials. Thus, the interstitial diffusion has only a contribution of interstitial migration, since one does not need to create the interstitial. In this case, the diffusivity of the small solved atoms can be much larger than the diffusivity of the (big) matrix atoms.

13.5 Diffusion in Amorphous Alloys

The most important difference between amorphous alloys and crystalline materials is that amorphous alloys do not have a regular crystalline lattice. The spatial distribution of atoms in amorphous bodies is more similar to that in liquids than in solids. Nevertheless, diffusivities in amorphous materials are much smaller than in liquids and, rather, comparable with those in corresponding solids [9, 10]. Amorphous materials are usually metastable. Thus, upon heating, they will crystallize at a certain temperature, referred to as the crystallization temperature, T_{cr}. In principle, as a consequence of their metastability, amorphous materials could crystallize at any temperature. At low temperatures, however, the rate of crystallization may become extremely small. The T_c value and the stability of amorphous alloys are thus largely controlled by their kinetics, i.e. by atomic diffusion.

We will now discuss the diffusion in classical amorphous metallic alloys. Usually, such alloys consist of noble and/or transition metals (like iron, cobalt, nickel, palladium or gold) and non-metals (like boron, carbon, phosphorus, silicon or germanium).Since amorphous materials do not have a unique structure as known from crystalline materials, their actual texture—and hence their intrinsic diffusivities—may most significantly depend on the manufacturing method materials. There may exist, as a consequence, many different amorphous atomic arrangements for one and the same alloy. Since some of them are, as a matter of course, more stable than others amorphous materials tend to change, upon heating, their structure toward a more stable (though still amorphous) one. Given the possibility of such phenomena of structural relaxation, it is frequently not easy to discriminate between effects of structural relaxation and genuine diffusion. Such disturbing influences can be largely excluded by prevenient annealing, e.g. at 0.96–0.98 T_{cr}, with subsequent diffusion experiments at temperatures below 0.9 T_{cr}.

Also in amorphous alloys the temperature dependence of the diffusivity is usually found to follow the Arrhenius law, Eq. 13.3. It is, indeed, a quite surprising observation! Recollect that the formation and migration of point defects has been identified as a prerequisite for diffusional motion in solids. From this one may immediately conclude that the activation energies of point defect formation and migration—and thus also the enthalpy of diffusion—have certain exact values because all positions in a crystalline lattice are equivalent! Amorphous materials, however, have an irregular structure. We would have to expect, therefore, a wide distribution of energies of defect formation and migration. Thus, it appears to be very paradoxical that the temperature dependence of the diffusivity in an amorphous alloy may be properly described by just one activation enthalpy. One should have in mind, however, that the observation of an Arrhenius dependence for diffusion in amorphous alloys may have another, simple explanation. Most diffusion studies in these materials have been performed in a relatively narrow temperature interval (of typically not more than 100 K). This restriction in the temperature range results from the conflict that the temperature, on the one hand side, must be high enough to allow the proper measurement of diffusion phenomena while, on the other hand, it must avoid getting too close to the crystallization temperature.

For diffusion in amorphous alloys several mechanisms have been proposed, with no consensus about the dominant mechanism. According to one suggestion, referred to as the quasi-vacancy mechanism, diffusion in amorphous alloys is assumed to follow a mechanism which is similar to the vacancy mechanism in crystals [11, 12]. One has, however, no real vacancies in an amorphous material. It accommodates instead an excess free volume, giving rise to the formation of a continuous spectrum of sizes of these free volume "clusters". They can be either smaller or larger than the atoms in an alloy. The density of an amorphous alloy is usually smaller than that of the respective crystalline alloy. The quasi-vacancy mechanism suggests that atoms can jump into such holes, just as they may jump into a vacancy in a crystalline lattice. We have to recollect, however, that in an amorphous structure each hole has to be associated with its own, specific activation energy of formation and migration!

A further suggestion is based on the assumption that the elementary process of diffusion in an amorphous structure is a cooperative displacement of a group of neighboring atoms. Such a step does, obviously, involve the movement of many atoms. The resulting activation enthalpy represents, thus, an averaged value over the entire group. This idea smoothly explains why there appears one single value for the activation enthalpy and why the activation energies for diffusion in non-metallic compounds and metals are comparable.

Diffusion of metal atoms within amorphous semi-conducting alloys is even more complicated. Thus several metals (including lithium, nickel, iron, copper and palladium) diffuse in amorphous silicon even slower than in crystalline silicon. We can explain this fact as follows. All these atoms diffuse in crystalline silicon via an interstitial mechanism. To the contrary, in amorphous silicon they can be trapped in quasi-vacancies. We remember that interstitial diffusion proceeds usually much faster than substitutional diffusion.

On the other hand, several metals (like gold, platinum or zirconium) diffuse in amorphous silicon faster than in crystalline silicon. For rationalizing such a behavior we have to recall that the atoms of gold, platinum or zirconium can occupy either lattice or interstitial sites in crystalline silicon. Diffusion of these atoms does include, therefore, periods of fast migration through the interstitials and of trapping.

Diffusion enhancement in amorphous silicon is thus a simple consequence of the fact that trapping in amorphous silicon occurs much less frequently than in crystalline silicon.

13.6 Diffusion in Polymers

The use of polymeric materials has really exploded during the last few decades. The world production per anno for polymers (equally, by volume or mass) has recently even overtaken that of metals! Diffusion in polymers may thus be regarded as even more important than diffusion in metals [13].

On penetrating into a polymer, "alien" small molecules can give rise to a conformational change of the polymer chains around them. Thus, the diffusion process includes both penetration of "alien" molecules and the conformational relaxation of the polymer chains. For a quantification of these two phenomena, the so-called Deborah number De has been introduced. It is defined as the ratio t_c/t_p between the time constants of conformational changes and particle propagation over the extension of the individual polymer molecules. If the Deborah number is small ($De < 0.1$) the characteristic time t_c of conformational relaxation is much shorter than the characteristic diffusion time t_p while for large Deborah numbers ($De > 10$), molecular structures remain essentially nearly unchanged during diffusive displacements along the extension of the individual polymeric chains. It is important to emphasize that in both cases guest penetration can be described by Fick's laws. In the intermediate case, i.e. for values t_c and t_p of similar orders of magnitude ($0.1 < De < 10$), the situation becomes much more complicated. Under such conditions, relaxation has to be in particular expected to facilitate the trapping of penetrating "alien" molecules. The total number of penetrating molecules has thus become a variable quantity which excludes the direct, unrestricted application of the Fick's laws.

Summing up, there is so far no generally accepted understanding of the mechanisms of metal diffusion in polymers. As an example of the still unsolved questions we refer to the remarkable observation that noble metals, which are frequently used for contacts in microelectronics, are found to diffuse slower than the gas molecules, irrespective of the fact that their sizes are comparable with or even smaller than the sizes of the molecules [14]. Two explanations of this phenomenon are under discussion. One is based on the assumption that the metal atoms diffuse as clusters which are considerably larger than single atoms. The second one implies that there exists a particularly strong interaction between the noble metals and the polymer molecules.

13.7 Diffusion During Severe Plastic Deformation

Over the last few decades, material sciences have emerged as an own, extremely promising field. Its development has most decisively been promoted by the invention of the so-called severe plastic deformation (SPD) methods [15–17]. SPD permits to strain a material in a confined space without giving it the possibility to fail. Figure 13.5 illustrates the operation principle of one of the most frequently used techniques for this purpose, the method of high pressure torsion (HPT). Here, a disk of the material to be strained is pressed between two anvils with a pressure of 4 up to 12 GPa. One anvil starts to rotate, so that the sample is strained by pure shear. Though the number of rotations can be very high (sometimes up to 100), the sample, obviously, cannot break. Torsion gives rise to the establishment of dynamic equilibrium between defect production and sample relaxation. Steady state is usually attained after 2 rotations so that the properties of the material do not change any more.

Under the given stress, an enormous amount of defects is produced in a material. Quite logically, the concentration of defects (vacancies, dislocations, interfaces etc.) cannot increase infinitely. It is due to this reason that relaxation sets in, giving rise to the formation of a steady state. This dynamic relaxation (or recovery) involves

Fig. 13.5 Scheme of the arrangement for high pressure torsion (HPT) experiments

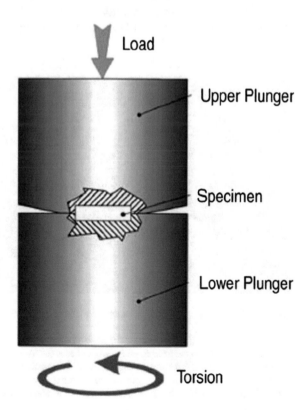

(sometimes very speedy) diffusion fluxes. The phases formed under steady-state conditions are, as a matter of course, different from those, which were present in the material before SPD treatment [18, 19]. In fact, composition and crystallographic structure of these phases can serve as a probe describing the (very quick and intensive) mass-transfer processes during SPD. The diffusion processes during SPD differ from those which take place in traditional materials technologies and are close to the equilibrium. The SPD-driven diffusion and diffusion-controlled phase transitions take place far from equilibrium. Their description and explanation is a real challenge for materials science.

SPD can drive different phase transitions in the materials under treatment [18, 19]. They include such diverse phenomena like the dissolution of phases, the synthesis of different allotropic modifications of elements (Fig. 13.6), the amorphization of crystalline phases, the decomposition of supersaturated solid solutions or dissolution of precipitates, the disordering of ordered phases and the nanocrystallization in amorphous matrices. The exploration of the manifold possibilities for the exploitation of these phenomena for the fabrication of materials with optimized performance characteristics is a hot topic of current research.

It can be observed that, after SPD, phases can be formed which usually appear after annealing at a certain (in general notably elevated) temperature. Such a temperature can be referred to as an effective temperature T_{eff}. The concept of effective temperatures has been originally suggested by George Martin for materials after severe neutron irradiation [19]. It does now appear that this concept is useful

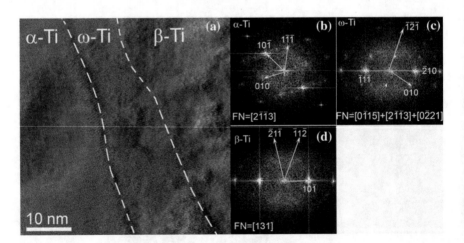

Fig. 13.6 HREM micrograph of an alloy of titanium containing 2 weight % of iron which had been annealed at 800 °C for 270 h, quenched and subjected to HPT under a pressure of 7 GPa and a total rotation of 36° (i.e. of around 1/10 of a full circle) with a rotational speed of 6° per second (**a**). Titanium is seen to occur in three different phases (α-, β- and ω-Ti), with ω-Ti resulting as the product of phase transformation during the HPT deformation. Evidence of the presence of these phases (and of even the relationship of their mutual orientation [20, 21]) is provided by Fast Fourier Transforms (FFT) (**b**–**d**) of the corresponding areas in the high resolution image (**a**)

and applicable also for the phenomenon of severe plastic deformation (SPD) [22, 23]. In both cases, the atomic movements driven by external actions (i.e. by irradiation or deformation) are accelerated in comparison with atomic movements during conventional thermal diffusion. The material has, obviously, been shifted into a state with a mobility which is otherwise only attained at a correspondingly increased (the "effective") temperature T_{eff}.

Severe plastic deformation at room temperature T_{SPD} can drive the phase transitions. This behavior is, obviously, caused by the high density of defects, namely a density of defects comparable with that at a notably increased temperature, T_{eff}. Some of the SPD-driven phase transformations require the long-range mass transfer, for other transitions only a small shift of constituent atoms is required. Usually, the sequence of these SPD-driven phase transformations is not easy to explain by the conventional bulk or even grain boundary diffusion at T_{SPD}, since T_{SPD} usually remains only slightly above room temperature.

For the phase transformations under SPD (like precipitation and dissolution of precipitates, amorphization, transitions between various "Hume-Rothery" phases etc. [24]) one usually requires a redistribution of components and, therefore, a certain mass transfer. Such SPD-driven redistribution of components proceeds extremely quickly, irrespective of the fact that it occurs at room temperature, without notable temperature increase. Thus, steady-state conditions during HPT are usually attained after not more than 2–5 min. As a measure of the SPD-driven mass-transfer one may use the concept of the effective temperature, defined by the requirement that, at this very temperature, the diffusivity under "normal" conditions (i.e. without SPD) would coincide with the diffusivity under HPT conditions. As a matter of course, this effective temperature usually exceeds the temperature of HPT treatment and measurement (T_{SPD}) [17, 24].

Thus, one can conclude that George Martin's idea [19], in its extended form, can be also applied for SPD-driven phase transitions. George Martin supposed [19] that the atomic displacements driven by the irradiation are similar to those which take place by thermal diffusion (like the jumps into a neighboring vacancy). In case of SPD such a suggestion is no longer valid. We suppose instead that, under similar SPD conditions, SPD-driven atomic movements are comparable in all alloys. The "natural" atomic movement, however, is quite different because the melting temperatures T_m of the considered materials are quite different. If T_{SPD} is almost the same (~300 K) and the diffusion coefficients for "natural" diffusion are low for alloys with high T_m atomic movements driven by SPD can be much larger than diffusion by atomic jumps, corresponding with large values of T_{eff} as predicted by G. Martin. If the melting temperature T_m of the considered material is low (as in the case of aluminum alloys), T_{eff} would be low. It can be even close to T_{SPD}, i.e. to room temperature [16].

13.8 Conclusions

Diffusion can control the rate of a wide range of important technological processes in materials manufacturing. These are also processes of materials modification, resulting in properties by which the materials may work properly—or fail. Examples include diffusion creep, sintering, pore formation and annihilation, grain boundary migration, grain growth, phase transformations, and precipitation. The present chapter introduces into the variety of phenomena associated with these processes. Though all of them, consistently, are based on thermal motion, i.e. on the random walk of the individual atoms of the material under study, the conditions under which movement occurs are manifold. As manifold are also the elementary steps by which, eventually, the overall rate of mass transfer is controlled. The spectrum of phenomena and processes thus considered ranges from the classical view on diffusion in metals up to such comparatively new topics as diffusion in amorphous alloys, polymers and during severe plastic deformation.

References

1. B. Bokstein, M. Mendelev, D. Srolovitz, *Thermodynamics and Kinetics in Materials Science* (Oxford University Press, 2005), 326 pp.
2. H. Mehrer, *Diffusion in Solids. Fundamentals, Methods, Materials, Diffusion-Controlled Processes* (Springer, Berlin, 2007)
3. A. Fick, Poggendorffs Ann. Phys. Chem. **94**, 59 (1855)
4. J. Fourier, *Theorie Analitique de la Chaleur* (Firmin-Didot pere et fils, Paris, 1822)
5. A. Einstein, Ann. Phys. **17**, 549 (1905)
6. M. von Smoluchowski, Ann. Phys. **326**, 756 (1906)
7. S. Arrhenius, Zt. für Phys. Chem. **4**, 226 (1889)
8. Y. Frenkel, Zt. für Physik **26**, 117 (1924)
9. F. Faupel, W. Frank, M.P. Macht, H. Mehrer, V. Naundorf, K. Rätzke, H.R. Schober, S.K. Sharma, H. Teichler, Rev. Mod. Phys. **75**, 237 (2003)
10. S.W. Basuki, A. Bartsch, F. Yang, K. Rätzke, A. Meyer, F. Faupel, Phys. Rev. Lett. **113**, 165901 (2014)
11. A. Bartsch, V. Zöllmer, K. Rätzke, A. Meyer, F. Faupel, J. Alloys Comp. **509**, S2 (2011)
12. W. Frank, J. Horváth, H. Kronmüller, Mater. Sci. Eng. **97**, 415 (1988)
13. J.-W. Rhim, H.-M. Park, C.-S. Ha, Progr. Polym. Sci. **38**, 1629 (2013)
14. F. Faupel, R. Willecke, A. Thran, Mater. Sci. Eng. R **22**, 1 (1998)
15. A.P. Zhilyaev, T.G. Langdon, Progr. Mater. Sci. **53**, 893 (2008)
16. R.Z. Valiev, T. Langdon, Progr. Mater. Sci. **51**, 881 (2006)
17. B.B. Straumal, A.R. Kilmametov, Yu. Ivanisenko, A.A. Mazilkin, O.A. Kogtenkova, L. Kurmanaeva, A. Korneva, P. Zięba, B. Baretzky, Int. J. Mater. Res. **106**, 657 (2015)
18. Y. Nasedkina, X. Sauvage, E.V. Bobruk, M.Y. Murashkin, R.Z. Valiev, N.A. Enikeev, J. Alloys Comp. **710**, 736 (2017)
19. G. Martin, Phys. Rev. B **30**, 1424 (1984)
20. D.R. Trinkle, R.G. Hennig, S.G. Srinivasan, D.M. Hatch, M.D. Jones, H.T. Stokes, R.C. Albers, J.W. Wilkins, Phys. Rev. Lett. **91**, 025701 (2003)
21. B.S. Hickman, J. Mater. Sci. **4**, 554 (1969)

22. B.B. Straumal, V. Pontikis, A.R. Kilmametov, A.A. Mazilkin, S.V. Dobatkin, B. Baretzky, Acta Mater. **122**, 60 (2017)
23. B.B. Straumal, A.R. Kilmametov, G.A. López, I. López-Ferreño, M.L. Nó, J. San Juan, H. Hahn, B. Baretzky, Acta Mater. **125**, 274 (2017)
24. B. Straumal, A. Korneva, P. Zięba, Arch. Civil Mech. Eng. **14**, 242 (2014)

Chapter 14
Spreading Innovations: Models, Designs and Research Directions

Albrecht Fritzsche

14.1 Introduction

Diffusion models describe local change processes that lead over time to a spread of particles or information in a topological space. Metric spaces are the most common examples of topological spaces, but there are other examples, too. Any kind of space in which the notion of proximity can be formed, mathematically addressed by the term 'neighbourhood', allows the application of diffusion models. This does not only include standard Euclidean spaces as they are frequently used in physics or geography, but also formal networks describing interconnected social or technical entities (see e.g. the contribution by Shekhtman et al. in this volume). Diffusion models have therefore not only proven to be quite useful in the natural sciences, but also in research on the connections between individual human behavior and the economic, cultural or technical development of a society as a whole. For example, they have helped to gain a better understanding of the way how the effects of technical inventions, scientific discoveries and artistic genius evolve in time and space and how society and economy are able to take advantage of it.

One of the first scientists who addressed this issue systematically was the French sociologist and psychologist Gabriel Tarde at the turn to the twentieth century [1]. Tarde, a contemporary and competitor of Émile Durkheim, subsumed the adoption of novelty among humans under the term imitation and looked specifically at the effects of blind obedience, explanation and training and their sequential combination. He made clear that the spread of innovations in society cannot be left to themselves. They have to be actively managed and require a lot of effort, which is widely neglected in simple histories that focus exclusively on the dates of discovery

A. Fritzsche (✉)
Institute of Information Management 1, Friedrich-Alexander University
Erlangen-Nuremberg, Nuremberg, Germany
e-mail: albrecht.fritzsche@fau.de

© Springer International Publishing AG 2018   277
A. Bunde et al. (eds.), *Diffusive Spreading in Nature, Technology and Society*,
https://doi.org/10.1007/978-3-319-67798-9_14

and invention, implicitly assuming that the results will spread more or less automatically in society.

Inspired by Tarde, the thesis of this contribution is that such management activities go further than the search for the most important factors of influence of the diffusion process. They can more likely be described as design efforts which organize the spread of novelty in a way that makes it possible to conceptualize it as a predictable diffusion process and exploit it accordingly. At the same time, the subject matter of the diffusion process must be considered by itself as a designed artefact, too (see also Chap. 15.4.1 on social construction). The notion of an innovation is a social construct that can be gained in different ways, adding further complexity to the discussion.

Every scientific discipline has its idiosyncrasies, and a book that discusses diffusion across many disciplinary boundaries therefore provides many opportunities for misunderstandings, caused by different ontological assumptions, epistemological interests or deviating nomenclature. In this chapter, for example, an artefact is understood in its literal sense as an object made by man, whereas physics uses the term to address systematic errors due to deficiencies in experimental procedure. Further problems arise from the limited comparability of the subject matter to which the diffusion models are applied. The intuition of a natural scientist is rather guided by phenomena like the diffusion of molecules in porous solids (Chap. 10) or the diffusion of plants and animals in their habitats (Chap. 3), which can be described with reference to conventional metric spaces by the Eqs. (2.1)–(2.18) in the chapter on "Spreading Fundamentals". Social scientists are confronted with a different kind of reality, in which proximity is not exclusively depending on physical distance, but also on personal acquaintance and technical connectedness. Individuals can accordingly feel closer to family members on a different continent than strangers in the next building.

The chapters by Brockmann and by Shekhtman et al. in this book discuss the notions of distance and network which provide the foundation for the understanding of space and proximity used in diffusion studies by the social sciences. It illustrates the wealth of information attainable from solely the topological structure of such networks: the paths existing between its entities. As it will be explained, society is built on numerous overlaying networks that connect individuals with one another, provided by different technologies, roles and relationships. The fact that these networks can be actively changed contributes largely to the specific way diffusion is treated on the following pages.

14.2 Diffusion Models in Innovation Research

14.2.1 Conceptual Approaches

The term technology does not describe a homogeneous entity. Technology rather has to be understood as an embodiment of any kind of instrumental action that occurs repetitively in the world [2, 3]. This definition of technology makes it possible to treat any kind of innovation as technical. At the same time, however, it creates the need for numerous different operational measures for the diffusion of innovations, depending on the given context. In many cases, such measures can be gained through sales figures for technical artefacts. This, for example, is the case in Griliches' seminal work on the economics of technical change [4]. For Gort and Klepper [5], "diffusion is defined as the spread in the number of producers engaged in manufacturing a new product" which would economically be described as the net entry rate in the market for a new product. Rogers, on the other hand, describes diffusion as "the process in which an innovation is communicated through certain channels over time among the members of a social system" [6], independently from any business operation. The spread of the internet requires yet another approach that takes the availability and, ideally, also the bandwidth of internet access into consideration [7]. In order to measure the actual adoption of a technology in daily routines, it is furthermore necessary to collect data on the frequency or intensity of usage.

In any case, innovations need a carrier medium to spread. This medium is provided by society, in terms of interrelated individuals, groups, corporations or other institutions that can be described as actors who hold certain information, are in possession of certain material goods, show certain behavior or have certain attitudes that entail certain decisions, which can then be empirically accessed. Regarding these actors, two important questions have to be asked: how are they connected and what resources do they have available to act? The connectedness determines the paths on which innovations can spread and thus induces a spatial structure on which diffusion can be observed. The availability of resources determines if and in what way the actors can make use of a technology, describing the capacity of the actors in their function as carriers of innovation.

The role of connectedness and resource availability for the diffusion of innovations is illustrated by the following examples:

- Innovative data processing algorithms can spread very quickly over the internet, if there is no further effort necessary for their installation. This effect is very well known from computer viruses. Their distribution is to a large extent a question of connectedness. One of the most popular ways of securing sensitive data is therefore to keep them isolated from the internet. Knowledge about new mathematical algorithms or construction methods in engineering spreads among sufficiently trained experts in a similar way.
- Expensive product innovations spread very slowly, even if many dealers keep them in stock. For example, this is currently the case for cars with electric engines, which need to be strongly subsidised to be sold. Connectedness does not matter, if people do not have the resources at their disposal to adopt them. Such resources do not only include financial means, but also qualification, time, space and the ability for habitual change.

As a general rule, one can say that connectedness matters most when innovations add novelty to an existing repertoire and are in this sense complementary to whatever is there already, whereas resource availability has to be considered whenever innovations entail a substitution process.

Inasmuch as the connections between the actors determine the paths on which innovations can travel and the available resources determine the potential for the adoption of an innovation, connectedness and resource availability both have to be considered in the design of a topology on which the diffusion of innovation is depicted as a formal process (illustrated by Brockmann in Chap. 19). Even if the transport infrastructure between the actors constitutes a small or an ultra-small network (see the contribution by Shekhtman et al. in Chap. 20), the actual travel times of innovations can be quite long, if their adoption requires an intensive substitution process. Furthermore, it is often necessary to take path dependencies during communication into account. Depending on the source of an innovation, actors can be more or less inclined to adopt it. Similar effects also have to be considered in the comparison of different communication channels such as internet blogs, e-mails, telephone calls, business gatherings or private meetings. Innovations, one can say, travel on very rough terrain and multi-dimensional surfaces.

Due to the increasing dynamics of technical, social and political development, the question of the durability of the change caused by the diffusion of innovations currently emerges as a new research topic for innovation studies. The switch to a new technology does not necessarily have to be permanent. Innovations often require a continuous flow of energy supply, consumer goods or regular expert maintenance. If the surrounding infrastructure breaks down, innovations can therefore disappear again. With the current discussion on the protection of critical social infrastructures against disruptive events, questions of technical robustness and resilience receive increased attention (e.g. [8]), and they are likely to become more important for innovation research in the future as well.

What can be learned for all this is that formal models to describe the diffusion of innovations have to be very specific about the subject matter they are concerned with and the social, technical and economic conditions under which the diffusion process is assumed to take place.

14.2.2 *Mathematical Models*

As a quantitative measure with reference to the diffusion of innovation, it stands to reason to consider the percentage of the target group who has adopted the innovation. Figure 14.1 shows this (in terms of the "fraction of the carrier medium that has adopted the innovation", f) schematically as a function of time.

Early studies found that the increase in innovations as shown in Fig. 14.1 followed the typical pattern of constrained exponential growth [4, 9], visualized by an S-shaped curve with asymptotic behavior at the outer limits and a central inflection point (see arrow in figure). The curve shown in Fig. 14.1 corresponds with the

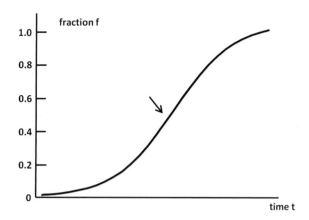

Fig. 14.1 Sample plot of constrained growth over time in a logistic function

so-called uptake curve indicating the relative number of molecules entering a nanoporous particle upon pressure increase in the surrounding atmosphere. It is illustrated in Chap. 10 that it is this type of information which over decades served as the main source of experimental evidence for the prediction of the diffusion characteristics in porous media, with all deficiencies of an "indirect" technique of measurement since evidence of such type of experiments concerns the effect of diffusion rather than the process of diffusion itself.

As this curve is characteristic for the logistic function (see Chap. 3, Sect. 3.4.1), Griliches [4] proposes that the diffusion of innovations be described by the according differential equation of the type

$$\frac{df}{dt} = af(1-f) \tag{14.1}$$

where f is the fraction of the carrier medium that has adopted the innovation, t is time, and a is a growth parameter.

This equation was introduced by Verhulst [10] in the discussion of limited population growth, without any data on the maximal population size that can actually be reached [11]. Innovation researchers are usually in a more comfortable situation, since they can estimate the maximal distribution of an innovation by the size of the current population or by referring to an older technology that is expected to be substituted by the innovation [12]. This approach is frequently used to forecast the progress of the diffusion of innovations, illustrated by the example of smartphones in Fig. 14.2.

Due to the conceptual challenges mentioned before, satisfactory explanations of this behaviour are hard to give. A common assumption is that differences in connectivity and resource availability cause adoption times t for innovations in society to follow a normal distribution $p(t)$. For each innovation, there are accordingly a few early adopters and laggards with exceptionally short or long adoption times, while the majority of the population stays within a smaller interval around the average adoption time (Fig. 14.3).

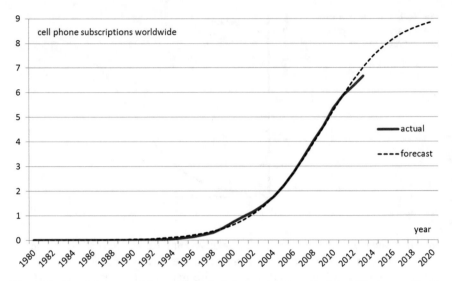

Fig. 14.2 Actual figures and forecast with logistic function for cell phone diffusion worldwide [13]

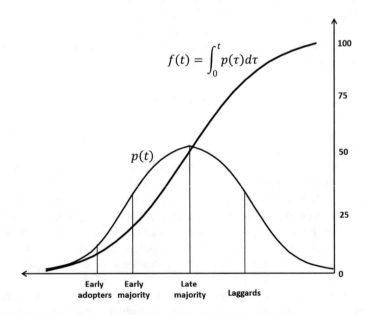

Fig. 14.3 Normal distribution of adoption characteristics in a population

Over the years, various modifications of this model have been suggested, in particular regarding the parameter a, which is not any more treated like a constant, but rather depending on changes in manufacturing and marketing [14]. In the course of the diffusion process, technology producers are expected to become more efficient, reduce prices and connect better to their audience, which accelerates the spread of the innovation.

While some data sets support this model, others raise questions about the general applicability of the logistics function to the diffusion of innovations. In a large survey on data sets about the diffusion of various technologies in different countries, Comin et al. [15] identify numerous cases in which logistic functions approximate the actual data very poorly and calculate unrealistic saturation times. While some of these findings may be caused by disruptive changes in the general setting (political change, economic crisis etc.), there is good reason to assume that there are also other internal dynamics at work which affect the diffusion process, in particular with respect to individual adoption behavior. These dynamics have become one of the major fields of study in innovation research.

14.3 Individual Adoption Behavior

14.3.1 Acceptance Models

Innovation research uses various different approaches to capture the causal relationships regarding the adoption behavior among social actors. A particularly high number of studies are based on the technology acceptance model, which looks at two different factors that influence the intention of an actor to use a technology [16]:

- the perceived ease of use refers to the complexity experienced by users in operating a technology and directing it to the outcome which they intend to achieve
- the perceived usefulness refers to the advantages that the users expect to result from applying the technology

These two factors reflect a distinction between costs and benefits of a technology in the ease of use expressing the effort necessary to handle it and the usefulness expressing the value generated by it. The perceived usefulness is by itself subject to various different influences, such as the quality of the output and its quality or the image of a technology in public and the social treatment of its users. Empirical evidence suggests that the perceived usefulness has a higher relevance for decision making process than the perceived ease of use [17]. As it seems, potential adopters expect a learning process over time that will make the technology easier to use in the future, while the perceived usefulness is considered as an attribute of the technology which cannot be influenced by them.

Another finding that has attracted a lot of attention during the last years is the contrast between personal acceptance and social acceptability of a technology. Public transportation may be taken as an example for a technology with higher social acceptability than personal acceptance: although most people agree that busses, subways and railroads are valuable means of transportation, many of them nevertheless prefer to drive by car for themselves. Smoking is an example of the opposite: despite all public concerns about it that they may share, many people still think that it is okay to have a cigarette for themselves.

In order to capture these differences, it is necessary to distinguish other factors influencing human behavior according to psychological theories [18, 19]:

- the personal attitude toward a certain act
- the social norms referring to its performance and outcome
- the perceived level of control over its execution

These factors evolve differently over time. They also react differently to specific forms of external interference.

The adoption behavior of individuals that provides the foundation for the spread of innovations must accordingly be considered as a result of a superposition of different cognitive processes. These processes are subjected to various influences which are unlikely to affect every person in society in the same way. With increasing social diversity, the carrier medium for the diffusion of innovations therefore becomes highly inhomogeneous. Even if the overall diffusion process indicates a high adoption rate, special focus groups might actually react differently. The automotive industry, for example, is lately confronted with the phenomenon that young people show significant differences in their adoption behavior from others. This fact remains invisible in general sales figures, since they only account for a small fraction of the market. Nevertheless, this phenomenon raises concerns about future sales opportunities [20, 21].

This is a rather unsatisfactory development for the manufacturer, since it indicates that the product does not find acceptance in an important part of society, no matter how successful it is elsewhere. As a consequence, the manufacturer is advised to action against this development. This, however, must be considered highly dangerous. Growth processes are known to react very sensitively to parameter change. In a diverse society, manufacturers have to expect chaotic reactions to change which are hard to predict or control. Many companies have therefore turned to strategies that relate innovation to specific target groups in the overall population which can be expected to show a more homogeneous behavior.

14.3.2 The Shifting Locus of Innovation

Figure 14.4 illustrates the development of product strategies in the automotive industry over the past decades, leading away from the idea of a single product that fits everyone's need towards a highly diversified family of different models which are designed according to specific application patterns that can be expected to meet the needs of certain social groups. In this example, the diversification is described in terms of the body shapes of the car. Diversification also proceeds with respect to

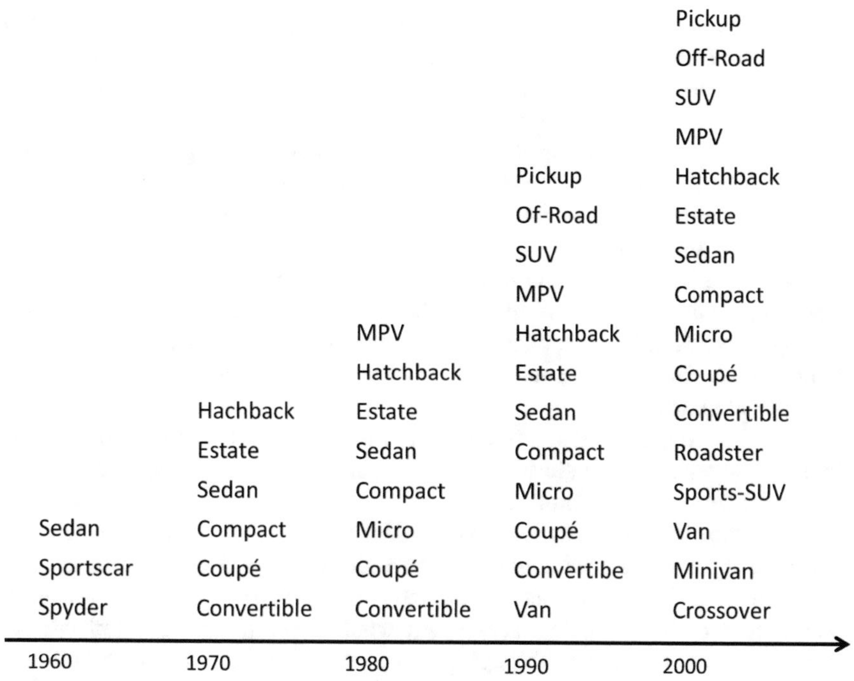

1960	1970	1980	1990	2000
				Pickup
				Off-Road
				SUV
				MPV
			Pickup	Hatchback
			Of-Road	Estate
			SUV	Sedan
			MPV	Compact
		MPV	Hatchback	Micro
		Hatchback	Estate	Coupé
	Hachback	Estate	Sedan	Convertible
	Estate	Sedan	Compact	Roadster
	Sedan	Compact	Micro	Sports-SUV
Sedan	Compact	Micro	Coupé	Van
Sportscar	Coupé	Coupé	Convertibe	Minivan
Spyder	Convertible	Convertible	Van	Crossover

Fig. 14.4 Increasing diversification in German automotive industry based on body shape [22]

such different aspects as engines, colors etc. Larger corporations such as General Motors or Volkswagen also diversify by differentiating brands according to specific lifestyles and personal values, from practicality over sports to comfort and luxury.

Product diversification allows companies to pursue different strategies to support the adoption of their products, depending on the respective target group. In addition to the technical features of the products, these strategies also address other aspects of business activity, including the pricing methods, the distribution network, and the communication channels to approach potential or existing customers. Companies can thus circumvent a large part of the complexity which they would have to face if they had to look at diffusion processes in the whole population. The separation of different target groups and the selection of different ways to approach them make it possible to differentiate separate diffusion processes on parts of the population which are, as carrier media of innovation, once again, largely more homogenous.

As a result, however, innovation also takes on a different quality. Although a larger notion of the term technology allows us to still think of innovation as technical change, this change is not focused on engineering solutions any more. Innovation now concerns the whole set of business operations that generate value for the customer. This is addressed in the current discussion on business model innovation [23, 24].

The shift towards value generation has various implications for the practice of modelling diffusion processes. There are now two different kinds of items which can be considered to spread: the overall business model and the offerings of the company that it contains. Business models spread among companies as a carrier medium; popular examples are leasing models, mobility packages, or flat rates in telecommunication. Offerings in the company remain more closely connected to the intuition of technical change; with the focus on value generation, however, the attention is drawn away from quantities of sales as means to make profit. What becomes more important is the control of the diffusion process that allows companies to plan the revenue and optimize the workload on their resources over time. In many respects, this also applies for larger attempts to spread innovations as they are undertaken by governments or other political institutions who want to ensure steady development.

Figure 14.5 provides a simplified visualization of this idea for a sequence of diffusion processes for single innovations appearing regularly over time, such as new model series that are produced.

Ideally, the diffusion of innovations should happen in a way that the capacities in manufacturing and logistics continue to have the same workload over time. Such conditions simplify the planning process and the operation of a company's facilities and reduce volatility in pricing. A company would accordingly use its influence on the diffusion of innovations through pricing, communication and distribution to ensure that the accumulated spread of subsequent innovations can be described by the simple and therefore easily manageable equation

$$\frac{df}{dt} = a \qquad (14.2)$$

where a is constant or at least increasing in rare steps or very slowly in comparison to f, if the production capacities are expanded. More likely than this expansion, however, is an increase in the prices for which innovations are sold, based on the

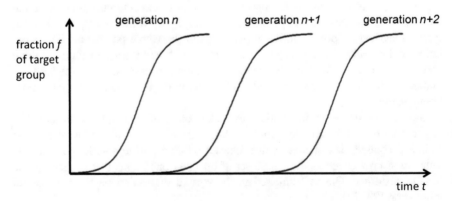

Fig. 14.5 Idealized sequence of innovations for optimal resource planning (see Fig. 14.1)

assumption that each new innovation will better meet the requirements of its target group and generate more value.

It seems reasonable to assume that innovation by value follows the pattern of novelty (at least inasmuch as the economic notion of value is concerned). In this sense, one could talk about a spread in terms of value which, excluding seasonality effects, follows the pattern of exponential growth, as many pricing schemes in innovation-driven industries illustrate. Monetary developments, however, lead back to the field of global phenomena in the whole population with all their complexity and require a wider investigation that goes beyond the boundaries of diffusion studies.

14.4 Diffusion and Co-creation

14.4.1 Platform Technology

So far, the adoption of innovations has been understood as a reactive process in which a new technology triggers certain behavior among actors according to their personal dispositions. This corresponds with the image of innovation as a rational problem-solving process (consisting of different stages with subsequent "control gates" to evaluate the success) in which the diffusion of innovations forms the last step. At this point, artefacts with a determinate function are already created and can now be introduced to the public (Fig. 14.6).

The shift towards individualized offerings can in many respects be interpreted as an expansion of the range that early process phases cover, because decisions about individual application patterns are already anticipated in the design phase and thus taken out of the hands of the adopters. With increasing data about usage behaviors, companies can further expand their reach into the personal lives of the adopters. At the same time, however, it can also be observed that the reach of the adopters also expands into the opposite direction with more opportunities to contribute to the design process.

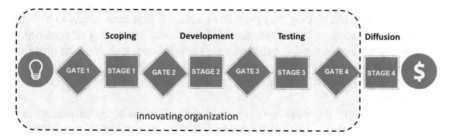

Fig. 14.6 Standardized "stage-gate"-innovation process with final diffusion [25]

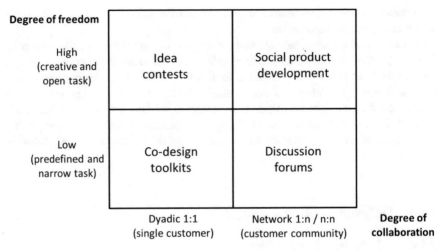

Fig. 14.7 Two-dimensional classification of contributions by participants in open innovation processes according to Piller and Ihl [26]

Figure 14.7 shows a classification of different contribution options that are offered during innovation processes. Popular examples include product configurators in which customers can choose from large lists of different options, contests in which participants can submit their own functional or aesthetic design, and numerous voting options and discussion groups on different aspects of innovation.

To allow such contributions from the user perspective, the technical architecture of the offerings in question must be modularized, so that different combinations and extensions become possible. In such cases, it seems doubtful whether the offerings that are brought to the market already constitute innovations or whether they just provide building blocks for further innovation activities which are executed by customers. In the latter sense, companies must be considered only to mediate innovation without accomplishing it themselves.

This is very prominently the case for many offerings in the field of information technology, such as smartphones or tablets, but also community platforms on the internet when they are stripped from further functions. They are widely celebrated as innovations, although they only provide operational platforms which, in order to generate value, have to be complemented through the installation of application software. When looking at the spread of such items in the population, one therefore has to ask to what extent this can be accounted for as a diffusion of innovation by itself and to what extent it rather has to be addressed as a spreading infrastructure for innovation.

Considering all this, there are apparently two different types of protagonists which nowadays have to be considered in the context of innovation: the engineer-innovator and the user-innovator. Both decide together about the meaning of an innovation in a communicative process (cf. [6]). While this process has previously been approached as a unidirectional transfer of matter and information, many researchers are nowadays interested in the bilateral exchange between engineer-innovators and user-innovators during the design and manufacturing processes. Since this exchange means that institutional boundaries are frequently crossed, it is customarily described as open innovation [27]. In an extreme form of open innovation, users might be able to propel the development of new technical solutions on their own, without any further involvement of companies [28].

Open innovation requires a fundamental change of perspective in the study of diffusion processes. Instead of assuming that there are predetermined points of origin from which innovations start to spread across the population, any kind of exchange between different actors in a societal network must now be expected to be a potential initiation point for innovations. While there is so far no systematic research agenda in this field, current research on open innovation can give a first impression of the directions that might be taken.

14.4.2 Innovation Incubators

While previous research on the diffusion of innovation allowed allusions to particle movement in various dimensions, open innovation rather seems to call for references in biology in order to give account of the continuous change of diffusion subjects and their carrier media, a general problem that is also addressed in Chap. 10 where in Sect. 10.6 guest molecules in porous materials are considered to undergo chemical reactions and where the last paragraph of Sect. 10.4 deals with guest-induced changes of the host material.

In that sense, a population and its practices of technology usage would be subjected to continuous change, caused by the replacement of its individual members over time with the possibility of mutation and recombination in every single case. Innovations could accordingly be assumed to originate and spread like successful genetic patterns, or, following a virological approach, like infectious diseases.

Similar to these references, it is interesting for research on diffusion from the perspective of open innovation to look for ways to anticipate, recognize and control "outbreaks" of innovation. Instead of fighting such "outbreaks", however, the ultimate goal of innovation research from a managerial perspective is obviously to provoke them and guide them into promising directions. In order to gain more transparency about the overall situation, innovation research requires scouting and scanning techniques which are able to identify occurrences of innovation. Such techniques have already been discussed for a long time in trend research. Currently, big data analysis adds further sophistication with advanced algorithms for pattern recognition.

Regarding the management of diffusion from the perspective of open innovation, research has been specifically attracted by the question how so-called incubators for open innovation can be set up and how they perform. Generally speaking, any kind of infrastructure that supports boundary-crossing interaction among innovators with

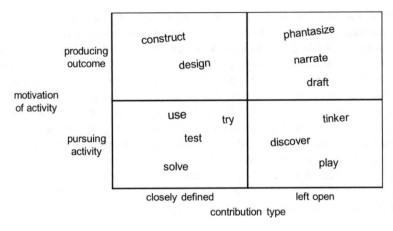

Fig. 14.8 Forms of collaboration in open innovation incubators, adapted from [29]

a positive effect on its outcome can be considered as an incubator for open innovation. This includes innovation communities and other platforms on the internet, but also physical spaces in which people come together. The types of interaction that are supported by the incubators can be quite diverse (Fig. 14.8). In particular, physical meeting spaces allow exchange in a large variety of ways. Among these spaces, two different kinds of incubators have lately been studied quite extensively: science parks and open innovation laboratories.

Science or technology parks are areas in which research institutions and companies with a strong focus on innovation are assembled to foster exchange and joint activities. Silicon Valley is usually considered as the archetype of such a park, although it exceeds most parks in size. In addition, most parks are intentionally planned by governmental organizations and strongly supported by different methods of financing: Science parks can thus be considered as public investments; research on science parks focusses on their performance on fostering innovation and the exchange between its inhabitants [30, 31]. Implicitly, science parks are assumed to be sources for innovation that can subsequently spread to other locations. However, this spread is expected to follow institutional structure. In that sense, innovation activities within the parks may be considered open, but the results are then redirected to conventional economic players.

Open innovation laboratories follow a different logic. They provide spaces for different people to come together for the purpose of problem solving and exploring novelty [32]. Such laboratories are usually established in central areas of larger cities or in the vicinity of universities or industrial districts where many people with higher education pass through. In order to use the equipment in the laboratories, visitor may have to pay a fee. Otherwise, there is no general entrance restriction. People join the activities in the laboratories whenever they want. Afterwards, they leave again and take their experiences with them to other places. Institutional actors can be involved in open innovation laboratories in different ways: as hosts of the facilities, organizer of events, or counterpart in the collaboration with other visitors [33]. Nevertheless, the interaction in the laboratories must be considered as a public exchange on innovation, since the visitors from the outside remain independent from internal company regulations and specifically designed contractual agreements.

While incubators in a biological or medical sense mainly serve the purpose of providing a hospitable environment for growth or reproduction, incubators for innovation can also influence the quality of the processes that are taking place by attracting certain people and providing special tools for innovation [34]. Some laboratories, for example Tech Shops, Fab Labs or Lego Stores, rely strongly on machinery to support the physical construction of new devices on site; others, like the Living Labs, the Fraunhofer Open Lab JOSEPHS or the Maker Faires focus their attention rather on the mode of social interaction [33]. Furthermore, the ratio between diverging, explorative activities and converging, exploitative activities is also different for each single laboratory concept. While some emphasize the construction of fully operative solutions, others give precedence to the clarification of goals and strategies in an open debate.

Each open laboratory accordingly defines its own constructive pattern of innovation. This does not only determine the possible outcomes; it also anticipates the path of the diffusion process, since it allows some persons or institutions to relate more easily to the innovations than others. First experiences during the last years suggest that explorative, discourse-oriented laboratories play an important role in social innovation that needs a high grade to public acceptance to spread; exploitative, engineering-oriented laboratories rather seem to serve as incubators for innovations that convince adopters by their technical function. These findings, however, have to be called preliminary. Until now, research has not had much time to study the impact of open innovation laboratories on broad range and there is still a lot to be learned in the future.

14.5 Conclusions

Diffusion models play an important role in innovation research—not only because of their descriptive capacity for the analysis of the processes that are taking place, but also because they provide the basis for the economic exploitation of the spread of novelty in society. Similar to different energetic potentials that initiate electric currents, an intentional design of the diffusion process of innovations to manage demand and supply can be used to gain revenue. Natural scientists create laboratory conditions to isolate certain effects from the environment; economic decision makers use the normative means of social organization to customize innovations with specific attributes for certain groups of people. This increases their control over the events that are taking place and enables them to focus their interferences on the effects that they intend to provoke. The design activities in the context of the diffusion of innovation therefore mostly take place on a detailed level, regarding individualized offerings for smaller groups of people, opposed to grand stories about technical progress that capture the long-term development of societies.

Grand stories on innovation rely on vague notions of objects: cars, planes, telephones—assuming that they remain the same during all the time that it takes for them to spread through society. With increasing detail, it becomes clear that technical devices and the conditions under which they are used change quite frequently; and in many respects, each of these changes can be referred to in terms of a small innovation, because it brings a new practice of using technology with it. For a long time, institutional centralization and formal standardization have made it possible to focus on the grand stories and neglect the details. Today, however, the situation has changed. Grand stories of innovation only continue to make sense where they refer to a platform technology: a solution that is by itself only an empty shell and requires further input to become meaningful in practice. As it turns out, this input is highly individual. Technology is customized with personal information which makes each single instance of a device different from all others. The study of diffusion processes for innovation consequentially become highly difficult.

Most companies have reacted to this difficulty by turning the focus towards the adoption of innovations among social groups with similar practices of technology usage and towards the design of solutions that are customized specifically for their needs and communicated accordingly. Comprehensive diffusion models for the whole population are traded in for a multitude of extremely simple diffusion models for many different artefacts and target groups, which avoid the effort necessary to address the complexity of general social dynamics. Mathematical models of diffusion as they are applied in industry remain accordingly comparably simple. At the same time, additional effort is created elsewhere. In order to cluster society in sufficiently homogeneous user groups that can then be addresses separately, it is necessary to collect more and more information about the users. Where companies are not able to do this, they integrate the users themselves in the design process in ways that allow them to organize themselves autonomously according to their interests. Users thus take over an active part in the creation of innovations, before it has reached maturity.

There is still a lot to learn about the consequences of opening up innovation procedures for user participation. Nevertheless, it seems clear that they will require a revision of the current diffusion models in innovation research. Inspiration can be drawn, for example, from biology. The image of a continuously changing population of individuals in which novelty can occur everywhere seems to provide a suitable background for research on open innovation, in particular where incubators are concerned that bring different people together under suitable conditions to foster innovation. Although first attempts into this direction have already been started, further conceptual and empirical work will be necessary to find out how much can actually be gained from it.

References

1. G. Tarde, *The Laws of Imitation* (Trans. E. C. Parsons) (Henry Holt, New York, 1903)
2. M. Weber, *Economy and Society* (University of California Press, Berkeley and Los Angeles, 1978)
3. C. Hubig, *Die Kunst des Möglichen I, Technikphilosophie als Reflexion der Medialität*, *transcript* (Bielefeld, 2006)
4. Z. Griliches, Hybrid Corn. Econometrica **25**, 501–522 (1957)
5. M. Gort, S. Klepper, Time paths in the diffusion of product innovations. Econ. J. **92**, 630–653 (1982)
6. E.M. Rogers, *Diffusion of Innovations*, 4th edn. (The Free Press, New York, 1995)
7. L. Andres, D. Cuberes, M. Diouf, T. Serebrisky, *Diffusion of the Internet, A Cross Country Analysis, Working Paper* (The World Bank, 2007)
8. M. Jalowski, A. Fritzsche, Ein Rahmenwerk zur Erfassung von IT-Sicherheit als Service-System, in *Proceedings of the MKWI, Ilmenau* (2016)
9. E. Mansfield, Econometrica **29**(4), 741–766 (1961)
10. P.-F. Verhulst, Notice sur la loi que la population poursuit dans son accroissement. Corresp. Math. Phys. **10**, 113–121 (1838)
11. N. Bacaer, *A Short History of Mathematical Population Dynamics* (Springer, London, 2011)
12. J.C. Fisher, R.H. Pry, Technol. Forecast. Soc. Chang. **2**, 75–88 (1971)
13. ITU, Mobile cellphone subscriptions, http://www.itu.int/en/ITU-D/Statistics/Pages/stat/default.aspx. Accessed 31 Jan 2016
14. S.C. Bhargava, Technol. Forecast. Soc. Chang. **49**, 27–33 (1995)
15. D. Comin, B. Hobijn, E. Rovito, J. Technol. Transf. **33**, 187–207 (2008)
16. F.D. Davis, *A Technology Acceptance Model for Empirically Testing New End-User Information Systems: Theory and Results* (Massachusetts Institute of Technology, Boston, 1986)
17. F.D. Davis, Perceived usefulness, perceived ease of use, and user acceptance of information technology. MIS Q. **13**(3), 319–340 (1989)
18. I. Ajzen, Organ. Behav. Hum. Decis. Process. **50**(2), 179–211 (1991)
19. V. Venkatesh, M.G. Morris, G.B. Davis, F.D. Davis, User acceptance of information technology: toward a unified view. MIS Q. **27**(3), 425–478 (2003)
20. S. Bratzel, Die junge Generation und das Automobil – Neue Kundenanforderungen an das Auto der Zukunft?, in *Automotive Management*, 2nd edn., ed. by B. Ebel, M.B. Hofer (Springer, Berlin, 2014), pp. 94–110
21. F. Dudenhöffer, Demographie und Innovation. ATZ - Automobiltechnische Zeitschrift **110** (1), 62–67 (2008)
22. H. Becker, *Auf Crashkurs, Automobilindustrie im globalen Verdrängungswettbewerb* (Springer, Berlin, 2005)
23. C. Zott, R.H. Amit, Lorenzo Massa, J. Manag. **37**(4), 1019–1042 (2011)
24. A. Osterwalder, Y. Pigneur, C.L. Tucci, Commun. AIS, **15**(5) (2005)
25. R.G. Cooper, Bus. Horiz. May–June 44–53 (1990)
26. F. Piller, C. Ihl, Co-Creation with customers, in *Leading Open Innovation*, ed. by A.S. Huff, K.M. Möslein, R. Reichwald (MIT Press, Cambridge, 2013), pp. 139–153
27. H.W. Chesbrough, *Open Innovation, The New Imperative for Creating and Profiting from Technology* (Harvard Business School Press, Boston, 2003)
28. E. von Hippel, *Democratizing Innovation* (MIT Press, Boston, 2005)
29. M. O'Hern, A. Rindfleisch, Rev. Market. Res. **6**, 94–106 (2008)
30. H. Löfsten, P. Lindelöf, Technovation **25**, 1025–1037 (2005)
31. C.-H. Yang, K. Motohashi, J.-R. Chen, Res. Policy **38**(1), 77–85 (2009)
32. A. Fritzsche, K.M. Möslein, Accelerating Scientific Research with Open Laboratories, *British Academy of Management Conference, Portsmouth* (2015)

33. A. Roth, A. Fritzsche, J. Jonas, K.M. Möslein, F. Danzinger, Interaktive Kunden als Herausforderung: Die Fallstudie, JOSEPHS® – Die Service-Manufaktur". HMD Praxis der Wirtschaftsinformatik **51**(6), 883–895 (2014)
34. K.M. Moeslein, A. Fritzsche, The evolution of strategic options, actors, tools and tensions in open innovation, in *Strategy and Communication for Innovation*, ed. by N. Pfeffermann, J. Gould (Springer, Heidelberg, 2017)

Chapter 15
The Spreading of Techno-visionary Futures

Armin Grunwald

15.1 Introduction

Visions are an established and perhaps, in some respect, necessary part of the scientific and technological communication. In general, they aim to create fascination and motivation among the public but also in science, increase public awareness on specific research fields, help motivating young people to choose science and technology as fields of education and career, and help gaining acceptance in the political system and in society for public funding. Visions are thus a major driver of the scientific and technological advance, as may be seen in the fields of spaceflight, nanotechnology, or synthetic biology. However, the spreading of scientific visions and their role in innovation processes is not well understood yet.

We have been witnessing a new wave of visionary and futuristic communication around science and technology in the last 15 years at the occasion of the so-called new and emerging sciences and technologies (NEST). Typical NEST areas are nanotechnology, converging technologies, synthetic biology, human enhancement, autonomous technologies, the different "omics" technologies, and climate engineering. The visions disseminated and debated in these fields show some specific characteristics which justifies speaking of them as *techno-visionary futures* [1]:

- techno-visionary futures refer to a more distant future, some decades ahead, and exhibit revolutionary aspects in terms of technology and in terms of culture, human behaviour, individual and social issues
- scientific and technological advances are regarded in a renewed techno-determinist fashion as by far the most important driving force in modern society (technology push perspective)

A. Grunwald (✉)
Institute for Technology Assessment and Systems Analysis (ITAS),
Karlsruhe Institute of Technology, Karlsruhe, Germany
e-mail: armin.grunwald@kit.edu

© Springer International Publishing AG 2018   295
A. Bunde et al. (eds.), *Diffusive Spreading in Nature, Technology and Society*,
https://doi.org/10.1007/978-3-319-67798-9_15

- the authors and promotors of techno-visionary futures are mostly scientists, science writers and science managers such as Eric Drexler and Ray Kurzweil; but also philosophers, non-governmental organizations and industry are developing and communicating techno-visionary futures
- high degrees of uncertainty are involved. As a rule, little if any knowledge is available about how the respective technology is likely to develop, about the products which such development may spawn and about the potential impact of using such products

The point of departure of this chapter is the observation that techno-visionary futures play a major role in the early stages of innovation (Sect. 15.2). In spite of only little if any knowledge about their feasibility, implications, and consequences being available in those early stages of research and development they might have a major impact on the course of the scientific research and technological development (Sect. 15.3). For example, they could heavily influence public debates and can possibly be crucial to public perception and attitudes by highlighting either chances or risks or by framing what is regarded chance and what is regarded risk. The possibly high impact of techno-visionary futures motivates the postulate to shed more light on the influential processes of their creation and spreading.

Techno-visionary futures are created and obviously have authors whose work initiates a process of spreading by communication and dissemination. While the spreading of these visions into scientific and societal debates and their impacts upon them can considerably influence the course of research and innovation, little is known about the processes and mechanisms of spreading of techno-visionary futures. We only can diagnose desiderates for future research and propose some ideas how this could be done. Answers to the question for the mechanisms of spreading and for the factors influencing or determining the impacts of techno-visionary futures are still not available (Sect. 15.4).

The analysis given in this Chapter makes use of some previous work of the author [2, 3] and builds on the highly reflected knowledge acquired in the previous decade about the role of technology futures and visions (e.g. [4–7]).

15.2 Techno-visionary Futures as Origins of Innovation Stories

In the past about 15 years, there has been a considerable increase in visionary communication on future technologies and their impacts on society. In particular, this has been and still is the case in the fields of nanotechnology [5, 8], human enhancement and the converging technologies [4, 9], synthetic biology and climate engineering. Visionary scientists and science managers have put forward far-ranging visions which have been disseminated by mass media and discussed in science and the humanities. These observations allow us to speak of an emergence of *techno-visionary sciences* in the past decades at the occasion of the occurrence of

NEST. NEST developments aim at providing *enabling technologies*. Their target is not to directly create products and innovations in specific areas of application. Rather they are open for a multitude of applications in greatly differing fields. It is this enabling character which opens up a huge space for techno-visionary ideas. This can be illustrated best at the occasion of nanotechnology.

Nanotechnology was early regarded an enabling technology [10]. Though there are some original nanotechnology products such as nanoparticles for medical applications, in most cases a nanotechnology component will be a small but decisive part of a much more complex product in a number of fields, such as energy technology, information and communication technology, or biotechnology. This mechanism that small devices from nanotechnology could lead to major, even disruptive innovation in certain application fields, makes predictions or valid scenarios of innovation pathways and product lines more or less impossible or restricts them to be speculative. This situation then opens up huge spaces for visionary innovation stories. A similar case is synthetic biology. In spite of the fact that it is predominantly laboratory research which raises fundamental questions far away from concrete application, there are great promises of some protagonists of synthetic biology to create artificial organisms, to produce biomass or novel materials. However, the feasibility and realization period of these techno-visionary futures are difficult if not impossible to assess. This is a general property of NEST: their "enabling" character is linked with a wealth of possible futures that are more or less speculative and very difficult to assess epistemologically.

As a thought experiment we imagine that in the year 2025 many nano-enabled products will be available at the marketplace. If a historian then would be interested in writing the history of nanotechnology he or she would go back to the origins of that debate and probably would be irritated. At the very beginning of nanotechnology there was a visionary statement of Nobel Laureate Feynman [11], followed by ideas of the futurist Drexler [12]. Spreading of these ideas led to an extensive global debate on a high diversity of techno-visionary futures related with nanotechnology. The prefix "nano" was frequently used as a synonym for good science and technology related with positive futures. The far-reaching expectations on nanotechnology were based on its potential to generate materials for completely new applications and to realize novel processes and systems as well as on the ability to target and fine-tune its properties by controlling its composition and structure down to molecular and atomic levels. Because of this, nanotechnology as an enabling technology was expected to trigger innovations in many areas of application and almost all branches of industry.

This situation changed completely in 2000. The initially positive visions of nanotechnology were transformed into horror scenarios based on precisely the same miniaturized technologies [13]—and these dystopian futures also spread quickly. The ambivalence of techno-visionary futures of nanotechnology became dramatically obvious [3, Chap. 5] and resulted in a scientific and public debate about its risks and chances. Both risks as well as chances were, at that stage of the debate in the early 2000s, related with visionary and more speculative expectations of fears around nanotechnology. It was Joy's [13] warnings about a post-human future world ruled by out-of-control nanotechnology which opened up this risk debate. Though it might look a bit crazy from today's perspective (perhaps even more from the perspective of the imagined historian) it spread quickly at the global level. Within months, people all over the world became familiar with concepts such as "grey goo", "nanobots", "cyborgs", and the dream of cybernetic immortality [3, Chap. 5].

Our imagined historian will probably find out that some years later such futuristic elements in public debate disappeared in favor of a more down to Earth risk debate [14]. Synthetic nanoparticles could spread, e.g., via emissions from production facilities or by the release of particles from everyday use of nano-based products, and they could end up in human bodies or in the environment and lead to adverse effects. This shift led to the debate on possible health and environmental risks of synthetic nanomaterials which is still ongoing [15]—and in contrast to the former debates on techno-visionary futures of nanotechnology this type of debate will probably maintain because the possibility of non-intended side-effects belongs to almost any technology [16].

This interesting history of nanotechnology appears typical for a so-called "hope, hype and fear" technology. Nanotechnology was believed to have the potential to solve global problems (hope), was associated with far-reaching visions of the future and with over-reaching expectations (hype), and because of its possible impacts that are difficult to foresee and even less to control they raise concerns no matter whether they are well founded or not (fear). At the beginning there was a powerful but speculative debate based on techno-visionary futures [3, Chap. 5] that were difficult to assess and led to a specific form of communication: high to extremely high expectations, on the one hand, but just as dramatic anxieties, on the other. After some years of intensive debates, however, the debate moved more and more down to Earth, and the techno-visionary futures as objects of debate were replaced by statements on toxicity. This story tells a lot about increasing understanding and thus of "appropriation" or "normalization" of nanotechnology [17]. It has become normality, like many other areas of technology, where we speak soberly about opportunities and risks without lapsing into the dramatics of salvation or apocalypse. This normalization is ultimately the result of the speculative debates in the early phase [3, Chap. 5.4] which gives some evidence to regard this more irritating debate part of the overall innovation process around nanotechnology. It started with the early visions by Feynman and Drexler mentioned above, passed the hope, hype & fear stadium and shifted then to a quite familiar type of debate on chances and risks concerning human health and the environment.

Our imagined historian of nanotechnology would uncover a lively and dynamic development at the early stage—and he or she obviously would ask for the mechanisms of this dynamics. It is a normalisation and appropriation story which includes the strong role of the early techno-visionary futures—and the fact that they lost relevance later on. The appropriation of nanotechnology, its transformation from a "fuzzy" and irritating field of development to a more or less normal one, would not be happened without those futures which might seem ridiculous from a today's perspective. The quick spreading of techno-visionary futures of nanotechnology among the relevant communities worldwide would probably be interesting to him or her, but also the disappearance of those visionary futures after some years of heavy debate. The lesson to be learned from this case is that techno-visionary futures shall be taken seriously even if they seem to be merely speculative.

15.3 The Power of Visionary Ideas

There are also systematic reasons for taking techno-visionary futures seriously. However, at a first glance this might be questioned. NEST developments are in early stages of development and still strongly rooted in basic research. It is not obvious that it makes sense at all to discuss the related techno-visionary futures with respect to stories of innovation. Should we not instead let scientists do their basic research until more consequential knowledge about innovation paths and possible product lines will be available? Aren't the positive and negative visions linked to them (see the example of nanotechnology introduced above) anything more than arbitrary and simple speculation? One could argue that many NEST debates are so speculative that they are hardly of any practical consequence. This exactly was the main criticism against the so-called speculative nanoethics [18]. It might accordingly be perhaps interesting in an only abstract philosophical and merely academic sense to discuss some obviously speculative questions related with techno-visionary futures (see Sect. 15.2 above for the field of nanotechnology). There might be some interest in circles of intellectuals or in the feuilletons of magazines. Yet, in view of the speculative nature of those questions, serious concern was expressed that the intellectual effort and the resources spent might be completely irrelevant in a practical sense. However, this argumentation has been proven misleading [19].

While techno-visionary futures ranging from high expectations to apocalyptic fears indeed are often more or less fictitious in content, such stories about possible futures can and often do have a real impact on scientific and public discussions [5]. Even pictures of the future lacking all facticity, being merely speculative, and probably never becoming reality, can influence debates, the formation of opinion, acceptance, and even decision making [4] with consequences in the real world in two ways at least [1]:

- Techno-visionary stories and images can change the way we perceive current and future developments of science and technology, just as they can change our picture of future societal constellations. Frequently, the societal and public debate about the opportunities and risks associated with new technology revolves around those stories, as has been the case in the field of nanotechnology and as is still the case in human enhancement and other NEST areas [3]. Visions and expectations motivate and fuel public debate because of the impact the related narratives may hold for everyday life and for the future of important areas of society, such as military, work, and health care—and this figure works independent of how realistic or speculative the futures under discussion are. Positive visions can contribute to fascination and public acceptance and also can attract creative young scientists to engage themselves there, just as negative visions and dystopias can cause concern and even mobilize resistance as was feared in particular in the early debate on nanotechnology [17].

• Techno-futures exert a particularly great influence on the scientific agenda which, as a consequence, partly determines what knowledge will be available and applicable in the future [20]. Directly or indirectly, they influence the views of researchers and, thus, ultimately also exert influence on political support and research funding. For example, even the speculative stories about improving human performance [9] quickly aroused great interest among policy makers and research funders [21]. Projections of future developments based on NEST expectations therefore might heavily influence decisions about the support and prioritization of scientific progress and the allocation of research funds, which then will have a real impact on further developments. The history of spaceflight is an impressive example for the power of visionary ideas from its origins in the 1920s on.

The common rationale behind both arguments is that the communication and dissemination of techno-visionary futures are interventions into ongoing communication, action, and decision-making. The spreading of those futures changes mindsets, convictions, beliefs and perceptions—and thus often has real impacts. The communication involving more or less speculative and visionary futures can exert real power.

Some practical experience gained in recent years support the diagnosis that policy makers are well aware of the factual power of techno-visionary communication. As an early example: A chapter about techno-visionary communication on human enhancement, converging technologies (nano-bio-info-cogno convergence; [9]), and other far-reaching visions compiled by the Office of Technology Assessment at the German Bundestag (TAB) was very well received by the Bundestag as part of a study on nanotechnology [22]. The authors came to the conclusion that this techno-visionary discourse played an important and to some extent new role in the governance of science and technology. Several policy makers and also experts in nanoscience and nanotechnologies communicated to the TAB team that they found the study's discussion of futuristic visions and description of the networks promoting them very useful. The TAB team's initial concerns that discussing these often far-fetched visions in a study which would become an official document of the parliament and an influential early publication on nanotechnology could cause irritations thus proved to be unfounded. Subsequently, TAB was requested to conduct several other projects on NEST fields: studies on the politics of converging technologies at the international level, on brain research, on pharmacological and technical interventions for improving performance, and on synthetic biology. Recently, the ceremony of the 25th anniversary of the foundation of TAB in 1990 was—upon request of members of parliament—dedicated to the issue of blurring the borderlines between humans and technology, e.g., by developments towards human enhancement and autonomous robots which also has a clear techno-visionary side.

This interest of policy makers in techno-visionary futures is also evident at the European level. NEST developments have been addressed by a fairly large number of projects as well as by other advisory activities such as the reflections on nanotechnology, synthetic biology, and ICT (information and communications technology) implants conducted by the European Group on Ethics in Science and New Technologies. The situation is much the same in the US which can be seen, for example, by considering the agenda of the Presidential Commission for the Study of Bioethical Issues which indicates indeed the factual power of techno-visionary narratives of possible futures and the high significance assigned to them by policy-makers. The factual significance and power of techno-visionary futures for the governance of science and in public debates are a strong argument in favor of the necessity of researching these futures and providing public and policy advice. Policy makers and society should know more about these positive or negative visions, their genesis and their spreading. The postulate to open up the "black box" of the creation and spreading of those futures is supported by calls for a more democratic governance of science and technology [23]. Its realization requires uncovering meanings, values, and interests hidden in the techno-visionary futures and enlightening the mechanisms of their spreading.

15.4 Creation and Spreading of Techno-visionary Futures

Techno-visionary futures are a phenomenon of communication in different areas such as science, science fiction, philosophy, literature, the arts, movies and public debate. My focus in this Chapter is on the lifetime of those futures: how do they come into being, what do they tell, how do they spread and which impacts do they have. Unfortunately, because of lack of knowledge about the underlying mechanisms the chapter remains at the stage of rising questions, offering patterns and identifying research directions to learn more about these issues.

15.4.1 The Social Construction of Futures

Obviously techno-visionary futures are socially constructed (following [1]). Their authors can be individual persons, such as the authors of science fiction novels, or collectives such as research institutes or participatory foresight processes. They always pursue specific purposes, for example, entertainment, supporting political decisions, sensitizing the public for problematic developments, mobilizing support for research, creating a vision for regional development, warning at an early stage about potential problems, creating fascination in the public etc. Constructing futures serves as a means in order to reach these goals. Creation and dissemination of techno-visionary futures are interventions into the real world and may have some impact—the intended ones but possibly also others.

Establishing statements about the future such as making predictions, simulating future developments, creating scenarios, formulating expectations and fears, setting goals and considering plans for their realization takes place in the medium of language. We need language for constructing futures which obviously always happens in a respective present time. Forecasters and visionary writers cannot break out of the present either, always making their predictions on the basis of present knowledge and present assessments involving also a lot of present assumptions. Societal futures can be neither logically deduced nor empirically investigated [1] but must rather be based on present knowledge, assessments, values, assumptions etc. Therefore, we must talk about possible futures in the plural, about alternative possibilities for imagining the future, and about the justification with which we can expect something to become real in the future. These are always present futures and not future presents [24]. If we talk, for instance, about cyborgs or far-reaching human enhancement which might become possible in the future, we are not talking about whether and how these developments will really occur but how we imagine them *today*—and these images differ greatly. Futures are thus something always contemporary and change with the changes in each present. In particular, techno-visionary futures are similar to living organisms having an origin but changing their shape during their lifetime and their spreading (see below).

Futures, regardless of whether they are forecasts, scenarios, plans, programs, visions, or speculative fears or expectations, are designed by authors following specific purposes, having certain interests and values in mind, based on specific diagnoses and pieces of knowledge. By creating techno-visionary futures their authors use a broad range of ingredients such as available knowledge, value judgments, suppositions and assumptions, some of them being explicit while others remain implicit. They also may include mere speculation and counterfactual assumptions (e.g. in the field of Science fiction literature and movies) but also visionary and utopian elements which do not contradict current knowledge (e.g. about natural laws) but are highly speculative and may serve as orientation, either to act towards their realization in case of positive futures or in order to prevent their occurrence in case of undesired, negative or dystopian views.

15.4.2 Providing Orientation by Assessing Futures

Scientific and technological progress leads to an increase of the options for human action. Whatever had been inaccessible to human intervention, whatever had to be accepted as not influenceable nature or as fate, becomes an object of technical manipulation or design. Emancipation from nature, from the traditions of the past, and from fate shows, however, another side of the coin: uncertainty, loss of orientation, and the necessity to be able to cope with the new freedoms by conscious

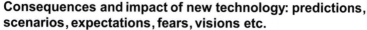

Consequences and impact of new technology: predictions, scenarios, expectations, fears, visions etc.

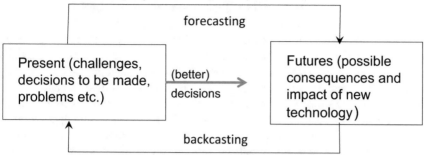

Orientations, modified problem perception, ideas about desirable and undesirable future societies, consequences for decisions to be made today, action strategies, measures, etc.

Fig. 15.1 The consequentialist mode of extracting orientation by processing a loop of forecasting and backcasting (*Source* [3], modified)

decisions. In this situation, which is characteristic for Modernity, the orientation needed for opinion formation and decision-making is drawn increasingly from debates about future developments, and less and less from existing traditions and values. The discourses on sustainable development, on the precautionary principle, on migration and the demographic change but also on NEST give evidence of this fact.

Within the familiar consequentialist approach pictures of the future (e.g. scenarios) are established in a foresight mode and assessed, e.g. with respect to their desirability or acceptability. Then conclusions are drawn in a back-casting mode for today's decision-making options taking into account the results of the assessment of the futures (see Fig. 15.1) which can be used to provide orientation, decision-support, and policy advice. Processing this loop shall add value to its input data. This expectation, however, is not realizable in case of completely diverging, arbitrary, speculative or heavily contested futures—that exactly is our case of the NEST debates involving the techno-visionary futures mentioned. But what can be done if there are no well-argued corridors of the envisaged future development or if proposed corridors are heavily contested as is in the case of NEST?

The hermeneutic approach developed for this constellation [1, 2], offers a completely different mechanism of providing orientation compared to what we normally expect from futures studies. In this approach the origins of the various futures must be considered. Visions of the future are social constructs, created and designed by people, groups, and organizations (see above). The variety or even divergence of visions of the future results from the consideration of controversial and divergent knowledge bases and disputed values during their creation: the divergence of futures mirrors the differences of contemporary positions, the diversity of attitudes, and reflects today's moral pluralism. Thus, uncovering these sources of diverging futures could tell us something about ourselves and today's society. Hermeneutic orientation implies a shift in perspective: instead of considering far futures and trying to extract orientation out of them, these stories of the future now are regarded as "social texts" of the present including potentially important content for today's debates. Thus, better understanding techno-visionary futures with respect to their content, diagnoses behind, values involved and ways of dissemination and spreading would be part of a self-enlightenment of contemporary debates. Instead of a senseless attempt to predict the future there is an opportunity to view the lively and controversial debates about NEST and other fields of science or technology not as anticipatory, prophetic, or quasi-prognostic talks of the future but as expressions of our present time. Therefore, there is a need to understand also the spreading of techno-visionary futures and their mechanisms.

15.4.3   The Spreading of Visions

In order to learn more about the spreading of visions we have to observe a specific property of this type of spreading. In contrast to the spreading of inert chemicals or electro-magnetic waves in vacuum the techno-visionary futures are in close exchange with their environment in two directions:

- The techno-visionary futures will have some impact on the real world as has been described in Sect. 15.3 above because they are interventions into ongoing communication and action. As interventions they might or could change the course of communication or mindsets of people. Interventions in diffusion from a corporate business perspective are also discussed in Chap. 14.
- Vice versa, the exchange of ideas on techno-visionary futures in related debates and controversies will usually not leave the content and meaning of those futures untouched. In contrary, this meaning will often be modified, it might take up new aspects or might be subject to changed accentuations. A dramatic example is the turnaround of the mostly positive ideas related with the so called "Molecular Assembler" proposed by Drexler [12] into a dystopian view on the future which might "no longer need us" [13].

Consequently, we see a co-evolution of the techno-visionary futures and their communicative and decision-making environment. It is a task of a continuous tracing and asking for better understanding to uncover the mechanisms of these co-evolutionary processes during the spreading of the visions. Techno-visionary futures do not just "travel" during spreading but will themselves be transformed. The hermeneutic circle (Fig. 15.2) provides a simple model of this type of spreading which affects and modifies the entities spreading.

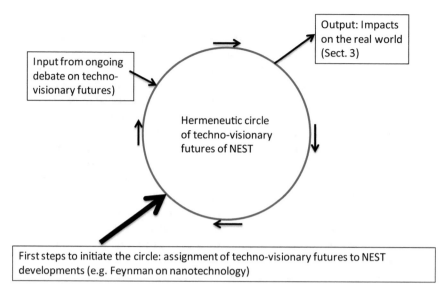

Fig. 15.2 The co-evolution of techno-visionary futures in NEST in a hermeneutic circle, including its stimulus (*Source* [3], modified)

First ideas of relating visionary futures with ongoing research and development constitute the initial stimulus which starts processing this circle [3]. Processing this circle constitutes of a series of communicative events by which the techno-visionary futures proposed are communicated, controversially discussed, supplemented or modified. Places like the innovation incubators described in Chap. 14.4.2 can play a role in this. The debate on nanotechnology [3, Chap. 5] is an excellent example for illustrating such a hermeneutic circle and its development over 10–15 years. For nanotechnology, Richard Feynman's famous lecture [12] or the book *Engines of Creation* [13] might have been such first steps or at least very early steps in starting the respective hermeneutic circle.

Figure 15.2 illustrates the high influence of the initializing stimulus. At the very beginning of the NEST debates the first facts are created for further communication and guidance. The meaning and framing given in the stimulus can decisively mold the ensuing debate (see Sect. 15.3) while the first framing can only be gradually modified by proposing alternative techno-visionary futures in subsequently processing the hermeneutic circle. This process can be self-reinforcing and lead, for example, to research funding being initiated, to massive investments being made in the affected field, and in this way to important real consequences for the agenda and the research process of science. Or the framework that was initially chosen might be challenged, strongly modified, or even changed into its opposite, leading to social resistance and rejection. Self-fulfilling as well as self-destroying narratives [25] about techno-visionary futures are the extremes of possible developments which could emerge out of the hermeneutic circle.

In processing the hermeneutic circle, techno-visionary futures are communicated via different channels, journals, networks, mass media, research applications, expert groups, ELSI (ethical, legal, and social implications) or TA (technology assessment, see [16]) projects on policy advice, etc. Some of them, finding no resonance, will "die" within these communication processes and quickly disappear again, while others will "survive" and motivate actors and groups to subscribe to or oppose the visions—in either case the story will continue. Only a few of the techno-visionary futures proposed will find an audience via the mass media and will therefore be able to achieve real impact for public debate and social perception or attitudes. Others may enter the political arena and result in political decisions, e.g. about research funding, and may disappear after having had big impact only. The history of spaceflight, for instance, is full of techno-visionary promises which regularly fail but nevertheless survive and attract further interest. The narratives of human settlements on the Mars or on artificial space stations belong to those persistent stories having impact without being realized or without even having a serious chance of realization.

Thus, it is evident that there are extremely different experiences with techno-visionary futures, with their spreading and impact. The vision of the molecular assembler [13] was among the motivating voices for the national nanotechnology initiative "Shaping the World Atom by Atom" [26] which was the first big funding program on nanotechnology. The narratives around climate engineering [3, Chap. 8], including some proposals being breath-taking with respect to the magnitude of human intervention into the global atmospheric system, did not reach a larger audience yet—neither in form of funding for research and development nor in public debates. Another interesting case is the revival of specific understandings after some time. In the 1970s there was a lively debate on artificial intelligence with high expectations and far-ranging techno-visionary futures following the establishment of computer sciences and cybernetics. These futures disappeared in subsequent decades but have been re-entering public and scientific debates over the last years. The normalization of today's or tomorrow's robots obviously has been prepared for by earlier debates on artificial intelligence and robots—and also by science fiction movies and literature which early took up ideas from that field. Stanley Kubrick's movie "2001: A Space Odyssey" (1968) thematizing the issue of power distribution between man and an intelligent machine is among the famous early examples which are still up-to date.

Thus we see different dynamics in different NEST fields with different techno-visionary futures influencing social debates and political decision-making. My conviction is that it would be worthwhile to better understand these dynamics including the biographies of the techno-visionary futures including the mechanisms of their spreading for ongoing and coming debates on NEST. Understanding must go beyond a mere description what happened but rather uncover the underlying mechanisms and dependencies.

The dynamic biographies of techno-visionary futures can be analyzed taking recent NEST developments as cases of study. This research could contribute to a deepened understanding of the dynamics of spreading dealing with issues of meaning of NEST but also of the creation and emergence of those future narratives. These could also be analyzed in an extended manner by examining their cultural and historical roots and philosophical backgrounds. Thus we can regard constructions of NEST techno-visionary futures as part of an ongoing communication process in science and at the interface of science and society in which specific assignments of meaning, e.g., the nanobots [12] or the chip in the brain, act as the necessary catalysts with their own individual biography or life cycle showing certain dynamics over time (see Fig. 15.2).

Biographies of techno-visionary futures as well as their dynamics are not well understood as yet [5]. The entire "life cycle" of techno-visionary futures, from their construction to dissemination, assessment, deliberation, and impact, thus raises a huge variety of research questions which can only be answered by giving interdisciplinary consideration to these aspects of spreading and impact. After some time usually some of the futures debated are sorted out, others might merge while only few "winners" remain and constitute a dominant understanding of the NEST under consideration. Again and again those developments happen in completely different fields such as nanotechnology, synthetic biology, care robots, or cyber-physical systems. A comparative analysis of the mechanisms of spreading and the conditions of "surviving" and having impact in the real world would probably shed some light on these processes and their dynamics. Investigating the emergence and dissemination of techno-visionary futures via different communication channels and its possible impact on decision-making in the policy arena and other arenas of public communication and debate involves empirical research and reconstructive understanding as well.

To answer these questions an interdisciplinary procedure employing various types of methods appears necessary. The empirical social sciences can contribute to clarifying the communication of techno-visionary futures by using media analyses or sociological discourse analysis and generate, for example, maps or models of the respective constellations of actors being involved in processing the hermeneutic circle (Fig. 15.2). Political science, especially the study of governance, can analyze the way in which techno-visionary futures can exert influence on political decision-making processes, such as via providing policy advice. In this way, a complete picture of the biography of the different techno-visionary future proposed including the mechanisms of their spreading can be created. It should include, for example, diffusion processes into different spheres of society, migrations of the techno-futures, related shifts in meaning and perception, consequences on, for example, social perception and political decision-making processes, and, if applicable, processes of the disappearance of the respective techno-future from the debate.

In view of the experience of the last 15 years it can be expected that in particular comparative research approaches to mechanisms of spreading hold the promise of new knowledge. These can, for example, compare the stories about the spreading of techno-visionary futures in specific NEST fields with one another, determine the common features and the differences, and ask about the causes and underlying mechanisms. For example, there seems to be an evident structural difference between the histories of nanotechnology [3, Chap. 5] and robotics [3, Chap. 6]. While nanotechnology initially appeared to be the disruptive technology par excellence causing many irritations in the initial stage of debate (Sect. 15.2) which had to be normalized through high effort, robots were normalized practically in advance by science fiction literature and films. Robots entered society in this way even before they came to exist in reality with the anticipated functions and meanings.

15.5 Conclusions

Techno-visionary futures have an important place in early stages of development, in particular in NEST fields. They can have major impact on the course of research and development, on public perception, research funding, and political decision-making. In spite of this high significance in the innovation process from early stages to later innovation paths, products, and services, only little is known about the ways and mechanisms of their spreading, about promoting factors and obstacles with regard to creating impacts in the real world. Thus, it seems desirable if not necessary to spend effort on shedding light on these processes of spreading and to conduct research aiming at enlightening these processes.

References

1. A. Grunwald, Technol. Innov. Stud. **9**(2), 21–23 (2013)
2. A. Grunwald, J. Responsib. Innov. **1**(3), 274–291 (2014)
3. A. Grunwald, *The Hermeneutic Side of Responsible Research and Innovation* (Wiley-ISE, 2017) (in press)
4. A. Grunwald, Futures **39**, 380–392 (2007)
5. C. Selin, Sociol. Compass **2**, 1878–1895 (2008)
6. C. Coenen, E. Simakova, Sci. Technol. Innov. Stud. **9**(2), 3–20 (2013)
7. A. Nordmann, J. Responsib. Innov. **1**(1), 87–98 (2014)
8. U. Fiedeler, C. Coenen, S.R. Davies, A. Ferrari (eds.), *Understanding Nanotechnology: Philosophy, Policy and Publics* (Akademische Verlagsgesellschaft, Heidelberg, 2010)
9. M.C. Roco, W.S. Bainbridge (eds.), *Converging Technologies for Improving Human Performance* (National Science Foundation, Arlington (VA), 2002)
10. T. Fleischer, Technikfolgenabschätzungen zur Nanotechnologie - Inhaltliche und konzeptionelle Überlegungen. Technikfolgenabschätzung - Theorie und Praxis **11**(3/4), 111–122 (2002)
11. R.P. Feynman, *Lecture Held at the Annual Meeting of the American Physical Society* (California Institute of Technology, 1959). http://www.zyvex.com/nanotech/feynman.html. Accessed 12 Dec 1959
12. K.E. Drexler, *Engines of Creation—The Coming Era of Nanotechnology* (Oxford University Press, Oxford, 1986)
13. B. Joy, Why the future does not need US. Wired Mag. **8**(04), 238–263 (2000)

14. A. Lösch, Visual dynamics: the defuturization of the popular 'nano-discourse' as an effect of increasing economization, in *Governing Future Technologies. Nanotechnology and the Rise of an Assessment Regime*, ed. M. Kaiser, M. Kurath, S. Maasen, C. Rehman-Sutter (Springer, Dordrecht, 2010)
15. J. Jahnel, Nanoethics **9**(3), 261–276 (2015)
16. A. Grunwald, *Technikfolgenabschätzung - eine Einführung* (Berlin, 2010)
17. A. Grunwald, Ten years of research on nanotechnology and society—outcomes and achievements, in *Quantum Engagements: Social Reflections of Nanoscience and Emerging Technologies*, ed. T.B. Zülsdorf, C. Coenen, A. Ferrari, U. Fiedeler, C. Milburn, M. Wienroth (AKA GmbH, Heidelberg, 2011), pp. 41–58
18. A. Nordmann, NanoEthics **1**(1), 31–46 (2007)
19. A. Grunwald, NanoEthics **4**(2), 91–101 (2010)
20. J.-P. Dupuy, Complexity and uncertainty: a prudential approach to nanotechnology, in *Nanoethics. The Ethical and Social Implications of Nanotechnology*, ed. by F. Allhoff, P. Lin, J. Moor, J. Weckert (Wiley, Hoboken (NJ), 2007), pp. 119–132
21. C. Coenen, M. Schuijff, M. Smits, P. Klaassen, L. Hennen, M. Rader, G. Wolbring, *Human Enhancement* (European Parliament, Brussels, 2009), http://www.itas.fzk.de/deu/lit/2009/coua09a.pdf
22. H. Paschen, C. Coenen, T. Fleischer, R. Grünwald, D. Oertel, C. Revermann, *Nanotechnologie: Forschung und Anwendungen* (Springer, Berlin, 2004)
23. K. Siune, E. Markus, M. Calloni, U. Felt, A. Gorski, A. Grunwald, A. Rip, V. de Semir, S. Wyatt, *Challenging Futures of Science in Society: Report of the MASIS Expert Group* (European Commission, Brussels, 2009)
24. N. Luhmann, Die Zukunft kann nicht beginnen: Temporalstrukturen der modernen Gesellschaft, in *Vor der Jahrtausendwende: Berichte zur Lage der Zukunft*, ed. by P. Sloterdijk (Suhrkamp, Frankfurt a. M., 1990)
25. R. Merton, The self-fulfilling prophecy. Antioch Rev. **8**(2), 193–210 (1948)
26. *National Nanotechnology Initiative*. (Washington, D.C., 1999)

Part IV
Society

Chapter 16
The Neolithic Transition: Diffusion of People or Diffusion of Culture?

Joaquim Fort

16.1 Introduction

The Neolithic transition is defined as the shift from hunting-gathering (Mesolithic) into farming and stockbreeding (Neolithic). The Neolithic arrived at about 8000 years Before Present (yr BP) from the Near East into Southeastern Europe. From there, it spread gradually westwards and northwards across Europe, until about 5000 yr BP. We know this from the radiocarbon dates of remains related to farming and stockbreeding that have been found in archeological sites. Europe is the continent for which more Neolithic sites per unit area have been dated so far than anywhere else in the world, and this is the reason why most models of the Neolithic transition have been applied to Europe. This spread of farming can be seen in Fig. 16.1, which is a recent interpolation of 918 early Neolithic sites [1].

Figure 16.1 shows at once that we are dealing with a gradual spread. Of course, there are some anomalously old/young regions (e.g., the patch inside the black rectangle in Fig. 16.1). Different interpolation methods yield some differences for small anomalous regions, but those of the size of that inside the rectangle in Fig. 16.1 and larger usually appear independently of the interpolation method used. Lemmen and Gronenborn (Chap. 17, this volume) rightly point out that some of such anomalous regions may be due to radiocarbon dating errors and/or other problems with the databases supplied by archeologists. This is certainly possible. However, in spite of the fact that most of such anomalously old/young regions are rather small and contain only a few sites, Lemmen and Gronenborn (Chap. 17, this

J. Fort (✉)
Complex Systems Laboratory and Physics Department, Universitat de Girona,
C/Ma. Aurèlia Capmany 61, 17071 Girona, Catalonia, Spain
e-mail: joaquim.fort@udg.edu

J. Fort
Catalan Institution for Research and Advanced Studies (ICREA),
Passeig Lluís Companys 23, 08010 Barcelona, Catalonia, Spain

© Springer International Publishing AG 2018
A. Bunde et al. (eds.), *Diffusive Spreading in Nature, Technology and Society*,
https://doi.org/10.1007/978-3-319-67798-9_16

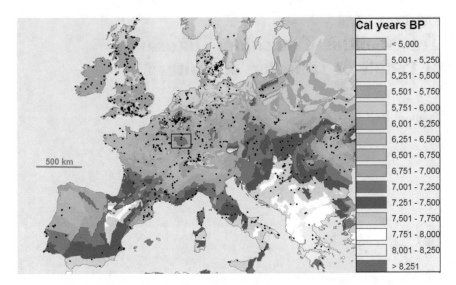

Fig. 16.1 Interpolation of 918 early Neolithic sites (circles). Each color corresponds to a 250-year interval. We see that the oldest sites are located in the southeast. Note also that farming propagated faster westwards than northwards. Moreover, slowdowns in the Alps and Northern continental Europe are clearly displayed. The patch inside the black rectangle is an example of an anomalously old region, as compared to its surroundings. Due to the paucity of sites, the contours are less detailed in some regions (e.g., upper right and lower left). This map was obtained by means of universal linear kriging interpolation. Adapted from Ref. [1]

volume) regard their presence as a 'pitfall' of the quantitative map in Fig. 16.1. They offer, instead, the qualitative map in their Fig. 17.1, which has no anomalous regions simply because it is just a drawing, not the result of any statistical interpolation. It is also important to note that, in order to totally avoid the presence of such anomalous regions in interpolation maps, several strong conditions would have to hold. A crucial one is that local geography should not have any effect, in the sense that all land should be equally attractive to farmers. Otherwise, it is reasonable to expect that farmers will sometimes move to more distant land even if there is nearer, less attractive land (which will not be settled by farmers until later on). The presence of rivers, mountains, different types of soils, etc., probably makes some areas more attractive for farmers than others. For this reason, the presence of anomalously old or young regions (such as the patch inside the black rectangle in Fig. 16.1) is probably unavoidable (even if we had a database totally free of errors and with all dates corresponding exactly to the earliest farming activity at each site). A very clear example are the Alps. These mountains cause an anomalously young region (as compared to its surroundings) in Fig. 16.1. It is observed independently not only of the interpolation method but also of the database used. Thus anomalous regions are not necessarily artifacts arising from limitations of the database and/or the interpolation technique. Nevertheless, some of them may certainly be artifacts, especially if they contain few or no sites and their presence depends on the interpolation method and/or database used. In practice, some such regions always seem to appear (i.e., whatever the database and the interpolation method that we use). But this should not be a problem after all. Smoothing techniques are well established in geographic analysis. They yield, with increasing coarse graining, maps with decreasing subtleness, where sufficiently small anomalous regions gradually disappear without substantially modifying the overall spread pattern (Fig. 16.2). This seems one reasonable solution to estimate local speed magnitudes (in kilometers per year) on a map based on quantitative, geostatistical methods [1], as we explain below. This is obviously impossible from drawings such as Fig. 17.1. Similarly to other researchers [2–4], the authors of Chap. 17 have compared the average speed implied by linear regressions according to the observed dates and to their mathematical model [5] (that model is briefly discussed at the end of this chapter).

Qualitative maps (such as that in Fig. 17.1) are certainly useful and interesting if we want to display the main features of the Neolithic spread. However, if we also want compare quantitatively the archaeological dates to the predictions of mathematical models we need quantitative maps, which are usually based on interpolation techniques (see, e.g., Fig. 16.1). An advantage of quantitative maps is that they are statistically justified. Another crucial advantage is that they make quantitative comparison to mathematical models possible [1]. This allows us to discuss a very important point, namely the implications concerning the mechanisms driving the spread of the Neolithic (diffusion of people versus diffusion of culture). Again, this seems impossible from drawings such as Fig. 17.1.

The spread of the Neolithic in Europe was clearly gradual, because as we move westwards and northwards, we find more and more recent dates (Figs. 16.1 and 17.1). This suggests that it may make sense to apply diffusive models to the spread of the Neolithic. A quantitative justification is the following. We know from Chap. 2 that diffusion equations provide large-scale descriptions of systems where there are, at the small scale, molecules or individuals following random walks (see Fig. 2.5). Does this scenario apply to the spread of the Neolithic? For the moment, assume a very simple model in which agriculture would have spread only due to the dispersal of farmers. Then each random walk is the trajectory obtained by joining, e.g., the birthplaces of an individual's parent, the individual in question, one of his/her children, and so on. Looking at Fig. 16.1, we can easily estimate that agriculture spread from Greece to the Balkans and Central Europe at a speed of roughly 1 km/year. Thus, assuming a generation time of about 32 year [6], farming spread about 32 km per generation. This is much less than the scale of Fig. 16.1 (3000 km or more). This comparison provides a quantitative justification for the use of diffusion equations in models of the Neolithic spread.

Ammerman and Cavalli-Sforza [7] were the first to apply a diffusive model to the spread of the Neolithic. They used Fisher's wave-of-advance model. In this model, the speed of the Neolithic front is given by Eq. (2.17),

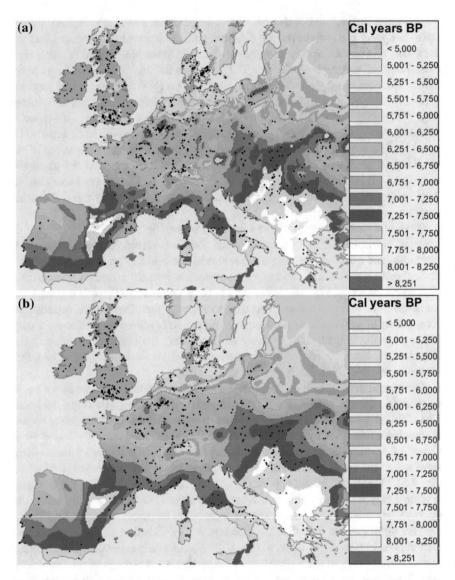

Fig. 16.2 Isochrones obtained by smoothing ("coarse graining") the map in Fig. 16.1 a single time (**a**), 10 times (**b**) and 20 times (**c**) (i.e. with 1, 10 and 20 iteration steps, where each step consists in replacing the date of each individual point of the map by the average of that date and those of the 8 surrounding points in a square grid). Note that anomalous regions (such as that inside the black rectangle in Fig. 16.1) gradually disappear. This is useful to perform quantitative estimates of local speed vectors and magnitudes (see Fig. 16.4b for the latter). Adapted from Ref. [1], Supp. Info. Appendix, Sect. S1

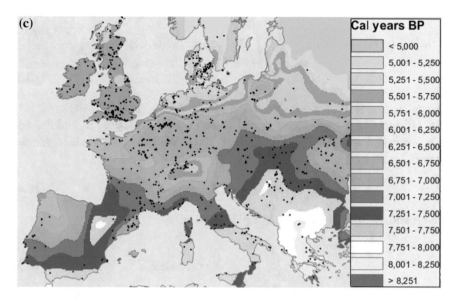

Fig. 16.2 (continued)

$$v_{Fisher} = 2\sqrt{D\alpha}, \tag{16.1}$$

where D is the diffusion coefficient and α the initial growth rate (i.e., the net reproduction rate at low population densities). This relation has already been introduced as Eq. (2.17) in Chap. 2. Following Ref. [2] we sketch, for the interested reader, the line of reasoning leading, eventually, to this relation.

Let $N(x, y, t)$ stand for the population density of Neolithic individuals (i.e., farmers), where x and y are Cartesian coordinates and t is the time. We assume that a well-defined time scale T between two successive migrations occurs. This model (to be improved in Sect. 16.3) is based on the assumption (see Ref. [8], Sect. 11.2) that, between the values t and $t + T$, we can add up the changes in the number of individuals in an area differential $ds = dx\, dy$ due to migrations (sub index m) and to population growth (sub index g),

$$[N(x, y, t+T) - N(x, y, t)]ds = [N(x, y, t+T) - N(x, y, t)]_m ds \\ + [N(x, y, t+T) - N(x, y, t)]_g ds. \tag{16.2}$$

Let Δ_x and Δ_y stand for the coordinate variations of a given individual during T. We introduce the dispersal kernel $\phi_N(\Delta_x, \Delta_y)$, defined such that $\phi_N(\Delta_x, \Delta_y)$ is the probability per unit area to move from $(x + \Delta_x, y + \Delta_y)$ at time t to (x, y) at time $t + T$. We can rewrite the parentheses in the first term on the right as

$$[N(x,y,t+T)-N(x,y,t)]_m = \int\limits_{-\infty}^{\infty} \int\limits_{-\infty}^{\infty} N_\Delta \phi_N d\Delta_x d\Delta_y - N(x,y,t) \approx \frac{\langle\Delta^2\rangle}{4}\left(\frac{\partial^2 N}{\partial x^2} + \frac{\partial^2 N}{\partial y^2}\right),$$

$$(16.3)$$

where N_Δ stands for $N(x+\Delta_x, y+\Delta_y, t)$, and ϕ_N for $\phi_N(\Delta_x, \Delta_y)$. In the last line in Eq. (16.3) we have performed a second-order Taylor expansion in Δ_x and Δ_y, and taken into account that $\int_{-\infty}^{\infty}\int_{-\infty}^{\infty}\phi_N d\Delta_x d\Delta_y = 1$. We have also assumed that the kernel is isotropic, i.e.,

$$\phi_N(\Delta_x, \Delta_y) = \phi_N(-\Delta_x, \Delta_y) = \phi_N(\Delta_x, -\Delta_y), \qquad (16.4)$$

and introduced the mean-squared displacement as

$$\langle\Delta^2\rangle = \int\limits_{-\infty}^{\infty} \int\limits_{-\infty}^{\infty} \Delta^2 \phi_N(\Delta_x, \Delta_y) d\Delta_x d\Delta_y, \qquad (16.5)$$

where $\Delta^2 = \Delta_x^2 + \Delta_y^2$. Note that Eq. (16.4) implies that $\langle\Delta_x\rangle = 0, \langle\Delta_y\rangle = 0$, $\langle\Delta_x\Delta_y\rangle = 0$ and $\langle\Delta_x^2\rangle = \langle\Delta_y^2\rangle$, which has been applied in the last step in Eq. (16.3). This is Einstein's approach to diffusion [9].

Finally we rewrite the parentheses in the last term in Eq. (16.2) as a Taylor expansion,

$$[N(x,y,t+T)-N(x,y,t)]_g = \left(TF(x,y,t) + \frac{T^2}{2}\frac{\partial F}{\partial t} + \ldots\right) \qquad (16.6)$$

where $F(x,y,t)$ is the change in population density per unit time, due to births and deaths.

Expanding the left-hand side of Eq. (16.2) up to first order and collecting terms, we arrive at Fisher's reaction-diffusion equation,

$$\frac{\partial N}{\partial t} = D\left(\frac{\partial^2 N}{\partial x^2} + \frac{\partial^2 N}{\partial y^2}\right) + F(x,y,t), \qquad (16.7)$$

where we have introduced the diffusion coefficient,

$$D = \frac{\langle\Delta^2\rangle}{4T}. \qquad (16.8)$$

which is the two-dimensional analogue of the one-dimensional Eq. (2.3). Concerning the net reproduction function $F(x,y,t)$, in Chap. 2 an example is presented such that

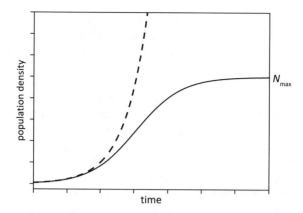

Fig. 16.3 Plots of population density N versus time t. The dashed line corresponds to exponential growth, $N = N_0 e^{\alpha t}$ (see the text below Eq. (16.9)), and the full line to logistic growth, $N = N_0 N_{max} e^{\alpha t} / (N_{max} + N_0 (e^{\alpha t} - 1))$ (see Eq. (16.12))

$$F(x, y, t) = \alpha N(x, y, t) \qquad (16.9)$$

(see the last term in Eq. (2.15)). This reproduction function corresponds to exponential growth, because without diffusion $(D = 0)$ Eq. (16.7) yields $N = N_0 \exp [\alpha t]$, with $N_0 = N(t = 0)$. Thus Eq. (16.9) is an example of interest, but the population density would never stop growing. A biologically more realistic case is the so-called logistic growth function,

$$F(x, y, t) = \alpha N(x, y, t) \left[1 - \frac{N(x, y, t)}{N_{max}} \right], \qquad (16.10)$$

were N_{max} is the saturation density, i.e. the population density at which net reproduction vanishes (note that $F(x, y, t) = 0$ if $N(x, y, t) = N_{max}$). The functions of exponential and logistic growth are compared in Fig. 16.3. A more detailed introduction into the formalism of logistic growth is provided by Sect. 3.4.1 of Chap. 3, with an example of the benefit of this reasoning on predicting the spreading of technological innovations given in Sect. 14.2.2 of Chap. 14.

Equation (16.7) with the logistic growth function (16.10) is called Fisher's equation. For our purposes here, we can consider the simple case in which all parameters $(D, \alpha$ and $N_{max})$ are independent of x, y and t. Travelling wave solutions (also called fronts or waves of advance) are defined as constant-shape solutions, i.e. those depending not on x, y and t separately but only on $z = r - vt$, where v is the front speed and $r = \sqrt{x^2 + y^2}$ the radial coordinate.

Kolmogorov et al. [10] showed that in Fisher's model, a front is formed and its speed is given by Eq. (16.1), assuming that initially the population density $N(x, y, t)$ has compact support. In practice, this assumption means that $N(x, y, t = 0) = 0$ everywhere except in a finite region. This is biologically realistic, in contrast to solutions such that $N(x, y, t = 0) \neq 0$ for all values of x, y $(-\infty < x < \infty,$ $-\infty < y < \infty)$. The latter solutions are not biologically realistic, because in practical applications we always want to analyze the spread of organisms that are initially present in a finite region of space.

Using variational methods, Aronson and Weinberger [11] also showed that the speed of front solutions to Fisher's equation is given by Eq. (16.1) (see Sect. IV.A in Ref. [12] for a simple derivation based on variational principles).

Importantly, Fisher's wave-of-advance speed (16.1) does not depend on N_{max}. Moreover, this speed is the same as for exponential growth (Eq. (16.9)), see Eq. (2.17). Thus the wave-of-advance speed is the same in both the logistic and the exponential models. However, their shape is different, because for exponential growth the population density keeps growing in time, whereas for logistic growth it stops growing at $N = N_{max}$ (see Fig. 16.3). Thus the waves of advance under logistic growth have the profile shown in Fig. 2.6, where we can see that the population density stops growing once $N = N_{max}$. In contrast, for exponential growth the population density keeps growing forever everywhere (see Ref. [13], Figs. 3.3 and 3.6).

Returning to the spread of the farming, Ammerman and Cavalli-Sforza [7] noted from archaeological dates that the speed of the Neolithic wave of advance was about 1 km/year. They next asked the following interesting question: what speed does Fisher's model [Eq. (16.1)] predict? In order to answer this, empirical values for $\langle \Delta^2 \rangle$ and T are needed to estimate D using Eq. (16.8). Additionally, an empirical value for α is needed to estimate the speed from Eq. (16.1). Ethnographic observations of preindustrial populations have measured the displacement of individuals and found the average for the mean-squared displacement per generation $\langle \Delta^2 \rangle = 1288$ km^2 [1, 14] and the mean generation time (defined as the age difference between a parent and his/her children) $T = 32$ year [6]. Thus we obtain from Eq. (16.8) $D = 10$ km^2/year. On the other hand, for populations which settle in empty space, $N \ll N_{max}$ and Eq. (16.10) reduces to (16.9), so that we can fit exponential curves (graphically, we can understand this because both curves in Fig. 3 overlap in the left-hand side). Ethnographic data yield the average exponent $\alpha = 0.028$ year^{-1} [14]. Using these values into Eq. (16.1) we estimate a front speed of about 1 km/year, which is similar to the speed obtained from the archeological observations. Indeed, as mentioned above, looking at Fig. 16.1 we can easily estimate that agriculture spread from Greece to the Balkans and Central Europe at a speed of roughly 1 km/year (more precise estimations with recent data, based on regression analysis [3] and geostatistical techniques [1], agree with this average). This agreement was first noted by Ammerman and Cavalli-Sforza [7, 4]. In this way, Ammerman and Cavalli-Sforza noted that diffusive models are useful not only because they make it possible to describe mathematically a major event in prehistory (the spread of agriculture), but also because they indicate a possible mechanism for it, namely the spread of people (i.e., of populations of farmers). They called this demic diffusion (from the Greek word *demos*, which means people). In contrast, most authors at the time advocated for the learning of farming by hunter-gatherers (i.e., for the spread of agriculture without substantial spread of people) [15]. The latter mechanism is called cultural diffusion.

16.2 First Improvement: Beyond the Second-Order Approximation

In the derivation of Eq. (16.7) we have performed Taylor expansions up to first order in time and second order in space. Without those approximations we obtain, instead of Eq. (16.7),

$$N(x,y,t+T) - N(x,y,t) = \int\limits_{-\infty}^{\infty} \int\limits_{-\infty}^{\infty} N_\Delta \phi_N d\Delta_x d\Delta_y - N(x,y,t) + R_T[N(x,y,t)]),$$

$$(16.11)$$

where the joint effects of reproduction and survival are, again, well-described by the solution to a logistic growth function, namely [8]

$$R_T[N(x,y,t)] = \frac{e^{\alpha T} N_{max} N(x,y,t)}{N_{max} + (e^{\alpha T} - 1)N(x,y,t)}. \tag{16.12}$$

When observed dispersal data are used, the kernel *per unit length* $\varphi_N(\Delta)$ is defined as the probability to disperse into a ring of radius Δ and width $d\Delta$, divided by $d\Delta$. If individuals of the population N have probabilities p_j to disperse at distances r_j ($j = 1, 2, ..., M$), we can write

$$\varphi_N(\Delta) = \sum_{j=1}^{M} p_j \delta^{(1)}(r_j), \tag{16.13}$$

where $\delta^{(1)}(r_j)$ is the 1D Dirac delta centered at r_j (i.e., a function that vanishes everywhere except at $\Delta = r_j$). Since the total probability must be one,

$$1 = \int\limits_{0}^{\infty} \varphi_N(\Delta) d\Delta, \tag{16.14}$$

and $\varphi_N(\Delta)$ is clearly a probability *per unit length*. In contrast, the kernel $\phi_N(\Delta_x, \Delta_y)$ in Eq. (16.11) is a probability *per unit area* (because it is multiplied by $d\Delta_x d\Delta_y$, which has units of area). The normalization condition for $\phi_N(\Delta_x, \Delta_y)$ is therefore

$$1 = \int\limits_{-\infty}^{\infty} \int\limits_{-\infty}^{\infty} \phi_N(\Delta_x, \Delta_y) d\Delta_x d\Delta_y = 2\pi \int\limits_{0}^{\infty} \phi_N(\Delta)\Delta d\Delta, \qquad (16.15)$$

where we have used polar coordinates $\Delta = \sqrt{\Delta_x^2 + \Delta_y^2}$, $\theta = tan^{-1}\frac{\Delta_y}{\Delta_x}$ and assumed the kernel is isotropic, $\phi_N(\Delta_x, \Delta_y) = \phi_N(\Delta)$. Comparing Eqs. (16.14) and (16.15), we see that the dispersal probability per unit length (i.e., into a ring of area $2\pi\Delta d\Delta$) $\varphi_N(\Delta)$ is related to that per unit area $\phi_N(\Delta)$ as [16]

$$\varphi_N(\Delta) = 2\pi\Delta\phi_N(\Delta) \qquad (16.16)$$

and Eq. (16.13) yields

$$\phi_N(\Delta) = \sum_{j=1}^{M} p_j \frac{\delta^{(1)}(r_j)}{2\pi\Delta}. \qquad (16.17)$$

For homogeneous parameter values, the speed will not depend on direction and can thus be more easily computed along the x-axis ($y=0$). Consider a coordinate frame $z = x - vt$ moving with the wave of advance (v is the front speed). The population density of farmers will be equal to its saturation density in regions where the Neolithic transition is over, and it will decay to zero in regions where few farmers have arrived. Thus we assume as usual the *ansatz* [16] $N(x, y, t) \approx N_0 \exp[-\lambda z] \to 0$ for $z \to \infty$ (with $\lambda > 0$). Then, assuming that the minimum speed is that of the front (which has been verified by numerical simulations), we obtain for the speed v of front solutions to Eq. (16.11) [14]

$$v_{NCohab} = \min_{\lambda > 0} \frac{\ln\left[(e^{\alpha T} - 1) \sum_{j=1}^{M} p_j I_0(\lambda r_j)\right]}{T\lambda}, \qquad (16.18)$$

where the sub index *NCohab* indicates that this is *not* a cohabitation model (see the next section), and $I_0(\lambda r_j)$ is the modified Bessel function of the first kind and order zero. In this model, the speed can be found by plotting the fraction in Eq. (16.18) as a function of λ and finding its minimum.

In Ref. [14] it has been shown that the differences in the front speed obtained from Eq. (16.13) and Fisher's approximation, Eq. (16.1), are up to 49% for human populations. So the effect of higher-order terms is not negligible.

16.3 Second Improvement: Cohabitation Equations

For human populations, newborn children cannot survive on their own. However, when they come on age they can move away from their parents. This point has led some authors to use an equation of the so-called cohabitation type, namely

$$N(x, y, t + T) = \int\limits_{-\infty}^{\infty} \int\limits_{-\infty}^{\infty} R_T[N_\Delta]\phi_N d\Delta_x d\Delta_y, \qquad (16.19)$$

where $R_T[N]$ is again given by Eq. (16.12). Then the speed of front solutions is [14, 17]

$$v_{Cohab} = \overset{min}{\underset{\lambda > 0}{}} \frac{\alpha T + \ln\left[\sum_{j=1}^{M} p_j I_0(\lambda r_j)\right]}{T\lambda}. \qquad (16.20)$$

The reason why Eq. (16.19) is more reasonable than Eq. (16.11) is that, clearly, Eq. (16.11) assumes that individuals born at (x, y) at time t (last-but-one term) will not move at all, i.e. they will all still be at (x, y) on coming of age (time $t + T$, left-hand side). Thus, for example, in the simple case in which all parents move, they will leave all of their children alone. Such an anthropologically unrealistic feature makes it clear that Eq. (16.11) is less accurate than Eq. (16.19). For additional derivations and figures showing that Eq. (16.11) is less realistic than the cohabitation Eq. (16.19), see especially Fig. 1 in Ref. [14], Fig. 17 in Ref. [16], and Ref. [17].

A more direct way to see the limitations of Fisher's speed (16.1) is to note that it yields $v_{Fisher} \to \infty$ for $\alpha \to \infty$. In contrast, numerical simulations have shown that the cohabitation speed (16.20) yields for $\alpha \to \infty$ the value $v_{Cohab} = r_{max}/T$, i.e. the maximum dispersal distance divided by the generation time [14, 18], which is physically reasonable. Moreover, the error of Fisher's speed (16.1) relative to Eq. (16.20) reaches 30% for realistic human kernels and parameter values [14]. This error is still larger when cultural diffusion is included [1] (next section).

16.4 Demic-Cultural Model

Up to now we have only considered equations with a single mechanism for the spread of the Neolithic, namely the dispersal of farmers (demic diffusion). But agriculture can be also learnt by hunter-gatherers (cultural diffusion). When this conversion of hunter-gatherers into farmers (cultural transmission) is taken into account, we might be tempted to generalize Eq. (16.19) into

$$N(x,y,t+T) = \int\limits_{-\infty}^{\infty} \int\limits_{-\infty}^{\infty} R_T[N_\Delta]\phi_N d\Delta_x d\Delta_y + \int\limits_{-\infty}^{\infty} \int\limits_{-\infty}^{\infty} c[N_\Delta, P_\Delta]\phi_N^{converts} d\Delta_x d\Delta_y,$$

$$(16.21)$$

where $P_\Delta = P(x+\Delta_x, y+\Delta_y)$ is the population density of hunter-gatherers at $(x+\Delta_x, y+\Delta_y)$. The cultural transmission function $c[\ldots]$ in Eq. (16.21) is due to the conversion of hunter-gatherers into farmers. Thus a similar equation for the population density of hunter-gatherers $P(x,y,t+T)$ could be proposed, with a minus sign in the last term. A recent derivation has found for the cultural transmission function $c[\ldots]$ (see Ref. [19], Eq. (1))

$$c[N(x,y,t), P(x,y,t)] = f \frac{N(x,y,t)P(x,y,t)}{N(x,y,t) + \gamma P(x,y,t)}, \tag{16.22}$$

where f and γ are cultural transmission parameters. The kernel $\phi_N^{converts}(\Delta_x, \Delta_y)$ in Eq. (16.22) is the dispersal kernel of hunter-gatherers that have been converted into farmers. Since they now behave as farmers, let us assume that this kernel is the same as $\phi_N(\Delta_x, \Delta_y)$. Then Eq. (16.22) becomes

$$N(x,y,t+T) = \int\limits_{-\infty}^{\infty} \int\limits_{-\infty}^{\infty} R_T[N_\Delta]\phi_N d\Delta_x d\Delta_y + \int\limits_{-\infty}^{\infty} \int\limits_{-\infty}^{\infty} f \frac{N_\Delta P_\Delta}{N_\Delta + \gamma P_\Delta} \phi_N d\Delta_x d\Delta_y.$$

$$(16.23)$$

A model of this kind was applied recently (see Eq. 5 in Ref. [19]). It is an approximation that may be valid in some regions (with mainly demic diffusion) but it cannot lead to a purely cultural model of Neolithic spread (because according to Eq. (16.23) there is no front propagation in the absence of demic diffusion, i.e. if $\phi_N(\Delta_x, \Delta_y) \neq 0$ only at vanishing distance, i.e. for $\Delta = (\Delta_x^2 + \Delta_y^2)^{1/2} = 0$). Thus we will here consider a more realistic model in two ways. Firstly we take into account that, according to ethnographic observations, hunter-gatherers can learn agriculture from farmers located some distance away [1]. Then Eq. (16.23) is generalized into

$$N(x,y,t+T) = \int\limits_{-\infty}^{\infty} \int\limits_{-\infty}^{\infty} R_T[N_\Delta]\phi_N d\Delta_x d\Delta_y$$
$$+ \int\limits_{-\infty}^{\infty} \int\limits_{-\infty}^{\infty} \phi_N d\Delta_x d\Delta_y \int\limits_{-\infty}^{\infty} \int\limits_{-\infty}^{\infty} \phi_p' d\Delta_x' d\Delta_y' f \frac{N_{\Delta+\Delta'} P_\Delta}{N_{\Delta+\Delta'} + \gamma P_\Delta},$$

$$(16.24)$$

where $N_{\Delta+\Delta'}$ stands for $N\left(x+\Delta_x+\Delta_x', y+\Delta_y+\Delta_y', t\right)$.

In practice, the cultural kernel $\phi'_P(\Delta'_x, \Delta'_y)$ (which is abbreviated as ϕ'_P in Eq. (16.24)) is a set of probabilities P_k for hunter-gatherers to learn agriculture from farmers living at distances $R_k = (\Delta_x'^2 + \Delta_y'^2)^{1/2}$, during a generation time T. This is similar to the fact, mentioned above Eq. (16.13), that in practice the demic kernel $\phi_N(\Delta_x, \Delta_y)$ is a set of probabilities p_j for farmers to disperse at distances $r_j = (\Delta_x^2 + \Delta_y^2)^{1/2}$, also during a generation time T.

Secondly we note that after a generation time T, reproduction will have led to new individuals not only in the population of farmers (first line in Eq. (16.24)) but also in the population of hunter-gatherers converted into farmers (second line in Eq. (16.24)). Thus we finally generalize Eq. (16.24) into

$$N(x, y, t+T) = \int_{-\infty}^{\infty} \int_{-\infty}^{\infty} R_T[N_\Delta] \phi_N d\Delta_x d\Delta_y$$
$$+ \int_{-\infty}^{\infty} \int_{-\infty}^{\infty} \phi_N d\Delta_x d\Delta_y \int_{-\infty}^{\infty} \int_{-\infty}^{\infty} \phi'_P d\Delta'_x d\Delta'_y R_T \left[f \frac{N_{\Delta+\Delta'} P_\Delta}{N_{\Delta+\Delta'} + \gamma P_\Delta} \right].$$

$$(16.25)$$

The speed of front solutions to Eq. (16.25) is [1]

$$v = \min_{\lambda > 0} \frac{\alpha T + \ln\left[\left(\sum_{j=1}^{M} p_j I_0(\lambda r_j) \right) \left(1 + C \left[\sum_{k=1}^{Q} P_k I_0(\lambda R_k) \right] \right) \right]}{T\lambda}, \qquad (16.26)$$

with $C = f/\gamma$. This reduced parameter C was called the intensity of cultural transmission [19] because, according to Eq. (16.22), $C = f/\gamma$ is the number of hunter-gatherers converted per farmer at the front leading edge (i.e. in regions such that $N \ll P$). Without cultural transmission ($C = 0$), the demic-cultural front speed, given by Eq. (16.26), reduces to the purely-demic speed, Eq. (16.20), as it should. With frequency-dependent cultural transmission, Eq. (16.22) is more complicated and the equations are longer, but the final results are exactly the same [1].

It is important to note that cultural transmission (the factor in brackets $[f \ldots]$ at the end of the second line in Eq. (16.25)) is applied in a term that also contains the effects of net reproduction (R_T) and dispersal (the kernel of farmers $\phi_N(\Delta_x, \Delta_y)$). Thus some hunter-gatherers will learn agriculture from farmers located a distance (Δ'_x, Δ'_y), and the children of those converted hunter-gatherers will possibly move a distance (Δ_x, Δ_y) (similarly to the children of farmers, first line). Therefore, some hunter-gatherers can learn agriculture from farmers and the next generation (i.e., the children) of those hunter-gatherers will be farmers. Such a conversion during a generation time is reported by ethnographic data [20] and implies that the individual acculturation process is not instantaneous but takes place within one generation time, which seems reasonable for a complex cultural trait as farming.

Finally, a purely cultural model means no demic diffusion. In this model, the front speed can be obtained from Eq. (16.26) without demic diffusion ($r_1 = 0$ and $p_1 = 1$), namely

$$v_C = \min_{\lambda > 0} \frac{\alpha T + \ln\left[1 + C\left(\sum_{k=1}^{Q} P_k I_0(\lambda R_k)\right)\right]}{T\lambda}, \tag{16.27}$$

where the sub index C stands for purely cultural diffusion. This is the purely cultural analogue to the purely demic speed given by Eq. (16.20). Both of them are, of course, cohabitation models.

16.5   Demic Versus Cultural Diffusion in the Spread of the Neolithic in Europe

What do the models above imply for the relative importance of demic and cultural diffusion in the spread of the Neolithic in different regions of Europe? Let us summarize a recent proposal [1], which is based on using the following realistic ranges for the parameters appearing in our equations.

The ranges for a_N and T that have been measured for preindustrial farming populations are $0.023 \, \text{year}^{-1} \leq a_N \leq 0.033 \, \text{year}^{-1}$ and $29 \, \text{year} \leq T \leq 35 \, \text{year}$ (see the *SI Appendix* to Ref. [19] for details).

The following 5 dispersal kernels $\phi_N(\Delta_x, \Delta_y)$ have been measured for preindustrial farming populations [14]. For each kernel we also give its purely demic speed range, as predicted by the cohabitation model, Eq. (16.20), with $a_N = 0.023 \, \text{year}^{-1}$ and $T = 35 \, \text{year}$ (slowest speed) or $a_N = 0.033 \, \text{year}^{-1}$ and $T = 29 \, \text{year}$ (fastest speed).

Population A (Gilishi 15): purely demic speed range 0.87–1.15 km/year.
Population B (Gilishi 25): purely demic speed range 0.92–1.21 km/year.
Population C (Shiri 15): purely demic speed range 1.14–1.48 km/year.
Population D (Yanomamö): purely demic speed range 1.12–1.48 km/year.
Population E (Issongos): purely demic speed range 0.68–0.92 km/year.

We see that demic diffusion predicts Neolithic front speeds of at least 0.68 km/year. Demic-cultural diffusion will be still faster. Thus it has been suggested that cultural diffusion is responsible for the Neolithic spread in regions with speeds below 0.68 km/year [1]. For simplicity, let us consider purely cultural diffusion, Eq. (16.27), although a short-range demic kernel can be also included (Sect. S6 in Ref. [1]). In order to estimate the speeds predicted by purely cultural diffusion, we need the following cultural parameters.

The cultural transmission intensity C from hunter-gathering to farming has been estimated from several case studies in Ref. [19] and the overall range is $1.0 \leq C \leq 10.9$.

The following 5 cultural kernels have been estimated from distances from hunter-gatherers camp locations to the villages of farmers, where the hunter-gatherers practice agriculture [1]. For each kernel, we also report the purely-cultural speed range obtained from Eq. (16.27) with $a_N = 0.023\,\text{year}^{-1}$, $T = 35\,\text{year}$ and $C = 1$ (slowest speed) or $a_N = 0.033\,\text{year}^{-1}$, $T = 29\,\text{year}$ and $C = 10.9$ (fastest speed).

Population 1 (Mbuti, band I): speed range 0.17–0.36 km/year.
Population 2 (Mbuti, band II): speed range 0.30–0.57 km/year.
Population 3 (Mbuti, band III): speed range 0.32–0.66 km/year.
Population 4 (Aka): speed range 0.09–0.19 km/year.
Population 5 (Baka): speed range 0.03–0.07 km/year.

Thus the purely cultural model yields 0.03–0.66 km/year. Note that this is slower than the purely demic speed range found above (0.68–0.92 km/year):

Finally, for the demic-cultural model, Eq. (16.26), the slowest speed is obviously 0.68 km/year (see the purely demic model above). The relevant result of the demic-cultural model is its fastest speed. This obviously corresponds to the strongest value observed for the intensity of cultural transmission ($C = 10.9$), the fastest cultural kernel (population 3), the fastest demic kernel (population C or D), the highest observed value of a_N ($0.033\,\text{year}^{-1}$) and the lowest observed value of T (29 year). Using these data in Eq. (16.25) we find that the fastest speed is obtained for the demic kernel of population D yielding 3.04 km/year.

In Fig. 16.4b, the color scale has been chosen so that the red color corresponds to the regions of purely cultural diffusion (0.03–0.66 km/year, from the purely cultural model above). The demic and demic-cultural models predict speeds above 0.68 km/year, and are thus too fast to be consistent with the archeological data in the red regions in Fig. 16.4b. This suggests that cultural diffusion explains the Neolithic transition in Northern Europe, as well as in the Alps and west of the Black Sea. The analysis of the areas where demic diffusion played a role is less straightforward, but it is possible to determine the regions where the speed was *mainly* demic (i.e. where the cultural effect was $< 50\%$) [1]. They correspond to the yellow regions in Fig. 16.4b. The regions where either demic or cultural diffusion could have dominated are the blue regions in Fig. 16.4b. The blue regions appear because we have used parameter ranges and several kernels (they would not appear if we had used a single value for each parameter, a single demic kernel, and a single cultural kernel). Finally, in the green regions in Fig. 16.4 the speed is too fast to agree with any of the three models in the present chapter, but in continental Europe those regions contain very few sites and will probably disappear using more complete databases (i.e., with more archeological sites).

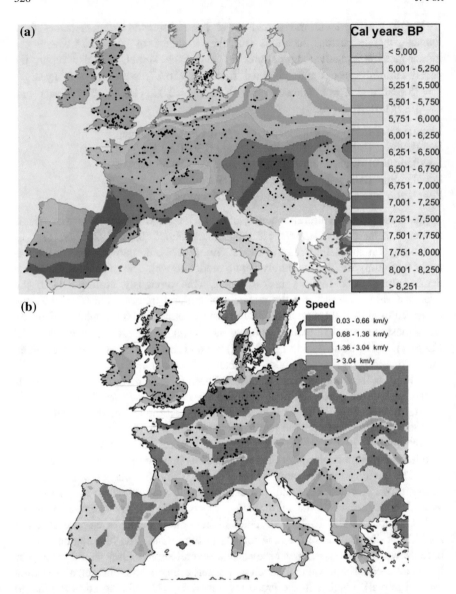

Fig. 16.4 Isochrones obtained by smoothing 40 times the map in Fig. 16.1 (**a**). Note that most anomalously old/recent areas have been disappeared. Smoothing 60 times yields almost the same map. (**b**) Displays the speed ranges obtained from (**a**). Closer isochrones correspond to slower speeds. Adapted from Ref. [1], Supp. Info. Appendix, Fig. S4

16.6 Conclusions

The models reviewed in this chapter suggest that the spread of the Neolithic in Europe was: (i) fast and mainly demic in the Balkans and Central Europe; (ii) slow and mainly cultural in Northern Europe, the Alpine region and west of the Black Sea (Fig. 16.4b) [1].

As seen in Fig. 16.4b, the process was fast (speeds above 0.68 km/year) in Greece, Italy, the Balkans, Hungary, Slovakia, Czechia and central Germany. This wide region includes a substantial part of the Linearbandkermic (LBK) culture in Central Europe. This is in agreement with the fact that the LBK is widely regarded as demic by archeologists. Also in agreement with our results, some archeologists have argued for the importance of demic diffusion in the Neolithic spread from the Aegean northwards and across the Balkans. On the other hand, our models suggest that farming populations did not spread much into Northern Europe, the Alps and West of the Black Sea (red color in Fig. 16.4b). In such regions, the transition was slow (speeds below 0.66 km/year) and, according to our models, not driven by demic or demic-cultural diffusion. Some archeologists have previously suggested that cultural diffusion played a strong role in the spread of the Neolithic in Northern Europe, the Alps and West of the Black sea. Note that these are precisely the mainly cultural diffusion regions according to our models (red color in Fig. 16.4b). For detailed archeological references on the importance of demic and cultural diffusion in different regions of Europe see, e.g., Sect. 3 in Ref. [1]. Ancient genetics also indicates that cultural diffusion was more important in Northern Europe [21], in agreement with our conclusions.

The slowness of cultural diffusion (as compared to demic diffusion) is due to the fact that, according to ethnographic observations, the distances appearing in the cultural kernel $\phi'_P(\Delta'_x, \Delta'_y)$ are substantially shorter than those appearing in the demic kernel $\phi_N(\Delta_x, \Delta_y)$ [1]. The intuitive reason may be that that agriculture is a difficult cultural trait to learn, and this leads to shorter cultural than demic diffusion distances. Note that the cultural distances are defined as those separating hunter-gatherers from the farmers who teach them how to farm. Indeed, according to ethnographic data, in the spread of farming cultural diffusion distances were short as compared to demic diffusion distances [1]. The latter are those along which the children of farmers disperse away from their parents. Such demic distances can obviously be larger than cultural distances, because the children of farmers have already learnt agriculture before leaving their parents.

Models similar to those summarized here have been applied to Paleolithic waves of advance [22], the extremely fast spread of the Neolithic in the Western Mediterranean [23], language substitution fronts [24], etc.

All models considered in this chapter operate with a minimum of parameters. Parameters in the demic model are, for instance, only the initial growth rate α, the generation T and the dispersal kernel which, in addition, have been estimated from ethnographic or archeological data. With such constraints one is able to largely avoid any unjustified bias in modeling which may easily occur by the use of too many parameters which, finally, degenerate to simple fitting parameters. For example, in some models of virus infection fronts, it was possible to reproduce the experimental front speeds if choosing several parameter values [25, 26]. However, this was not possible for realistic parameter values [26, 27]. Later, more refined models reproduced the data without choosing any parameter values [28]. We have to be aware, however, that one may quite reasonably introduce much larger parameter sets as demonstrated in Chap. 17. However, their values (see the caption to Table 17.2) become questionable with the lack of possibilities for their determination from reliable, independent sources (see also p. 3462 in Ref. [5], where 8 parameter values are chosen to replicate the observed spread rate, etc., rather than from independent data). Optimum strategies will notably change with changes in data accessibility and in the course of exchange between the scientists in the various disciplines involved in the problem.

References

1. J. Fort, J.R. Soc, Interface **12**, 20150166 (2015)
2. J. Fort, V. Méndez, Phys. Rev. Lett. **82**, 867 (1999)
3. R. Pinhasi, J. Fort, A.J. Ammerman, PLoS Biol. **3**(e410), 2220 (2005)
4. A.J. Ammerman, L.L. Cavalli-Sforza, *The Neolithic Transition and the Genetics of Populations in Europe* (Princeton University Press, Princeton, 1984)
5. C. Lemmen, D. Gronenborn, K.W. Wirtz, J. Arch. Sci. **38**, 3459 (2011)
6. J. Fort, D. Jana, J.M. Humet, Phys. Rev. E **70**, 031913 (2004)
7. A.J. Ammerman, L.L. Cavalli-Sforza, in *The Explanation of Culture Change: Models in Prehistory*, ed. by C. Renfrew (Duckworth, London, 1973), pp. 343–357
8. J.D. Murray, *Mathematical Biology*, vol. 1 (Springer, Berlin, 2001)
9. A. Einstein, *Investigations on the Theory of Brownian Movement* (Dover, New York, 1956)
10. A.N. Kolmogorov, I.G. Petrovsky, N. Piskunov, Bull. Univ. Moscow Ser. Int. A **1**, 1 (1937)
11. D.G. Aronson, H.F. Weinberger, Adv. Math. **30**, 33 (1978)
12. V. Méndez, J. Fort, J. Farjas, Phys. Rev. E **60**, 5231 (1999) (Sect. IV.A)
13. N. Shigesada, K. Kawasaki, *Biological Invasions: Theory and Practice* (Oxford University Press, Oxford, 1997)
14. N. Isern, J. Fort, J. Pérez-Losada, J. Stat. Mech. **2008**, P10012 (2008)
15. M.S. Edmonson, Curr. Anthropol. **2**, 71 (1961)
16. J. Fort, T. Pujol, Rep. Prog. Phys. **71**, 086001 (2008)
17. J. Fort, J. Pérez-Losada, N. Isern, Phys. Rev. E **76**, 031913 (2007)
18. J. Fort, J. Pérez-Losada, J.J. Suñol, J.M. Massaneda, L. Escoda, New J. Phys. **10**, 043045 (2008)
19. J. Fort, Proc. Natl. Acad. Sci. U.S.A. **109**, 18669 (2012)
20. J.D. Early, T.N. Headland, *Population Dynamics of a Philippine Rain Forest People. The San Ildefonso Agta* (University Press of Florida, Gainesville, 1998)
21. E.R. Jones, G. Zarina, V. Moiseyev, E. Lightfoot, P.R. Nigst, A. Manica, R. Pinhasi, D.G. Bradley, Curr. Biol. **27**, 576–582 (2017)
22. J. Fort, T. Pujol, L.L. Cavalli-Sforza, Cambr. Archaeol. J. **14**, 53 (2004)

23. N. Isern, J. Zilhao, J. Fort, A.J. Ammerman, Proc. Natl. Acad. Sci. U.S.A. **114**, 897 (2017)
24. N. Isern, J. Fort, J. R. Soc. Interface **11**, 20140028 (2014)
25. J. Yin, J.S. McCaskill, Biophys. J. **61**, 1540 (1992)
26. L. You, J. Yin, J. Theor. Biol. **200**, 365 (1999)
27. J. Fort, J. Theor. Biol. **214**, 515 (2002)
28. J. Fort, V. Méndez, Phys. Rev. Lett. **89**, 178101 (2002)

Chapter 17
The Diffusion of Humans and Cultures in the Course of the Spread of Farming

Carsten Lemmen and Detlef Gronenborn

17.1 Introduction

The most profound change in the relationship between humans and their environment was the introduction of agriculture and pastoralism. With this millennia-lasting economic shift from simple food acquisition to complex food production humankind paved the way for its grand transitional process from mobile groups to sedentary villages, towns and ultimately cities, and from egalitarian bands to chiefdoms and lastly states. Given this enormous historic impetus, Gordon Childe coined the term "Neolithic Revolution" [1] almost a hundred years ago.

The first experiments towards agriculture began with the end of the Glacial period about 10,000 years ago in the so called Fertile Crescent [2]. They were followed by other endeavors in various locations both in the Americas and in Afroeurasia. Today farming has spread to all but the most secluded or marginal environments of the planet [3]. Cultivation of plants and animals on the global scale appears to have changed energy and material flows—like greenhouse gas emissions—so fundamentally, that the term "early anthropocene" has been proposed for the era following the Mid-Holocene [4].

Possible reasons for the emergence of farming during the relatively confined period between the Early and Mid-Holocene in locations independent of each other are continuously being debated [2]. Once these inventions were in place, however, they immediately become visible in the archeological and paleoenvironmental records. From then on we can trace the spatial expansion of the newly domesticated plants and animals, the spatial expansion of a life style based on these domesticates,

C. Lemmen (✉)
Institute of Coastal Research, Helmholtz-Zentrum Geesthacht, Geesthacht, Germany
e-mail: carsten.lemmen@hzg.de

D. Gronenborn
Romano-German Central Museum and Johannes Gutenberg University of Mainz,
Mainz, Germany
e-mail: gronenborn@rgzm.de

© Springer International Publishing AG 2018
A. Bunde et al. (eds.), *Diffusive Spreading in Nature, Technology and Society*,
https://doi.org/10.1007/978-3-319-67798-9_17

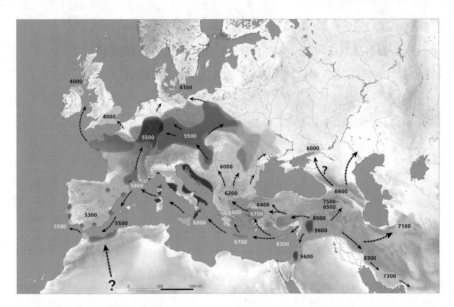

Fig. 17.1 Overview of the study area and the archeologically visible expansion of farming. Figure by Detlef Gronenborn (Romano-German Central Museum, Mainz, Germany). Reprinted with permission, licensed under CC BY 4.0

and the induced changes in land cover [5, 6]. From such empirically derived data the characteristic condensed map of the spread of farming into Western Eurasia is produced (Fig. 17.1).

The local changes introduced spatial differences in knowledge, labor, technology, materials, population density, and—more indirectly—social structure and political organization, amongst others [7, 8]. Consequently, the dynamics occurring along such spatial gradients may be modeled as a diffusive process. In Chap. 2, Fick's first law was introduced, which describes that the average flux across a spatial boundary is proportional to the difference of concentration across this boundary (Chap. 2, Eq. (2.6)). Each of the local inventions would then spread outward from its respective point of origin. Indeed, these spatiotemporal gradients have been observed in ceramics [1], radiocarbon dates [5, 9], domesticates [10, 11], land use change [12, 13], and the genetic composition of paleopopulations [14–17].

For an understanding of the expansion process, it appears appropriate to apply a diffusive model. Broadly, these numerical modeling approaches can be categorized in correlative, continuous and discrete, as specified further below. Common to all approaches is the comparison to collections of radiocarbon data that show the apparent wave of advance [18] of the transition to farming. However, these data sets differ in entry density and data quality. Often they disregard local and regional specifics and research gaps, or dating uncertainties. Thus, most of these data bases may only be used on a very general, broad scale. One of the pitfalls of using irregularly spaced or irregularly documented radiocarbon data becomes evident from the map generated by Fort (this volume, Chap. 16): while the general east-west and south-north trends are well represented, some areas appear as having undergone anomalously early transitions to farming.

Correlative models compare the timing of the transition (or other archeologically visible frontiers) with the distance from one or more points of origin. These are among the earliest models proposed, such as those by Clark [5] or by Ammerman and Cavalli Sforza [18]. These models have been used to roughly estimate the front propagation speed of the introduction of agriculture into Europe, and the original speed of around 1 km per year has not been substantially refined until today.

Continuous models predict at each location within the specified domain the transition time as the solution of a differential equation, mostly of a Fisher–Skellam type, in relation to the distance from one or more points of origin. Often, this distance is not only the geometric distance but also factors in geography and topography, including ease of migration. The prediction from the continuous model is compared to the archeologically visible frontier [19]. This is the approach taken by Fort (this volume, Chap. 16) who compares the wave-front propagation of different models for the transition from a hunting and gathering economy to a farming economy in Europe with the spatiotemporal pattern of the earliest radiocarbon dates locally associated with farming.

Discrete models are often realized as agent-based models (see also Sect. 2.5 of Chap. 2), with geographic areas (or their populations) representing the "agents", and rules that describe the interaction, especially the diffusion properties, between them. They also predict for each geographic area the transition time, but not as an analytic, but rather as an emergent property of the system. We here introduce, as an example, a discrete agent-based and gradient-adaptive model, referred to as the "Global Land Use and technological Evolution Simulator" (GLUES).

The chapter starts with introducing into the special features of this simulation model, notably into the set of local characteristic variables ("traits", see Sect. 17.2.1). They are exploited for characterizing the given state of the "agent" (i.e. of the population under consideration) and decide, simultaneously, about the further development of the system. Section 17.2.2 introduces into the general formalism of evolution. Correlations between the different traits, their evolution and local population growth are considered in Sect. 17.2.3, with a summarizing discussion of the various types of flux and diffusion provided in Sects. 17.2.4 and 17.2.5. Section 17.3 illustrates the surplus of information attainable by data analysis via GLUES, including clear quantitative assessment between demic and cultural diffusion (i.e. between the movement of people or ideas) as a function of time and space.

17.2 The Agent-Based Gradient-Adaptive Model GLUES

We employ the Global Land Use and technological Evolution Simulator (GLUES [20, 21])—a numerical model of prehistoric innovation, demography, and subsistence economy based on interacting geographic populations as agents and

Fig. 17.2 Regions
constituting the set of agents
in the simulation (shown for
Western Eurasia and North
Africa) in 685 globally
distributed regions)

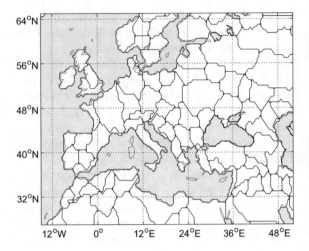

gradient adaptive trait dynamics to describe local evolution. There are currently
685 regions representing the "cells" of agent-based models (Fig. 17.2), together
with interaction rules that describe diffusion of people, material and information
between these regions. The "agent" is the population living within a region. Its state
is described by its density and by a number of characteristic features, referred to
as traits. They result as averages over the considered population. Such averages are
referred to as "aggregated traits". GLUES operates with three different traits which
shall be described in more detail below.

The "numerical model" is able to hindcast the regional transitions to agropastoralism and the diffusion of people and innovations across the world for the time
span between approximately 8000 BCE (before the common era) and 1500 CE. It
has been successfully compared to radiocarbon data for Europe [21], Eastern North
America [22], and South Asia [23].

Regions are generated from ecozone clusters that have been derived to represent
homogeneous net primary productivity (E_{NPP}) based on a 3000 BCE $1° \times 1°$ paleo-
productivity estimate; this estimate was derived from a climatologically downscaled
dynamic paleovegetation simulation [20]. By using E_{NPP}, many of the environmental
factors taken into account by other expansion or predictive models, such as altitude,
latitude, rainfall, or temperature [10, 24] are implicitly considered.

17.2.1 Local Characteristic Variables

The model as displayed in Fig. 17.2 is, at any instant of time, completely described by
knowledge of (i) population (B), (ii) associated traits (X) and (iii) environmental conditions (E) of each individual region. In a series of different applications (covering a
time span of close to 10,000 years in Europe, America and Asia, it has turned out that

Table 17.1 Characteristic traits used in the Gradient Adaptive Dynamics formulation of GLUES; a full table of symbols used is available as Table 17.2

Characteristic trait	Symbol	Quantification	Typical range
Technology efficiency	T	Factor of efficiency gain over Mesolithic	0.9–15
Economic diversity	N	Richness of economic agropastoral strategies	0.1–8
Agropastoral share	C	Fraction of activities in agropastoralism	0–1

overall developments could be satisfactorily described by introducing three different types of "traits", i.e. of characteristic features characterizing the productivity-related intellectual level of the population. They all result as sociocultural averages and may, in short, be referred to as "technological efficiency" (T), "share of agropastoral activities" (C) and "economic diversity" (N), as summarized by Table 17.1. In detail, the various traits can be characterized as follows:

1. Technology T is a trait which describes the efficiency for enhancing biological growth rates, or diminishing mortality. It is represented by the efficiency of food procurement—related to both foraging and farming—and improvements in health care. In particular, technology as a model describes the availability of tools, weapons, and transport or storage facilities, and includes institutional aspects related to work organization and knowledge management. These are often synergistic: the technical and societal skill of writing as a means for cultural storage and administration, with the latter acting as an organizational lubricant for food procurement and its optimal allocation in space and among social groups. Quantitative measure of T is the (estimated) efficiency gain over Mesolithic technology.
2. Economic diversity N resolves the number of different agropastoral economies available to a regional population. This trait is closely tied to regional vegetation resources and climate constraints. A larger economic diversity offering different niches for agricultural or pastoral practices enhances the reliability of subsistence and the efficacy in exploiting heterogeneous landscapes.
3. The third model variable C represents the share of farming and herding activities, encompassing both animal husbandry and plant cultivation. It describes the allocation of energy, time, or manpower to agropastoralism with respect to the total food sector; this is the only variable that is directly comparable to data from the archeological record.

17.2.2 Adaptive Dynamics

The entities of the "local characteristic variables" as introduced in Sect. 17.2.1 are subject to continuous variation. This process is controlled by the current state of these variables. Dynamics of evolution is thus immediately recognized as a function of the given evolutionary stage (represented by the set T, N, C) of traits (also referred to as the "food production system") and the population density. This concept of "adaptive dynamics" is related to E. Boserup's observation that "The close relationship which exists today between population density and food production system is the result of two long-existing processes of adaptation. On the one hand, population density has adapted to the natural conditions for food production [...]; on the other hand, food supply systems have adapted to changes in population density." [25, 26].

Mathematically, this conceptual model is implemented in the so-called Gradient Adaptive Dynamics (GAD) approach: Whenever traits can be related to growth rate, then an approach known as adaptive dynamics can be applied to generate the equations for the temporal change of traits, the so-called evolution equations. This adaptive dynamics goes back to earlier work by Fisher in the 1930s [27] and the field of genetics. When genetically encoded traits influence the fitness of individuals, the prevalence of the genes encoding this phenotype changes. Adaptive dynamics describes the change of the probability of the trait in the population by considering its mutation rate and its fitness gradient, i.e., the marginal benefit of changes in the trait for the (reproductive) fitness of the individual.

To ecological systems, this metaphor was first applied by Wirtz and Eckhardt in 1996 [28], and to cultural traits by Wirtz and Lemmen in 2003 [20]. In this translation, the genetically motivated term mutation rate was replaced by the ecologically observable variability of a trait. Because many traits are usually involved in (socio)ecological applications (here T, N, C), the term Gradient Adaptive Dynamics was introduced to emphasize the usage of the growth-rate gradient of the vector of traits. Here, we explain the published equations in an updated and consistent form.

In a local population B composed of n sub-population members $\iota \in \{1 \ldots n\}$, each member with relative contribution B_ι/B, characteristic traits X_ι, and time-dependent environmental condition $E_\iota(t)$, has a relative growth rate r_ι

$$r_\iota = \frac{1}{B_\iota} \cdot \frac{dB_\iota}{dt} = r_\iota\left(X_\iota, E_\iota(t)\right). \tag{17.1}$$

This equation is often formulated in terms of the population density $P = B/A$, where A is the area populated by B:

$$r_\iota = \frac{1}{P_\iota} \cdot \frac{dP_\iota}{dt} = r_\iota\left(X_\iota, E_\iota(t)\right) \tag{17.2}$$

$$\text{and} \quad \sum_\iota^n (P_\iota/P) = 1.$$

Fig. 17.3 The adaptive dynamics of a characteristic trait X in a fitness landscape $r(X, E(t))$ is described by the width of the trait distribution (σ_X) and the marginal benefit that a small change in X has on the growth rate r. Modified from [29]

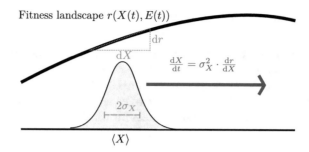

The mean of a quantity X over all individuals ι is calculated as

$$\langle X \rangle = \sum_{\iota=1}^{n} \frac{X_\iota P_\iota}{P}. \tag{17.3}$$

The adaptive dynamics rooted in genetics assumes that mutation errors are only relevant at cell duplication, and not during cell growth. Translated to the ecological entity population this restriction enforces that all traits X_ι of a member of this population are stable during the lifetime of this member: $\frac{d}{dt}X_\iota = 0$ for all X_ι. Changes in the aggregated traits $\langle X \rangle$ are a result of frequency selection (the number of members carrying a specific characteristic trait increases or decreases as a result of selection) only. Differentiating Eq. (17.3) with respect to time and considering $\frac{d}{dt}X_\iota = 0$, gives

$$\frac{d \langle X \rangle}{dt} = \sum_{\iota=1}^{n} \frac{\partial \frac{X_\iota P_\iota}{P}}{\partial t}, \tag{17.4}$$

which can be further simplified to

$$\frac{dX}{dt} = \sigma_X^2 \cdot \frac{\partial r(X)}{\partial X}, \tag{17.5}$$

where $\sigma_X^2 = \langle (X - \langle X \rangle)^2 \rangle$ denotes the variance of X. The angular brackets around $\langle X \rangle$ have been left out for better readability. Figure 17.3 is provided for illustrating the essence of Eq. (17.5): A given distribution in trait (curve around $\langle X \rangle$ of width σ_X) is seen to evolve by being shifted into the direction of increasing "fitness landscape", i.e. into the direction giving rise to higher growth rates.

17.2.3 Local Population Growth

Key to adaptive dynamics is the formulation of the growth rate as a function of all characteristic traits. Once this dependence is specified, the evolution equations for X are generated automatically from Eq. (17.5).

The relative growth rate r of an agent population may obviously be noted as the difference of gain and loss rates for which we use the shorthand notation

$$r = r_{gain} - r_{loss}. \tag{17.6}$$

We are now going to illustrate how gain and loss is reasonably correlated with the characteristic traits and environmental conditions. Corresponding with the high degree of complexity of the system under consideration, gain and loss are subject to quite a substantial number of parameters. Following the introduction in the various "traits" by Table 17.1, a complete overview of these parameters is provided by Table 17.2.

On the way towards quantitating the gain r_{gain} it is useful to introduce a quantity which describes a community's effectiveness in generating consumable food and secondary products. This quantity is referred to as the "subsistence intensity". It is dimensionless and scaled such that a value of unity expresses the mean subsistence intensity of a hunter-gatherer society equipped with tools typical for the mature Mesolithic. With Table 17.1, mature Mesolithic is seen to be characterized by traits $T = 1$ and $C \approx 0$. We note that the first term on the right-hand side of Eq. (17.7), used for quantifying the subsistence intensity s

$$s = (1 - C) \cdot \sqrt{T} + C \cdot N \cdot T \cdot E_{TLI}, \tag{17.7}$$

does exactly reflect this condition by becoming equal to unity, with the second term disappearing for a hunter-gatherer society. The second term, the "agropastoral part", is assumed to increase linearly with N and T: The more economies (N) there are, the better are sub-regional scaled niches utilized and the more reliable returns are generated when annual weather conditions are variable; the higher the technology level (T), the better the efficiency of using natural resources (by definition of T). While a variety of techniques can steeply increase harvests of domesticated species, analogous benefits for foraging productivity are less pronounced, giving rise to a less than linear dependence of the hunting-gathering calorie procurement on T, which is taken into account by considering, in the first term on the right of Eq. (17.7), only the square root of T.

With the parameter E_{TLI} we have introduced an additional temperature constraint on agricultural productivity which considers that cold temperature could only moderately be overcome by Neolithic technologies. While E_{TLI} is thus set unity at low latitudes, it approaches zero at permafrost conditions. The domestication process is represented by N, which is the number of realized agropastoral economies. We link N to natural resources by expressing it as the fraction f of potentially available

Table 17.2 Symbols and variables used in the text and equations. A useful parameter set (see e.g. [23]) is $\mu = \rho = 0.004\ a^{-1}$, $\omega = 0.04$, $\gamma = 0.12$, $\delta_T = 0.025$, $\varsigma_{demic} = 0.002$, $\varsigma_{info} = 0.2$, $\delta_N = 0.9$; and initial values for $P_0 = 0.01$, $T_0 = 1.0$, $N_0 = 0.8$, and $C_0 = 0.04$

Symbol	Description	Unit	Typical range
P	Population density	km^{-2}	>0
X	Growth-influencing trait		>0
T	Technology trait		>0
N	Economic trait		>0
C	Labor allocation trait		0–1
t	Time	a	9500–1000 BCE
r	Specific growth rate	a^{-1}	
f	Economy availability		0–1
E	Environmental constraints		
E_{TLI}	Temperature limitation		0–1
E_{PAE}	Potentially available economies		0–1
E_{FEP}	Food extraction potential		0–1
$\langle \cdot \rangle$	Mean/first moment of $\cdot$		
σ^2	Variance		>0
ς	Diffusion parameter		>0
s	Subsistence intensity		
ω	Administration parameter		
γ	Exploitation parameter		
μ	Fertility rate	a^{-1}	
ρ	Mortality rate	a^{-1}	

economies (E_{PAE}) by specifying $N = f \cdot E_{PAE}$, where the latter corresponds to the richness in domesticable animal or plant species within a specific region.

The increase in s may be accompanied by processes which tend to mitigate rather than to enhance fertility. This concerns, in particular, the overexploitation of natural resources which is taken account of by multiplying s by a factor $(E_{FEP} - \gamma\sqrt{TP})$, where E_{FEP} is introduced as a measure of the multitude of natural resources and γ stands for a suitably chosen scaling parameter. With a second factor $(1 - \omega T)$ one takes account of the so-called organizational losses, which emerge when people neither farm nor hunt: Construction, maintenance, and administration draw a small fraction of the workforce away from food production.

Summing up, the overall gain may be noted as

$$r_{gain} = \mu \cdot (E_{\text{FEP}} - \gamma \sqrt{TP}) \cdot (1 - \omega T) \cdot s \qquad (17.8)$$

where μ has been introduced as a scaling parameter referred to as the gain coefficient or fertility rate. For the loss term one applies the standard ecological form

$$r_{loss} = \rho \cdot P \cdot e^{-T/T_{\text{lit}}}, \qquad (17.9)$$

modeled on the crowding effect (also known as ecological capacity), implying proportionality between loss and population density. It is mediated by technologies (T) which mitigate, for example, losses due to disease, where $T_{\text{lit}} = 12$ proved to serve as a good health standard. The scaling parameter ρ is the equivalent of the fertility rate μ in Eq. (17.8) and referred to as the loss coefficient or mortality rate.

17.2.4   Spatial Diffusion Model

Information, material and people are implied to exchange between the various regions by fluxes which may be modeled by the Fickian diffusion equation (Chap. 2, Eqs. (2.6–2.9)), where the discrete region arrangement and the locally varying diffusivity coefficient D_{ik} have to be taken account of. By adopting the notation of the continuity equation, Eq. (2.8) the change of any characteristic trait X_i in a region i due to diffusion from/to all regions $k \in \mathcal{N}_i$ in its neighborhood $\mathcal{N}_i$ with neighbor distance Δx_{ik} may, thus, be noted as

$$\frac{\Delta X_i}{\Delta t} = \sum_{k \in \mathcal{N}_i} -j_{ik}/\Delta x_{ik}, \qquad (17.10)$$

with $j_{ik} = -D_{ik}\Delta X_{ik}/\Delta x_{ik}$ constituting the diffusive flux between i and k (Fig. 17.4). In this formulation of the diffusive flux we easily recognize the structure of Fick's

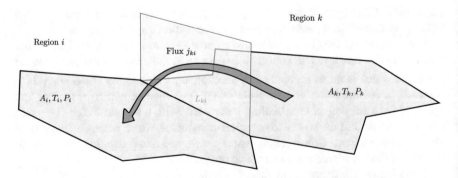

Fig. 17.4   Schematic representation of interregional exchange in GLUES

first law, Eq. (2.6), with the concentration c replaced by the characteristic trait X_i under study and, correspondingly, the particle flux j replaced by the flux of traits, driven by their "gradient" $\Delta X_{ik}/\Delta x_{ik}$ between regions i and k. Insertion into Eq. (17.10) yields

$$\frac{\Delta X_i}{\Delta t} = \sum_{k \in \mathcal{N}_i} D_{ik} \Delta X_{ik} \Delta x_{ik}^{-2}. \tag{17.11}$$

This equation can be reformulated [30] as

$$\frac{\Delta X_i}{\Delta t} = \varsigma \sum_{k \in \mathcal{N}_i} f_{ik} \Delta X_{ik} \tag{17.12}$$

with $f_{ik} = D_{ik} \Delta x_{ik}^{-2} \varsigma^{-1}$, where ς is a global diffusion property characterizing the underlying process (see below) and f_{ik} collects all regionally varying spatial and social diffusive aspects.

The social factor in the formulation of f_{ik} is the difference between two regions' influences, where influence is defined as the product of population density P and technology T, scaled by the average influence of regions i, k. The geographic factor is the conductance between the two regions, which is constructed from the common boundary length L_{ik} divided by the mean area of the regions $\sqrt{A_i A_k}$. Non-neighbor regions have no common boundary, and hence have zero conductance. To connect across the Strait of Gibraltar, the English Channel, and the Bosporus, the respective conductances were calculated as if narrow land bridges connected them. No additional account is made for increased conductivity along rivers [31].

17.2.5 Three Types of Diffusion

Three types of diffusion are distinguished: (1) demic diffusion, i.e. the migration of people, (2) the hitchhiking of traits with migrants, and (3) cultural diffusion, i.e. the information exchange of characteristic traits.

Demic diffusion is the mass-balanced migration of people between different regions. The diffusion equation (17.12) is applied to the number of inhabitants $B_i = P_i A_i$ in each region i.

$$\left.\frac{dB_i}{dt}\right|_{\text{demic}} = \begin{cases} \varsigma_{\text{demic}} \sum_{j \in \mathcal{N}_i} f_{ij}(B_j - B_i), & r \geq 0 \\ 0 & \text{otherwise} \end{cases} \tag{17.13}$$

The free parameter ς_{demic} has to be determined from comparison to data. The parameter estimation based on the European dataset by Pinhasi [30] and the typical front speed extracted from this dataset yields $\varsigma_{\text{demic}} = 0.002$ (see [32] for parameter estimation). We impose an additional restriction to migration by requiring positive growth rate $r_i \geq 0$, i.e. favorable living conditions, in the receiving region i.

Hitchhiking traits: Whenever people move in a demic process, they carry along their traits to the receiving region. Changes in trait are proportional to the number of immigrants (proportional to B_j) and inversely proportional to the number of original inhabitants B_i

$$\frac{dX_i}{dt}\bigg|_{demic} = \begin{cases} \varsigma_{demic} \sum_{j \in \mathcal{N}_i} f_{ij} X_j \frac{B_j}{B_i}, & r_i \geq 0 \\ 0 & \text{otherwise} \end{cases} \tag{17.14}$$

Information exchange: Traits do not decrease when they are exported. Thus, only the positive contribution from the diffusion equation Eq. (17.12) is considered:

$$\frac{dX_i}{dt}\bigg|_{info} = \varsigma_{info} \sum_{j \in \mathcal{N}_i, f_{ij} > 0} f_{ij} \cdot (X_j - X_i) \tag{17.15}$$

The diffusion parameter was estimated to be $\varsigma_{info} = 0.2$ in a reference scenario [21]. Despite the formal similarity of Eqs. (17.14) and (17.15) suggesting a mere factor $\varsigma_{demic}/\varsigma_{info}$ as the difference, the processes are rather different: migration is mass-conserving, information exchange is not (note the summation of only positive f_{ij} for information exchange) and migration is hindered by bad living conditions, information exchange is not.

17.3   Model Applications to Diffusion Questions

Two questions have been addressed with GLUES that are specific to diffusion. First and foremost, the wave front propagation speed was diagnosed from the model with respect to both demic and cultural diffusion [21]. For a mixed demic and cultural diffusion scenario, the authors found a wave front propagation speed of 0.81 km a^{-1} radiating outward of an assumed center near Beirut (Lebanon) in the European dataset, somewhat faster than the speed diagnosed from radiocarbon data (0.72 km a^{-1} [30]). Both in the radiocarbon data and the model simulation, however, there is large scatter from the linear time-distance relationship, with a lower than average propagation speed in the Levante before 7000 BCE, and with higher than average propagation speed with the expansion of the Linearbandkeramik (LBK) in the sixth millennium BCE.

It was also found, that there is a regionally heterogeneous contribution of demic and cultural diffusion, and of local innovation in the simulated transition to agropastoralism. While either diffusion mechanism is necessary for a good reconstruction of the emergence of farming, the major contribution to local increases in T or C is local innovation. Diffusion (its contribution is in many regions around 20% to the change in an effective variable) seems to have been a necessary trigger to local invention.

Not only is the contribution of diffusive processes heterogeneous in space, but it also varies in time. This was shown by studying the interregional exchange fluxes in the transition to farming for Eurasia with GLUES [32]. Most Eurasian regions exhibited an equal proportion of demic and cultural diffusion events when integrated over time, with the exception of some mountainous regions (Alps, Himalayas), where demic diffusion is probably overestimated by the model: the higher populations in the surrounding regions may lead to a constant influx of people into the enclosed and sparsely inhabited mountain region.

When time is considered, however, it appears that diffusion from the Fertile Crescent is predominantly demic before 4900 BCE, and cultural thereafter; that east of the Black Sea, diffusion is demic until 4200 BCE, and cultural from 4000 BCE. The expansion of Southeastern and Anatolian agropastoralism northward is predominantly cultural at 5500 BCE, and predominantly demic 500 years later. At 5000 BCE, it is demic west of the Black Sea and cultural east of the Black Sea; at 4500 BCE, demic processes again take over part of the eastern Black Sea northward expansion. This underlines that "Previous attempts to prove either demic or cultural diffusion processes as solely responsible [. . .] seem too short-fetched, when the spatial and temporal interference of cultural and diffusive processes might have left a complex imprint on the genetic, linguistic and artefactual record" [32].

Unlike in many other models, the diffusion coefficient D here is an emergent property, that varies in space and time, and that varies among all neighbors of each region. The diffusion coefficient varies between zero and 7 km^2 a^{-1}; Fig. 17.5 shows the topology of the interregional connections in Europe and their maximum diffusion coefficients. Maximum diffusion is highest on the Balkan and within Italy (up to 4 km^2 a^{-1}), it is one order of magnitude lower for all of Northern Europe. This shows the importance of the Balkans as a central hub for the diffusion of Neolithic technology, people, and ideas; there seem to have been main routes for Neolithic diffusion across the Central Balkan, along Adriatic coastlines, or, to a lesser extent, up the Rhône valley.

The diffusion coefficient D seems first and foremost to match the migration rate of populations of ultimately Anatolian/Near Eastern ancestry into and within Europe. On a continental scale this rate should have been higher in Southeastern Europe and possibly in Italy, equally along the Rhône. This is supported by recent archeological and archeogenetic data, at least for Southeastern and Central Europe [16, 17]. Therefore, it is to be assumed that the proportion of non-indigenous populations should have been highest in these areas. Towards the north the spread of these immigrant Neolithic populations was halted until about 4000 BCE, after which farming spread further across the northern and northwestern European continent as well as to the British Isles. This stagnation pattern is visible from archeological evidence [13, 33] and represented in model simulations [34]. Towards the continental west the evidence for a lesser proportion of allochtonous cultural traits in the archeological record of farming societies has continuously been interpreted as an increase in indigenous populations within these societies; therefore the rate of immigrants

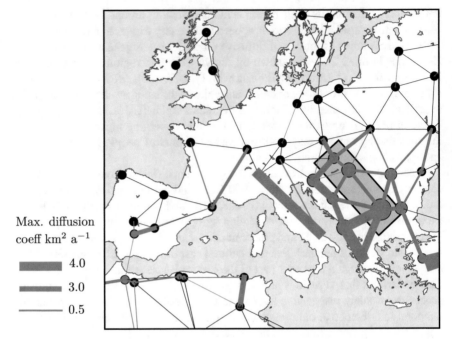

Fig. 17.5 Topology of European regional connections and maximum diffusion coefficient for each region. Circles represent geographic centers of regions, red circles highlight regions with large maximum influence; the size of the highlighted connections represents the maximum diffusion coefficient between two adjacent region. Shading indicates the three regions analysed in Fig. 17.6, labelled Bulgaria, Serbia, Hungary (from south to north)

should have been lower. This has at least been suggested by archeology [35]; recent genetic studies have shown, however, that the influx of a population of ultimately Balkanic/Anatolian origin seems also to have been strong in the Paris Basin and Eastern France [36].

While the simulated Neolithic transition is reasonably well reflected on the continental scale, the model skill in representing the individual regional spatial expansions varies. For example, the particular geographic expansion of the LBK in Central Europe occurs too late and is too small in extent towards the Paris Basin. On the other hand, the timing of the arrival of the Neolithic in the Balkans, in Southern Spain, or in Northern Europe is well represented [34].

For three selected regions along the Central Balkan diffusion main route (highlighted in Fig. 17.5) we analysed the temporal evolution of their diffusion coefficients (Fig. 17.6). A similar pattern is visible in all three regions and all diffusion coefficients: D starts at zero, then rapidly rises to a marked peak and slowly decays asymptotically to an intermediate value. This behavior is a consequence of the local influence and its difference to adjacent regions. Initially the influence difference is zero, because all regions have similar technology and population. As soon as one

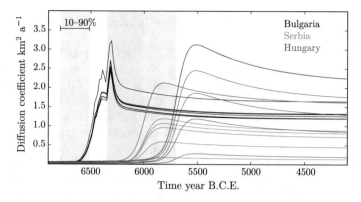

Fig. 17.6 Time evolution of the diffusion coefficient for three selected regions of Central Europe (Bulgaria, Serbia, Hungary; shown in black, red, and blue, respectively). For each of these three regions, the family of trajectories represents the region's diffusion coefficients with respect to each of its adjacent regions. The shaded bars indicate the time interval of a regional transition to agropastoralism in the simulation (10–90% of C)

region innovates (or receives via diffusion technology and population from one of its neighbors), population and technology increase, and so does the influence difference to all other neighbors. With an increase in influence and diffusion coefficient, demic and cultural diffusion to neighbors decrease the influence differences. Relative proportions among the diffusion coefficients of one region to all its neighbors are constant and attributed to the geographical setting.

The time evolution of the diffusion coefficient plotted in Fig. 17.6 reflects the population statistics for advancing Neolithic technology: Early farming appears to be associated to a rapid increase in population, this on a supra-regional scale [37, 38]. At the regional level, the diffusion coefficient lags the onset of farming by several hundred years. This lag is also empirically reflected in the data set of the Western LBK [39]. Any pioneering farming society seems to have followed more or less the same general population trajectory with a gradual increase over several centuries, followed by a sudden rise-and-decline. The causes for this general pattern are yet unclear, but may have to be sought more in social behavior patterns rather than purely economic or environmental determinants [40].

17.4   Conclusions

It has been long evident, that the Neolithic "Revolution" is not a single event, but heterogeneous in space and time. Statistical models for understanding the diffusion processes, however, have so far assumed that a physical model of Fickian diffusion can be applied to the pattern of the emergence of farming and pastoralism using constant diffusion coefficients. Relaxing this constraint, and reformulating the diffusivity as a function of influence differences between regions, demonstrates how diffusivity varies in space and time.

When results using this variable correlation coefficient (D) are compared to empirical archeological data, they represent the dynamics on a continental scale and on the regional scale for many regions well, but not for all: The impetus of the Neolithic in Greece and the Balkans is well represented, also in Southeastern Central Europe. The emergence and the expansion of the Central European LBK shows, however, a too early expansion in the model, whereas the stagnation following the initial expansion is again very well represented.

Divergence between the mathematical model and the empirical findings provided by archeology is unsurprising and expected, because human societies behave in much more complex ways than are described in the highly aggregated and simplified model. Individuals may have chosen to act independent of the social and environmental context and against rational maximization of benefits. Rather than perfectly capturing each regional diffusion event, the mathematical model serves as a null hypothesis which is broadly consistent with the archeologically reconstructed picture, and against which individual decisions can be assessed. In this respect, the simple model helps to disentangle in complex histories general forcing agents and individual choices.

The numerical model and necessary datasets have been publicly released under an open source license. The code is available from SourceForge (https://sf.net/p/glues/).

References

1. V.G. Childe, *Dawn of European Civilization*, 1st edn. (Routledge [reprinted 2005], 1925), p. 256
2. N. Goring-Morris, A. Belfer-Cohen, Current Anthropology, vol. 52 (2011)
3. G. Barker, *The Agricultural Revolution in Prehistory: Why Did Foragers Become Farmers?* (Oxford University Press, Oxford, United Kingdom, 2006), p. 616
4. W.F. Ruddiman, D.Q. Fuller, J.E. Kutzbach, P.C. Tzedakis, J.O. Kaplan, E.C. Ellis, S.J. Vavrus, C.N. Roberts, R. Fyfe, F. He, C. Lemmen, J. Woodbridge, Rev. Geophys. **54**, 93 (2016)
5. J.G.D. Clark, Proc. Prehist. Soc. **31**, 57 (1965)
6. M.-J. Gaillard, S. Sugita, F. Mazier, A.-K. Trondman, A. Broström, T. Hickler, J.O. Kaplan, E. Kjellström, U. Kokfelt, P. Kuneš, C. Lemmen, P. Miller, J. Olofsson, A. Poska, M. Rundgren, B. Smith, G. Strandberg, R. Fyfe, A.B. Nielsen, T. Alenius, L. Balakauskas, L. Barnekow, H.J.B. Birks, A. Bjune, L. Björkman, T. Giesecke, K. Hjelle, L. Kalnina, M. Kangur, W.O. van der Knaap, T. Koff, P. Lagerås, M. Latałowa, M. Leydet, J. Lechterbeck, M. Lindbladh, B. Odgaard, S. Peglar, U. Segerström, H. von Stedingk, H. Seppä, Clim. Past **6**, 483 (2010)
7. H.J. Zahid, E. Robinson, R.L. Kelly, Proc. Natl. Acad. Sci. 201517650 (2015)
8. B.G. Purzycki, C. Apicella, Q.D. Atkinson, E. Cohen, R.A. McNamara, A.K. Willard, D. Xygalatas, A. Norenzayan, J. Henrich, Nature **530**, 327 (2016)
9. J.-P. Bocquet-Appel, S. Naji, M. Van der Linden, J. Kozlowski, J. Archaeol. Sci. **39**, 531 (2012)
10. F. Silva, J. Steele, K. Gibbs, P. Jordan, Radiocarbon **56**, 723 (2014)
11. K. Manning, S.S. Downey, S. Colledge, J. Conolly, B. Stopp, K. Dobney, S. Shennan, Antiquity **87**, 1046 (2013)
12. A. Bogaard, *Neolithic Farming in Central Europe : An Archaeobotanical Study of Crop Husbandry Practices* (Routledge, 2004), p. 210

13. W. Schier, Praehistorische Zeitschrift **84**, 15 (2009)
14. F.-X. Ricaut, Adv. Anthropol. **02**, 14 (2012)
15. W. Haak, I. Lazaridis, N. Patterson, N. Rohland, S. Mallick, B. Llamas, G. Brandt, S. Nordenfelt, E. Harney, K. Stewardson, Q. Fu, A. Mittnik, E. Bánffy, C. Economou, M. Francken, S. Friederich, R.G. Pena, F. Hallgren, V. Khartanovich, A. Khokhlov, M. Kunst, P. Kuznetsov, H. Meller, O. Mochalov, V. Moiseyev, N. Nicklisch, S.L. Pichler, R. Risch, M.A. Rojo Guerra, C. Roth, A. Szécsényi-Nagy, J. Wahl, M. Meyer, J. Krause, D. Brown, D. Anthony, A. Cooper, K.W. Alt, D. Reich, Nature **522**, 207 (2015)
16. G. Brandt, A. Szécsényi-Nagy, C. Roth, K.W. Alt, W. Haak, J. Hum. Evol. **79**, 73 (2015)
17. Z. Hofmanová, S. Kreutzer, G. Hellenthal, C. Sell, Y. Diekmann, D. Díez del Molino, L. van Dorp, S. López, A. Kousathanas, V. Link, K. Kirsanow, L.M. Cassidy, R. Martiniano, M. Strobel, A. Scheu, K. Kotsakis, P. Halstead, S. Triantaphyllou, N. Kyparissi-Apostolika, D.-C. Urem-Kotsou, C. Ziota, F. Adaktylou, S. Gopalan, D.M. Bobo, L. Winkelbach, J. Blöcher, M. Unterländer, C. Leuenberger, Ç. Çilingiroglu, B. Horejs, F. Gerritsen, S. Shennan, D.G. Bradley, M. Currat, K. Veeramah, D. Wegmann, M.G. Thomas, C. Papageorgopoulou, J. Burger, Proc. Natl. Acad. Sci. USA (2016)
18. A. Ammerman, L. Cavalli Sforza, *The Neolithic Transition and the Population Genetics of Europe* (Princeton University, Princeton, NJ, 1984)
19. J. Fort, Proc. Natl. Acad. Sci. USA **109**, 18669 (2012)
20. K.W. Wirtz, C. Lemmen, Clim. Change **59**, 333 (2003)
21. C. Lemmen, D. Gronenborn, K.W. Wirtz, J. Archaeol. Sci. **38**, 3459 (2011)
22. C. Lemmen, Archaeol. Ethnol. Anthropol. Eurasia **41**, 48 (2013)
23. C. Lemmen, A. Khan, in *Climates, Landscapes, and Civilizations*, ed. by L. Giosan, D.Q. Fuller, K. Nicoll, R.K. Flad, P.D. Clift (American Geophysical Union, Washington, 2012), pp. 107–114
24. B. Arıkan, J. Archaeol. Sci. **43**, 38 (2014)
25. E. Boserup, *Population and Technological Change* (University of Chicago Press, 1981), p. 255
26. C Lemmen, in *Ester Boserup's Legacy on Sustainability: Orientations for Contemporary Research*, ed. by M. Fischer-Kowalski, A. Reenberg, A. Schaffartzik, A. Mayer (Springer, Vienna, 2014), pp. 87–97
27. R.A. Fisher, *The Genetical Theory of Natural Selection* (At the Clarendon, Oxford, 1930), p. 308
28. K.W. Wirtz, B. Eckhardt, Ecol. Modell. **92**, 33 (1996)
29. K.W. Wirtz, Modellierung von Anpassungsvorgängen in der belebten Natur. Ph.D. thesis, Universität Kassel, 1996
30. R. Pinhasi, J. Fort, A.J. Ammerman, Public Libr. Sci. Biol. **3** (2005)
31. K. Davison, P.M. Dolukhanov, G.R. Sarson, A. Shukurov, J. Archaeol. Sci. **33**, 641 (2006)
32. C. Lemmen, Documenta Praehistorica **XLII**, 93 (2015)
33. A.W.R. Whittle, F.M.A. Healy, A. Bayliss, *Gathering Time: Dating the Early Neolithic Enclosures of Southern Britain and Ireland* (Oxbow Books, 2011)
34. C. Lemmen, K.W. Wirtz, J. Archaeol. Sci. **51**, 65 (2014)
35. C. Jeunesse, S. Van Willigen, *The Spread of the Neolithic to Central Europe* (Mainz, Germany, 2010), pp. 569–605
36. M. Rivollat, H. Réveillas, F. Mendisco, M.-H. Pemonge, P. Justeau, C. Couture, P. Lefranc, C. Féliu, M.-F. Deguilloux, Am. J. Phys. Anthropol. **161**, 522 (2016)
37. J.-P. Bocquet-Appel, Science **333**, 560 (2011)
38. S.J. Shennan, S.S. Downey, A. Timpson, K. Edinborough, S. Colledge, T. Kerig, K. Manning, M.G. Thomas, Nat. Commun. **4**, 1 (2013)
39. D. Gronenborn, H.-C. Strien, S. Dietrich, F. Sirocko, J. Archaeol. Sci. **51**, 73 (2014)
40. D. Gronenborn, H.-C. Strien, C. Lemmen, Quat. Int. **54**, 446 (2017)

Chapter 18
Modeling Language Shift

Anne Kandler and Roman Unger

18.1 Introduction

Languages behave similarly to living species [1, 2] (see also Chap. 3). They display diversity, differentiate in space and time, emerge and disappear. While processes of differentiation happen at a relatively slow rate with a typical timescale of the order of 1,000 years to evolve into different languages (e.g. [3, 4]) language extinction takes place at a substantially faster rate [5]. Language birth and extinction are natural processes that have taken place ever since language came into existence but the current linguistic extinction rate is immense; it even exceeds the rate of loss of biodiversity (e.g. [6–8]). It is estimated that half of the world's languages existing today will disappear in the 21st century [7]. Serious concerns over the loss of linguistic diversity, seen as a benchmark for overall cultural diversity, have driven governments and international organizations to actively engage in the conservation of endangered languages [6, 9, 10].

Most recent language extinction events are caused by language shift rather than the extinction of the population speaking this language [5]. Language shift is defined as the process where members of a community in which more than one language is spoken abandon their original vernacular language in favor of another. Knowledge of a language can selectively facilitate and inhibit interaction, enable social contracts and cooperative exchange and give access to accumulated and linguistically encoded knowledge [11]. Therefore in language contact situations people are confronted with choices about which language to speak. Now in the

A. Kandler (✉)
Department of Human Behavior, Ecology and Culture, Max Planck Institute
for Evolutionary Anthropology, Deutscher Platz 6, 04103 Leipzig, Germany
e-mail: anne_kandler@eva.mpg.de

R. Unger
Technische Universität Chemnitz, Straße der Nationen 62, 09111 Chemnitz, Germany
e-mail: roman.unger@mathematik.tu-chemnitz.de

© Springer International Publishing AG 2018
A. Bunde et al. (eds.), *Diffusive Spreading in Nature, Technology and Society*,
https://doi.org/10.1007/978-3-319-67798-9_18

course of globalization and of recent trends for urbanization and long-distance economic migration, interactions between groups speaking different languages have increased and so has the need for a common language of communication. Language shift is initiated by the decision to abandon a more local or less prestigious language, typically because the target of the shift is a language seen as more modern, useful or giving access to greater social mobility and economic opportunities [1, 5, 12]. But crucially language shift is not caused by cultural selection acting on particular features of a language but by people shifting between two languages because of their perceived benefits [13].

A language dies with its last speaker. In other words, a language is not a self-sustaining entity; it can only exist when there is a community to speak and transmit it [14]. The number of speakers can therefore be interpreted as a measure of the 'health' of the language. In mathematical terms we can consider the process of language shift as a competition where two or more languages compete for speakers. Modeling the competition dynamics between interacting species has a long tradition in the ecological literature (see e.g. [15]) and based on the similarity between the ecological and linguistic situations a number of mathematical approaches have been proposed to describe the temporal and spatial dynamics of language shift. These approaches can potentially contribute to a better understanding of the process of language shift by identifying the demographic, socio-economic, cultural and/or linguistic processes that are needed to explain observed patterns of real-world language shift scenarios. The formal analysis of those mathematical models can then inform about the long-term outcome of language shift provided the competition environment stays unchanged or changes according to the scenario assumed in the model. We stress that even if a model can replicate past demographic trajectories accurately the validity of prediction about the future course of language shift depends on the validity of the assumptions about the future state of the competition environment. Additionally, mathematical models can be used as an artificial experiment. They can give an indication about what needs to be changed in order to alter the language shift dynamics. In this context, models can provide useful information about the potential success of different intervention strategies, in particular they can inform about the total intervention strength that is needed to achieve a desired goal, e.g. the stabilization of the bilingual population group.

In this chapter we provide a brief overview of the recent literature on modeling language shift with special emphasis on spatial dynamics. Further, we introduce a diffusion-reaction approach and illustrate its usefulness for questions related to revitalization efforts. Further, we apply this framework to the English-Gaelic language shift situation in Western Scotland and demonstrate what kind of information can be obtained from mathematical modeling efforts.

18.2 Modeling Approaches

Research into mathematical modeling of the dynamics of language shift has gained momentum with the seminal paper by Abrams and Strogatz [16]. However, already before its publication a number of modeling approaches had been published on this subject (e.g. [17, 18]). In this section we briefly introduce the Abrams and Strogatz model [16] and some of its generalization with particular focus on spatial dynamics. We stress that we do not provide an exhaustive literature review but concentrate on diffusion-reaction approaches to language shift.

18.2.1 Abrams and Strogatz Model

The modeling framework proposed by Abrams and Strogatz [16] assumes that two mutually unintelligible languages A and B compete for a fixed number of potential speakers. The time-dependent variables $n_i, i = 1, 2$ describe the relative frequencies of monolingual speakers of languages A and B, respectively (each individual speaks either language A or B). In other words, the variables n_i describe the fractions of speakers of either language in the population. Further, population size is assumed to be constant and therefore it holds $n_1 + n_2 = 1$. The dynamics of language shift is governed by the following differential equation

$$\frac{dn_1}{dt} = n_2 P_{12}(n_1, s) - n_1 P_{21}(n_2, 1 - s). \tag{18.1}$$

The variable s describes the perceived relative status of language A on a scale from 0 to 1 and reflects the social and economic opportunities afforded to its speakers. The status of language B is given by $1 - s$. The term P_{12} denotes the probability that an individual speaking language B converts to speaking language A. (For consistency purposes we assume in the remainder of this chapter that the first index of shift probabilities and coefficients indicates the target of the shift process while the second index describes the source. Further, index 1 stands for language A and index 2 for language B.) This probability is assumed to be frequency- and status-dependent and is given by the power law

$$P_{12}(n_1, s) = m n_1^a s. \tag{18.2}$$

The exponent a models the level of resistance of monolingual speakers to change their language [19], and the coefficient m controls the peak rate at which speakers of language B shift to language A. Similarly, P_{21} denotes the probability that an individual speaking language A converts to speaking language B

$$P_{21}(n_2, 1-s) = mn_2^a(1-s). \tag{18.3}$$

Consequently, the higher the status of a language and the higher the number of its speakers, the more speakers the language will recruit per time unit.

Model (18.1) predicts that one language (depending on status and initial frequencies) will always go extinct over time. Abrams and Strogatz [16] fitted their model to time series data describing several language shift situations (including the English-Gaelic shift in the county Sutherland and the Spanish-Quechua shift in Huanuco) and were able to accurately describe the observed temporal language shift dynamics.

Based on these results model (18.1) appeared to be a promising approach to model language shift. Nevertheless, the framework rests on assumptions which have been deemed unrealistic. In particular, model (18.1) assumes that (i) languages are fixed, (ii) the population is highly connected with no spatial or social structure, (iii) all speakers are considered monolingual, (iv) population size is assumed to be constant and (v) the use and usefulness of both competing languages is the same in all social contexts. Subsequent modeling approaches generalized the Abrams and Strogatz model (18.1) by addressing one or more of these shortcomings.

18.2.2   Generalizations

Generalizations of the modeling idea by Abrams and Strogatz [16] can be broadly divided into differential equation-based approaches and simulation-based approaches (for a review of this literature see e.g. [20–22]). While some approaches focused on a more detailed description of the demographic properties of the interacting population groups others focused on a more realistic description of the process of language shift. In the following we concentrate on equation-based modeling frameworks with a particular emphasis on spatial dynamics. Nevertheless, we note that a large amount of research has been devoted to simulation-based approaches and produced crucial insights into the process of language shift (e.g. [23–28]).

It is widely known that the introduction of spatial dynamics can lead to the emergence of qualitative changes in dynamical patterns, in particular it can change the interaction dynamics between populations (e.g. [15, 29, 30]). To explore this fact, Patriarca and Leppänen [31] introduced spatial dependence into model (18.1) by formulating a diffusion-reaction system of the form (see also Eq. (2.19) in Chap. 2)

$$\begin{aligned}
\frac{\partial c_1}{\partial t} &= D_1 \Delta c_1 + c_2 P_{12}(n_1, s) - c_1 P_{21}(n_2, 1-s) \\
\frac{\partial c_2}{\partial t} &= D_2 \Delta c_2 + c_1 P_{21}(n_2, 1-s) - c_2 P_{12}(n_1, s).
\end{aligned} \tag{18.4}$$

The space- and time-dependent variables c_i describe the absolute frequencies of the two population groups speaking language A or B at location x and time t, i.e. the variables c_i stand for the total number of speakers of both languages in the population. As above, the time-dependent variables n_i denote the relative frequencies of both languages in the population. Spatial dispersal is modeled by the diffusion components $D_i \Delta c_i$. The diffusion coefficients D_i are measures of the spatial mobility of both population groups. The shift probabilities P_{12} and P_{21} are given by Eqs. (18.2) and (18.3).

The analysis of model (18.4) revealed that if two languages of different status exist initially in the same spatial domain, only extinction states are stable equilibria. But if the initial "home ranges" of the two language groups speaking languages A and B are spatially separated and if the shift probabilities P_{12} and P_{21} in those home ranges depend only on the relative frequencies of languages A and B in the local populations then the two languages will exclude each other in their home ranges but coexist globally (see also [21] for additional analyses). Summarizing, competition under spatial structure can generate novel results: since competitive exclusion depends on initial conditions, both equilbria are (locally) possible.

Kandler and Steele [32] expanded model (18.4) by allowing for population growth (see also Chap. 2, Eqs. (2.15) and (2.18) for an introduction to logistic growth processes). They analyzed the system

$$\frac{\partial c_1}{\partial t} = D_1 \Delta c_1 + \alpha_1 \left(1 - \frac{c_1}{K - c_2} \right) + k c_1 c_2$$
$$\frac{\partial c_2}{\partial t} = D_2 \Delta c_2 + \alpha_2 \left(1 - \frac{c_2}{K - c_1} \right) - k c_1 c_2$$

(18.5)

where population growth is model by the logistic growth terms $\alpha_i \left(1 - \frac{c_i}{K - c_j} \right), i \neq j$.

The coefficients α_i express the internal growth rates of both population groups and K defines the carrying capacity, i.e. the upper limit of the population size regardless of the language spoken at location x any time t. This naturally leads to the condition $c_1 + c_2 \leq K, \forall t, x$. The coefficient k represents the shift coefficient and, analogously to model (18.1), expresses the difference of social and economic opportunities afforded to the speakers of both languages. Model (18.5) predicts that coexistence between two languages of different status is not possible, even when initially spatially separated. For $k > 0$ language A will always prevail in competition (and similarly language B for $k < 0$) and the spatial dynamics of the extinction process shows a travelling wave-like pattern. Coexistence between the two languages, however, is possible when their status differences change between spatial regions. This can be implemented in model (18.5) by allowing the shift coefficient $k = k(x)$ to be space-dependent.

Walters [33] considered a system similar to (18.5) but assumed separate carrying capacities for both population groups. Global stability analysis indicated that, subject to appropriate parameter constraints, extinction and coexistence states might be stable, depending on the initial number of speakers of both languages.

Patriarca and Heinsalu [34] presented a diffusion-reaction framework with an additional advection term and analyzed the influence of external factors not related to the cultural transmission process on the language shift dynamics. They showed that the initial distribution of the population groups speaking languages A and B, geographic boundaries as well as spatial inhomogeneities strongly affect the dynamics of language shift.

Fort and Perez-Losada [19] used an integral formulation and described population dispersal through a dispersal kernel ansatz, which also accounted for a cohabitation effect (defined as joint dispersal of new-borns with their parents). The diffusion kernel describes the probability distribution of different migration distances and allows for the analysis of non-local dispersal patterns. Based on this model they estimated the speed with which a novel language spreads into a region. They applied their model to the English-Welsh language shift in Wales and showed that the predicted front speed coincided reasonably well with the observed speed. Further, Fort and Perez-Losada [19] concluded that the dynamics of language shift is more sensitive to linguistic parameters (i.e. the model parameters controlling the strength of the shift process) than to reproductive and dispersal parameters.

Isern and Fort [35] described the dynamics of language shift by the following one-dimensional diffusion-reaction approach

$$
\frac{\partial c_1}{\partial t} = D \frac{\partial^2 c_1}{\partial x^2} + \alpha c_1 \left(1 - \frac{c_1 + c_2}{K}\right) + \frac{k}{(c_1 + c_2)^{\eta + \lambda - 1}} c_1^{\eta} c_2^{\lambda}
$$
$$
\frac{\partial c_2}{\partial t} = D \frac{\partial^2 c_2}{\partial x^2} + \alpha c_2 \left(1 - \frac{c_1 + c_2}{K}\right) - \frac{k}{(c_1 + c_2)^{\eta + \lambda - 1}} c_1^{\eta} c_2^{\lambda}.
$$

(18.6)

Spatial dispersal is again modeled by a diffusion process and D denotes the diffusion coefficient. The coefficient α describes the growth rate, $k (> 0)$ is a time-scaling parameter and the coefficients η, $\lambda \geq 1$ control the status differences of both languages. In model (18.6) extinction of the language B is inevitable. As in [20] the spatial extinction dynamics showed a travelling wave-like pattern. Isern and Fort [35] derived estimations of the front speeds and applied the model to a number of historical case studies of language shift. They showed that the fit of model (18.6) is comparable to the fit of the original Abrams and Strogatz model (18.1).

Zhang and Gong [36] pointed out that all modeling approaches mentioned above rely on model fitting procedures, especially for determining the status s of a language and the shift coefficients, when applied to real world data. This implies that the temporal (and spatial) resolution of the observed frequency data needs to be sufficiently high so that reliable estimates of the model parameters can be obtained. To circumvent this problem they proposed a more mechanistic approach. They assumed that the temporal language shift dynamics can be described by the following system

$$\frac{dc_1}{dt} = \alpha_1 c_1 \left(1 - \frac{c_1}{K_1} - k_1 \frac{c_2}{K_2} \right)$$
$$\frac{dc_2}{dt} = \alpha_2 c_2 \left(1 - \frac{c_2}{K_2} - k_2 \frac{c_2}{K_1} \right)$$

(18.7)

where K_i describes the maximum population sizes of the monolingual population groups speaking languages A or B. Crucially, Zhang and Gong [36] determined the values of α_i and k_i externally through the so called language diffusion principle (based on Fourier's law of heat conduction) and the language inheritance principle. They applied model (18.7) to a number of case studies and concluded that historical shift trajectories could be well replicated.

All modeling approaches discussed so far have been mainly concerned with a more realistic description of reproduction and dispersal properties of the population groups speaking different languages but assumed that language shift happens instantaneously. Monolingual speakers of one language convert directly to be monolingual speakers of another language. Below we briefly introduce how the concept of bilingualism can be incorporated in models of language shift.

Mira and Paredes [37] generalized the original Abrams and Strogatz model (18.1) by adding a third, bilingual population group whose frequency is denoted by n_3. This means individuals can be either monolingual in language A or B, or bilingual in both languages. Population size is still assumed to be constant and therefore it holds $n_1 + n_2 + n_3 = 1$. Further, they introduced the parameter κ, which measures the likelihood that two monolingual speakers of the languages A and B can communicate with each other ($\kappa = 0$ means that both languages are mutually unintelligible and therefore no communication is possible; an increase in κ signals increasing similarities between the languages) and proposed the following system of differential equations

$$\frac{dn_1}{dt} = (n_2 + n_3)P_{12}(n_1, s) - n_1(P_{21}(n_2, 1-s) + P_{31}(n_2, 1-s))$$
$$\frac{dn_2}{dt} = (n_1 + n_3)P_{21}(n_2, 1-s) - n_2(P_{12}(n_1, s) + P_{32}(n_1, s)).$$

(18.8)

Similarly to the relations (18.2) and (18.3) of the Abrams and Strogatz approach (18.1) the shift probabilities are defined by

$$P_{21}(n_2, 1-s) = m(1-\kappa)(1-n_1)^a(1-s), P_{31}(n_2, 1-s) = m\kappa(1-n_1)^a(1-s),$$
$$P_{12}(n_1, s) = m(1-\kappa)(1-n_2)^a s, P_{32}(n_1, s) = m\kappa(1-n_2)^a s,$$
$$P_{13}(n_1, s) = P_{12}(n_1, s) \text{ and } P_{23}(n_2, 1-s) = P_{21}(n_2, 1-s).$$

This implies that the probability for speakers of, for example, language A to adopt language B is divided between the probability of becoming bilingual, P_{31}, and the probability of becoming monolingual in language B, P_{21}. The more similar the languages, i.e. the closer κ is to 1, the more likely individuals become bilingual (instead of abandoning their mother tongue) [37]. Mira and Paredes [37] showed that coexistence between languages of different status is possible given they are sufficiently similar to each other. Importantly, the low-status language is maintained in the bilingual population. Model (18.8) predicts that in the long term only the bilingual and monolingual population group speaking the high-status language can coexist. They fitted their model to time series data collected for the language shift between Castillian Spanish and Galician and found a good coincidence.

Minett and Wang [38] also analyzed the effect of a bilingual strategy on the language shift dynamics but considered mutually unintelligible languages A and B. They introduced a bilingual population group, as above its frequency is denoted by n_3, and assumed that language shift cannot happen directly but must involve a transitional bilingual state, i.e. once individuals have acquired a language they cannot lose it over their life time. Further, they assumed a constant population size (which implies $n_1 + n_2 + n_3 = 1$) and formulated the following system of differential equations

$$
\begin{aligned}
\frac{dn_1}{dt} &= \mu m_{13} s (1 - n_1 - n_2) n_1^a - (1 - \mu) m_{31} (1 - s) n_1 n_2^a \\
\frac{dn_2}{dt} &= \mu m_{23} (1 - s)(1 - n_1 - n_2) n_2^a - (1 - \mu) m_{32} s n_2 n_1^a.
\end{aligned}
\tag{18.9}
$$

The variables m_{ij} define the peak attractiveness of state i on individuals in state j (As before state 1 stands for monolingual language A, state 2 for monolingual in language B and state 3 for bilingual). The coefficient μ describes the mortality rate at which adults are replaced by children. The first terms in both equations of model (18.9) describe the dynamics of vertical transmission. It is assumed that children of monolingual parents necessarily acquire the language of their parents but children to bilingual parents can adopt either or both of the competing language [38]. In contrast, the second terms describe the dynamics of horizontal transmission and therefore the process of becoming bilingual. Model (18.9) predicts the extinction of one language: the possibility of bilingualism alone cannot produce coexistence. Minett and Wang [38] went on and explored how coexistence could be engineered externally. They found that raising the status of the endangered language whenever its frequency falls below a certain threshold together with isolations of the two languages by encouraging monolingual education of children can result in coexistence between all three population groups.

Parshad et al. [39] investigated the effect of a hybrid Hinglish code-switching population group on the competition between English and Hindi in India using a three-population diffusion-reaction system (Hinglish stands for the macaronic hybrid use of English and South Asian languages.). They found that coexistence between a Hindi-English bilingual group, a Hindi monolingual group and a Hinglish group is possible and argued that this might be the most realistic outcome.

18.3 Diffusion-Reaction Models with Bilingual Transition State

In this section we describe the modeling approaches taken in [11, 20]. We assume the existence of two mutually unintelligible languages A and B in a bounded, two-dimensional domain G, which compete for speakers. The time- and space-dependent variables $c_1(t,x)$ and $c_2(t,x)$ describes the frequencies of the two monolingual population groups (speaking languages A and B, respectively) and $c_3(t,x)$ the frequency of the bilingual population group at time t and location $x \in G$. Being bilingual in this context simply means being proficient in both languages.

18.3.1 Basic Model

We start our analysis by describing the properties of the process of language shift when its temporal and spatial dynamics is solely driven by the frequencies of the three population groups, the demographic and cultural attributes of these groups and the benefits both languages convey to their speakers. In the following, we assume that the temporal changes of the frequencies c_i of the three population groups are determined by

- Spatial spread processes, $D_i \Delta c_i$ with $i = 1, 2, 3$,
- Processes of biological and cultural reproduction, $\alpha_i c_i \left(1 - \frac{c_i}{K - c_j - c_k} \right)$ with $i, j, k = 1, 2, 3$ and $i \neq j \neq k$ and
- Processes of language shift, $k_{ij} c_i c_j$ with $i, j = 1, 2, 3$ and $i \neq j$.

In more detail, the spatial mobility of individuals of each population group is modeled by the diffusion terms $D_i \Delta c_i$. The diffusion coefficients D_i are a measure of the scale of spatial interactions within the different groups. Spatial dispersal therefore has only a local dimension (for a comprehensive review of the application of diffusion processes to human dispersal see [22]).

The logistic growth terms $\alpha_i c_i (1 - c_i/(K - c_j - c_k))$ model biological and cultural reproduction in each population group whereby the coefficients α_i express the growth rates. The variable K defines the upper limit of the population size regardless of the language spoken at any time t and location x. This naturally leads to the condition $c_1 + c_2 + c_3 \leq K, \forall t, x$.

Language shift is modeled by frequency-dependent shift terms $k_{ij} c_i c_j$. The coefficients $k_{ij}, i, j = 1, 2, i \neq j$ quantify the pressure one language puts on monolingual speakers of the other language. But it is assumed that language shift cannot happen by directly shifting from speaking one language only to speaking another language only but must involve a bilingual transition state. Therefore the consequences of the exerted pressure are that monolingual speakers become bilingual, and the coefficients k_{12} and k_{21} can be interpreted as the rate at which monolingual speakers become bilingual due to the attractiveness or status of the other language. Similarly to [38], we define $k_{12} = \tilde{k}_{12}s$ and $k_{21} = \tilde{k}_{21}(1 - s)$ where the variable s describes the status difference between the two languages ranging from 0 to 1. The status of a language, very simplistically, quantifies the social, cultural, economic or political opportunities afforded to its speakers [16]. The coefficients $\tilde{k}_{12}$ and $\tilde{k}_{21}$ indicate how strong monolinguals respond to those status differences. Similarly, the coefficients k_{13} and k_{23} quantify the rate at which bilinguals become monolinguals. This transition back to monolingualism can e.g. be associated with bilingual parents who choose to raise their children in one language only. Again we assume $k_{13} = \tilde{k}_{13}s$ and $k_{23} = \tilde{k}_{23}(1 - s)$ and the coefficients $\tilde{k}_{13}$ and $\tilde{k}_{23}$ indicate how strong bilinguals respond to the status differences of the two competing languages.

These assumptions lead to the following diffusion-reaction system (where, as mentioned above, $c_1(t, x)$ and $c_2(t, x)$ describes the frequencies of the two monolingual population groups speaking languages A and B, respectively and $c_3(t, x)$ the frequency of the bilingual population group at time t and location $x \in G$)

$$\frac{\partial c_1}{\partial t} = D_1 \Delta c_1 + \alpha_1 c_1 \left(1 - \frac{c_1}{K - c_2 - c_3} \right) - k_{21} c_2 c_1 + k_{13} c_3 c_1$$

$$\frac{\partial c_2}{\partial t} = D_2 \Delta c_2 + \alpha_2 c_2 \left(1 - \frac{c_2}{K - c_1 - c_3} \right) - k_{12} c_1 c_2 + k_{23} c_3 c_2$$

$$\frac{\partial c_3}{\partial t} = D_3 \Delta c_3 + \alpha_3 c_3 \left(1 - \frac{c_3}{K - c_1 - c_2} \right) + (k_{21} + k_{12}) c_2 c_1 - (k_{13} c_1 + k_{23} c_2) c_3$$

$$(18.10)$$

with the boundary conditions $\partial c_i / \partial n = 0, x \in \partial G, i = 1, 2, 3$ (where $\partial / \partial n$ describes the outer normal derivation). These conditions imply that no spatial spread is possible beyond the boundary ∂G.

Summarizing, the outflow of speakers from the two monolingual population groups is governed by the status difference of the two competing languages, the propensity of the groups to respond to those differences and the frequency of the other monolingual population group (see Fig. 18.1 for an illustration). It holds: the higher the status of a language the lower is the shift towards the bilingual group but the higher the frequency of the other monolingual population group the higher is the shift towards the bilingual group.

In the following we investigate the dynamics of language shift between a high-status and a low-status language, i.e. it has to hold $k_{21} < k_{12}$ and $k_{13} > k_{23}$. In this situation language A is considered more advantageous or high-status as monolinguals of language A are less likely to become bilinguals and bilinguals are more likely to become monolingual in language A.

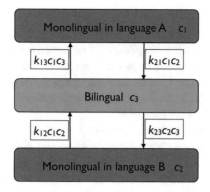

Fig. 18.1 Schematic illustration of the shift dynamics assumed in model (18.10). The dynamic depends on the frequencies c_i of the different population groups and the benefits both languages convey to their speakers. The coefficients $k_{ij}, i, j = 1, 2, i \neq j$ quantify the pressure one language puts on monolingual speakers of the other language and can be interpreted as the rate at which monolingual speakers become bilingual. Similarly, the coefficients k_{13} and k_{23} quantify the rate at which bilinguals become monolinguals

18.3.1.1   Role of Spatial Dispersal

The only stable equilibria of model (18.10) are the extinction states $(K, 0, 0)$ and $(0, K, 0)$. Depending on the status difference between the two competing languages and the demographic and cultural attributes of the three population groups, especially their initial frequency distributions, one language will acquire all speakers over time. Importantly, that does not have to be the high-status language.

To illustrate this point in more detail we consider the following situation. Initially the low-status language B is spoken by the vast majority of the population situated in a two-dimensional domain G but the high-status language A has entered the population in a small region G' where both languages are in direct contact. For simplicity we assume a uniform initial frequency distribution of monolingual speakers of language A in G'. Now the shift dynamics of model (18.10) results in the emergence of the bilingual population group in the language contact zone. If all population groups show similar demographic properties (i.e. possess similar growth rates α_i and sufficiently small diffusion coefficients D_i) the monolingual population group speaking the high-status language A and the bilingual group grow in frequency in the region G' causing in turn the frequency monolingual population group speaking the low-status language B to decline in this region. Crucially, model (18.10) assumes that bilingualism facilitates the communication between two monolingual population groups and therefore the bilingual group will disappear in a local area soon after languages B has disappeared from there. Local diffusion causes a steady expansion of language A in a travelling wave-like manner. With time the contact and mixing zone between the two languages is shifted toward the edges of the domain with extinction of the population group speaking language B followed by bilingual group as the long-term outcome.

However, if the 'invading' population group speaking language A shows a high spatial mobility (i.e. high diffusion coefficient D_1) the shift dynamics can be reversed. The greater mobility of speakers of language A causes a dramatic dilution of the initial frequency of speakers in the domain G'. If the intrinsic growth rate α_1 is not able to compensate for this loss the frequency-dependent dynamics predominate, leading to the extinction of the population group speaking language A and subsequently of the bilingual group.

Figure 18.2 illustrates the basins of attraction of the extinction states $(K, 0, 0)$ (all parameter combinations above the curves) and $(0, K, 0)$ (all parameter combinations below the curves) for the parameter s and the initial frequency of the high status language in the domain G' for different values of $D_1 = 10^{-5}, 10^{-4}, 10^{-3}$. We observe a nonlinear relationship and as expected the outcome of language shift is strongly influenced by the status difference of the two competing language. If the monolingual population group speaking language A shows a larger mobility then language A need to have a higher status or the 'invading' population group must be more frequent to nevertheless attract all speakers in the domain G.

Summarizing, as expected the difference in status between the competing languages influences the dynamics of language shift greatly: the higher the status of a language the higher its chances to dominate the shift scenario. But also the dispersal behavior and the initial distribution of the different population groups play an important role. If the low-status language is sufficiently established in the population, it can prevail in competition with a higher-status language.

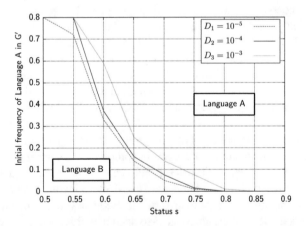

Fig. 18.2 Long-term outcome of language shift in region G, initially populated by monolingual speakers of language B with a certain number of monolingual speakers of language A concentrated in a sub-region G', as a function of the initial size ('frequency') of the population group speaking language A situated in G', the value of the status variable s and the spatial mobility of the speakers (denoted by D_i). The parameter combinations below the curves lead to populations speaking language A only $(K, 0, 0)$ while the combinations below the curves lead to populations speaking language B only $(0, K, 0)$. Despite the status disadvantage, language B can outcompete language A in situations where there are not enough speakers of language A initially and they disperse too quickly

Model (18.10) describes spatial dispersal by the locally acting Laplace operator Δ. This implies that speakers interact only with their local neighborhood. This however, might not be realistic for all episodes of human dispersal and it has been argued that dispersal distances are better approximated by long-range dispersal kernels (e.g. [40, 41]). This can be included in model (18.10) by replacing the diffusion terms $D_i \Delta c_i$ with the integral formulation

$$\lambda_i \left[\int_G c_i(t, x+\delta)\phi_i(\delta)d\delta - c_i(t, x) \right] \tag{18.11}$$

where the kernel functions $\phi_i(\delta)$ define the probability distributions of the dispersal length δ for each population group and the coefficients λ_i are a measure of the dispersal rates (see e.g. [42, 43]) for a detailed analysis of such dispersal models and [19] for an application to language shift). If researchers possess data to estimate the kernel functions $\phi_i(\delta)$ describing the dispersal distances of the different population groups reliably then the integral formulation (18.11) should be preferred to the diffusion formulation of model (18.10) as we have seen that the spatial dispersal behavior can qualitatively and quantitatively change the dynamics of language shift (see [20] for a more detailed analysis of the effects on different diffusion kernels on the language shift dynamics).

18.3.1.2   Role of Bilingualism

As already mentioned, model (18.10) assumes that bilingualism facilitates the communication between two monolingual (and spatially separated) population groups. Now if one monolingual group has gone extinct and there are no cultural or external reasons to still use this language then the bilingual strategy is not needed anymore for communication purposes and goes extinct as a consequence. Therefore the bilingual strategy cannot be maintained in the population in homogeneous environments (expressed by constant model parameters).

Nevertheless the existence of the bilingual population group influences the dynamics of language shift greatly. Firstly (and not surprisingly) it slows down the process of language extinction. Secondly, it can allow the low-status language to successfully prevent the spread of the high-status language if it is sufficiently established in the considered domain. To see this we analyze model (18.10) under the assumption that there is no bilingual population, i.e. $c_3(t, x) = 0, \forall t, x$. In this situation only the extinction state $(K, 0, 0)$ is stable, or in other words, the low-status language will always go extinct over time [32].

18.3.1.3 Coexistence

Coexistence between two different languages can be achieved in model (18.10) when their status differences vary in different spatial regions. In other words, if the status variable s, and consequently the shift coefficients k_{12}, k_{21}, k_{13} and k_{23} are space-dependent with each language being the preferred medium of communication in its own 'home range' then the two languages still outcompete each other in their 'home ranges' but coexist globally.

18.3.2 Diglossia Model

Model (18.10) describes language shift in a single social domain, and assumes that the shift dynamics is solely governed by the frequencies of the three population groups, the demographic and cultural attributes of these groups and the status difference of the two competing languages. We have seen that in this situation the loss of linguistic diversity is inevitable: the extinction of a monolingual population group is followed by the extinction of the bilingual group. But reality is likely to be more complex. For instance, the benefit a language conveys to its speakers might be different in different social domains. Additionally, the aim of intervention strategies is usually not to reverse the outcome of language shift as the high-status language provides its speakers with additional benefits compared to the low-status language (e.g. participation in higher education or 'global' business) but to strengthen the survival chances of the endangered language by creating social and cultural domains where the low-status language is still used and potentially even the preferred medium of communication [44].

To consider the effects of the existences of segregated and complementary sociolinguistic domains, in each of which both languages are differentially preferred as medium of communication, model (18.10) has been generalized based on a simplified concept of diglossia [11]. Diglossia, in the strict sense, refers to situations where the mother tongue of the community is used in everyday (low status) settings, but another language (or another form of the vernacular language) is used in certain high status domains [45, 46]. We assume that in the majority of social domains the shift mechanisms of model (18.10) apply but there exist some restricted social domains in which the balance of competitive advantage between the two languages differs from that which drives the main shift process. The temporal and spatial dynamics of the process of language shift is now determined by

$$\frac{\partial c_1}{\partial t} = D_1 \Delta c_1 + \alpha_1 c_1 \left(1 - \frac{c_1}{K - c_2 - c_3} \right) - k_{21} c_2 c_1 + k_{13} c_3 c_1 - w_1 c_1$$

$$\frac{\partial c_2}{\partial t} = D_2 \Delta c_2 + \alpha_2 c_2 \left(1 - \frac{c_2}{K - c_1 - c_3} \right) - k_{12} c_1 c_2 + k_{23} c_3 c_2 - w_2 c_2$$

$$\frac{\partial c_3}{\partial t} = D_3 \Delta c_3 + \alpha_3 c_3 \left(1 - \frac{c_3}{K - c_1 - c_2} \right) + (k_{21} + k_{12}) c_2 c_1 - (k_{13} c_1 + k_{23} c_2) c_3 + w_1 c_1 + w_2 c_2$$

$$(18.12)$$

where the coefficients w_1 and w_2 quantify the pressure to participate in social domains where the other language is the preferred medium of communication. The main shift dynamics is still frequency- and status-dependent as described in model (18.10) but additionally we assume that if monolinguals of the low-status language want to participate in domains where the high-status language is required (such as higher education or 'global' businesses), they need to learn that language and become bilingual as a consequence. This is modeled by the term $w_2 c_2$. Similarly, if monolinguals of the high-status language want to participate in domains where the low-status language is required (such as small 'local' businesses or administrations), they also need to learn that language. This dynamics is modeled by the term $w_1 c_1$. Importantly, the strengths of both shift terms $w_1 c_1$ and $w_2 c_2$ do not depend on the frequencies of the other language but only on the (social, cultural, economic or political) pressure to participate in the associated domains. As long as $w_1 > 0$, meaning as long as there is a need for speaking the low-status language in at least some (relevant) domains, the bilingual population will persist, however, the monolingual population group speaking language B will go extinct nevertheless. Consequently, the language shift dynamics described by model (18.12) is characterized by the extinction of the monolingual population group speaking the endangered language and the maintenance of the bilingual strategy in the population. But this result is conditioned on the existence of domains where the low-status language is the preferred medium of communication (e.g. $w_1 > 0$). Therefore intervention strategies aimed at creating those domains (e.g. through legislation that requires the use of the endangered local language in a specific set of contexts) affect the magnitude of the coefficient w_1.

Summarizing, the existence of social domains of competition, which differ in the competitive advantage between the two languages may allow for coexistence. Importantly, however, the endangered language is only maintained through the bilingual population group.

18.4   The Gaelic-English Language Shift

In this section we illustrate the results of the analysis of the language shift scenario between English and Scottish Gaelic in Scotland (see [11] for an additional analysis of the Welsh situation). By late mediaeval times, Gaelic was the main language of the Scottish Highlands and western islands, with Scots (descended from the Old Northumbrian dialect of Old English) and English prevailing in the Lowlands. This division appears to have been reinforced by a contrast between these two regions in their social structure, marriage and migration patterns. The breakdown of the geographical 'niche' for Scottish Gaelic is closely linked to the English political and economic dominance (and the subsequent interference with the Highlands' political and economic systems). Drastic demographic changes (the eighteenth-nineteenth century 'Highland clearances') and the establishment of English as the language of education and advancement were associated with increasing rates of Gaelic-to-English language shift [47]. The late stages of this shift process can be reconstructed from census records (see electronic supplementary material in [11]). It must be noted that historical census data on language use will include 'noise' owing to inaccurate answers (for instance, owing to the perceived social status implications of self-classification into a particular category), and to changes in the phrasing of the questions in successive censuses. The first census to enumerate Gaelic speakers was that of 1881, but only from 1891 were data gathered separately on numbers of Gaelic monolinguals and Gaelic-English bilinguals (in all cases, among those aged 3 years or older). After 1961, no data were collected for Gaelic monolinguals, as these were assumed to be approaching extinction. From 1891 until 1971, the census enumerations were collated and analyzed on the basis of the old county divisions. The Highland counties Argyll, Inverness, Ross and Cromarty, and Sutherland are seen as the 'core land' of the Gaelic language ('*Gaidhealtachd*'): in 1891, 73 per cent of all Scotland's Gaelic speakers were located among the 8 per cent of Scotland's population that lived in these 'Highland Counties', covering the mainland Highlands and the Western Isles. From 1981 onwards, these counties were subsumed into new administrative units.

Generally, we observe a sharp decline of the number of monolingual Gaelic and bilingual speakers in the period between 1891–2001. Areas where Gaelic is still spoken by at least 50% of the population are pushed towards the Western Islands over time (see Fig. 18.3) and these empirical travelling wave-like patterns partly motivated the application of the diffusion-reaction framework. The absolute numbers of Gaelic speakers in Scotland have declined through this period, from about 250,000 in the 1891 census of Scotland to about 65,000 in the 2001 census. Of these, the majority has always been bilingual in Gaelic and English, with the last census record of Gaelic monolinguals finding fewer than 1000 still alive in 1961.

18.4.1   Basic Model

We start by applying the basic model (18.10) to this shift scenario. In particular, we are interested in exploring how well our model can describe the observed trajectories of the three population groups over time and in different spatial location. Figure 18.4 (solid lines) shows the change in the proportions of monolingual English and Gaelic speakers and bilinguals for the counties of Argyll, Inverness,

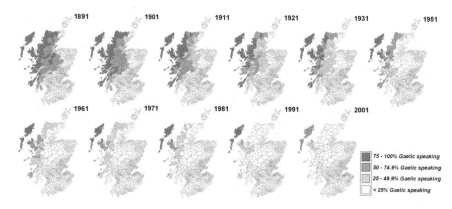

Fig. 18.3 Percentages of Gaelic speakers (mono- and bilingual) in Western Scotland in successive census years, 1891–2001. Data for civil parishes: 1891–1971 from Withers ([48], pp. 227–234); 1981 from Withers ([49], p. 40); 1991–2001 from General Register Office for Scotland ([50], Table 3). Redrawn from [11]

Ross and Cromarty, and Sutherland during the time period 1891–1971 (as mentioned above due to changes to the county division data for these four counties is only available within this time frame).

We fitted model (18.10) to these census data whereby the growth rates α_i and the diffusion coefficients D_i are estimated from demographic data and only the shift coefficients k_{12}, k_{21}, k_{13} and k_{23} are free to vary. To avoid over-fitting we firstly restrict ourselves to the parameter constellations $k_{21} = k_{23}$ and $k_{12} = k_{13}$, i.e. we assume that, for example, language A exerts the same pressure on the population group speaking language B and on the bilingual population group (see Fig. 18.1). Figure 18.4 (dotted lines) illustrate that model (18.10) with $k_{21} = k_{23}$ and $k_{12} = k_{13}$ and values as shown in Table 18.1 captures the general shift dynamics well. It is obvious that Gaelic is not able to attract speakers; the outflow from the English monolingual population group to the bilingual group is zero for all counties. However, we also observe a systematic overestimation of the monolingual Gaelic population group and underestimation of the bilingual group.

In the next step we allow for parameter constellations with $k_{21} \neq k_{23}$ and $k_{12} \neq k_{13}$. With these additional degrees of freedom we (unsurprisingly) obtain a better fit between model (18.10) and the data (Fig. 18.4, dashed lines). Interestingly, the improved fit is almost entirely generated by an increase in the coefficient k_{12}, and consequently by an increased shift from the Gaelic monolingual population to the bilingual population. This can be interpreted as evidence that the shift is mainly driven be the desire to learn English and not by the desire to abandon Gaelic. However, we have shown in Sect. 18.3 that the extinction of the Gaelic monolingual population group and the bilingual group is inevitable over time.

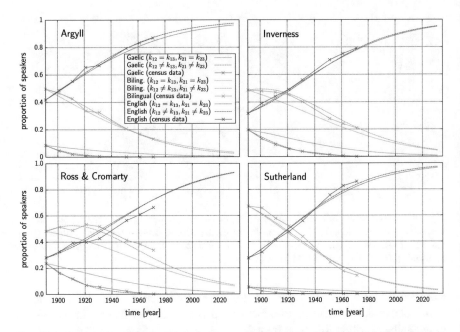

Fig. 18.4 Frequencies of the three population groups (monolingual Gaelic: red, bilingual: green, monolingual English: blue) in the four Scottish Highland counties (**a** Argyll, **b** Inverness, **c** Ross and Cromarty, **d** Sutherland). Empirical data (solid lines) and theoretical predictions of model (18.10) with $k_{21} = k_{23}$ and $k_{12} = k_{13}$ (dotted lines) and $k_{21} \neq k_{23}$ and $k_{12} \neq k_{13}$ (dashed lines). Shift parameters are given in Table 18.1 and describe rates per year. Redrawn from [11]

Table 18.1 Fitted shift coefficients of model (18.1) under the assumptions $k_{12} = k_{13}$ and $k_{21} = k_{23}$ (top two lines) and $k_{12} \neq k_{13}$ and $k_{21} \neq k_{23}$ (bottom four lines). All coefficients indicate rates per year

	Scottish Highlands	Argyll	Inverness	Ross and Cromarty	Sutherland
$k_{12} = k_{13}$	0.03	0.03	0.035	0.03	0.035
$k_{21} = k_{23}$	0	0	0	0	0
k_{12}	0.07	0.115	0.1	0.12	0.075
k_{13}	0.025	0.03	0.03	0.025	0.035
k_{21}	0	0.005	0	0.005	0
k_{23}	0	0	0	0.005	0

18.4.2 Diglossia Model

The prediction of model (18.10) about the future of Scottish Gaelic is of course only valid if the 'competition environment' stays unchanged. However, the Scottish government has started to implement a number of revitalization measures. Recent efforts have included the establishing of Gaelic-medium pre-school and primary school units [51] and the development of Gaelic-medium broadcasting [47]. In 2005, the Gaelic Language (Scotland) Act was passed by the Scottish Parliament, providing a planning framework for a number of additional shift-reversal measures, while Comhairle nan Eilean Siar, the Western Isles Council, has adopted Gaelic as its primary language.

In this section we use the diglossia model (18.12) and ask the question of how strong do those intervention strategies need to be in order to at least maintain the overall bilingual population group at the level of year 2001. To do so we fitted model (18.10) to the accumulated numbers of monolingual Gaelic, monolingual English and bilingual speakers of the Scottish highlands in the time interval 1891–2001. Based on the estimated coefficients k_{12}, k_{21}, k_{13} and k_{23} (see Table 18.1, second column for their values) we then applied model (18.12) and asked how large the coefficient w_1 needs to be so that the frequency of the bilingual population, c_3, stays constant over time. We note that by 2009 the frequency of the monolingual Gaelic population group has already approached zero and therefore the shift term $w_2 c_2$ can be neglected.

The results of this analysis are summarized in Fig. 18.5. Firstly, we observe that model (18.10) replicates the language shift dynamics very well (cf. solid lines for the empirical data and dashed lines for the model prediction). The estimated values of the shift coefficients are shown in the second column of Table 18.1. Now $w_1 = 0.0031$ (whereby w_1 is a rate per year) is sufficient to prevent the further decline of the bilingual population group. This implies that roughly 860 English speakers have to become bilingual every year based on a Highland population of about 315,000 individuals. However, the coexistence between the bilingual and the

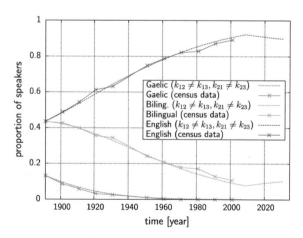

Fig. 18.5 Frequencies of the three population groups (monolingual Gaelic: red, bilingual: green, monolingual English: blue) in the Scottish Highlands. Empirical data (solid lines) and theoretical predictions of model (18.10) until 2009 and model (18.12) after 2009 (dotted lines). Shift parameters for model (18.10) are given in Table 18.1, model (18.12) is parameterized with the same parameters and $w_1 = 0.0031$ and $w_2 = 0$. All coefficients describe rates per year. Redrawn from [11]

English-speaking population groups depends in this case entirely on the existence of (potentially externally engineered) sociolinguistic domains where Gaelic is the preferred medium of communication. Intervention strategies may prove much more successful if the rate of intergenerational transmission of the bilingual strategy could be increased as well. Thus, for example, the number of English monolinguals required to learn Gaelic each year could drop down to roughly 440 if the rate of intergenerational transmission of Gaelic at home could be increased (i.e. k_{12} changes from 0.025 to 0.0125). This means that beside the 440 English speakers who become bilingual, roughly 340 more children who live in bilingual households would have to be raised in both languages to prevent a further decline of the bilingual population group. These numbers indicate that an increase in the rate of intergenerational transmission is a highly effective language maintenance strategy, although one that is also harder to achieve in practice.

18.5   Discussion

The rapid increase in the rate of language extinction, witnessed over the last 100 years, is mainly caused by the process of language shift. From a phylogenetic point of view language shift can be seen as a process of shifting between different branches of the phylogenetic tree (or in other words, as a process of selective cultural migration, see e.g. [52]). Frequent instances of language shift potentially result in divergences between linguistic and genetic trees and therefore cast some doubt on the demographic assumption of tree building approaches that the linguistic tree is also representative of the bifurcating population history (see [53] for a discussion). Consequently understanding the process of language shift and identifying its main drivers is a crucial step towards understanding the general process of language extinction.

   In this chapter we discussed how mathematical modeling can contribute to this task. Considering language shift as competition between two languages (of possibly different status) for speakers allows us to make use of a well-developed theory describing the interactions between different species under limited resources (e.g. [15]). However, it is crucial to note that despite the similarities between ecological and linguistic competition dynamics there are linguistic phenomena, which have no equivalent in the ecological situation. Consequently existing modeling frameworks have to be adopted to include important cultural concepts such as e.g. bilingualism.

   We have demonstrated that if two languages of different status compete in a single social domain and the shift dynamics is solely governed by the frequencies of the three population groups (i.e. the two monolingual and one bilingual groups), the demographic and cultural attributes of these groups and the status difference of the two languages, then the extinction of one language and therefore the loss of linguistic diversity is the only long-term outcome. In this situation bilingualism is a temporary transition state and not a stable long-term outcome securing the maintenance of the endangered (i.e. low-frequency) language.

Our analyses of the basic language shift model (18.10) showed that demographic and cultural factors influence the dynamics of language shift greatly. For instance, a change in the dispersal behavior of a population group alone can change the outcome of language shift (see Sect. 18.3.1.1). Consequently, in order to realistically describe and predict the spatial and temporal dynamics of language change, the demographic properties of the different population groups have to be summarized appropriately.

But how well does this model reflect linguistic reality? And given the increasing conservation effort by governments and international organization how can linguistic diversity be maintained? So far the analysis has been on the assumption that both languages compete in a single social domain (or equivalently both languages possess the same properties in all social domains). However, this does not have to be the case. In order to allow for the differential use of both languages in different social context the concept of diglossia was included into the modeling framework. This means, superimposed on the basic shift dynamics described in model (18.10), there is an additional demand for the endangered language as the preferred medium of communication in some restricted sociolinguistic domain. This additional dynamics creates a flow from the monolingual population groups speaking the high-status language to the bilingual group whereby the strength of this flow is controlled by the model parameter w_1. The demand for both languages, each in its own preferred domain allows bilingualism to be a stable final state and we find a wider range of possible stable extinction and coexistence states depending on the strength of the various in- and out-flows between the three population groups.

We applied the modeling frameworks to the English-Gaelic shift situation in Western Scotland. Firstly, Fig. 18.4 revealed that the basic shift dynamics as described in model (18.10) is able to replicate the past demographic trajectories of the language shift scenario. We then used the values of the parameters, which produced this close fit between model and data in the diglossia model (18.12). The model predicted that roughly 860 English monolingual have to become bilingual each year, is needed to maintain the bilingual population group at the level of the year 2009. This number could, however, drop down to roughly 440 if 340 more children who live in bilingual households would be raised in both languages, which points to the crucial importance of the intergenerational transmission of the bilingual strategy in conservation efforts.

18.6 Conclusions

We believe that mathematical modeling can provide meaningful indicators for the potential success or failure of certain language intervention strategies. Those strategies usually do not attempt to reverse language shift completely as there are good reasons why speakers abandon a language in favor of another (mainly because speakers receive an economic gain from switching). But they do aim at creating stable bilingualism by developing or preserving essential social domains in which the endangered language is the preferred or only acceptable medium of communication. In our modeling framework this means that successful intervention strategies affect the strength of the flow between the monolingual population group speaking the high-status language and the bilingual group and therefore the magnitude of the model parameter w_1. This in turn allows us to ask the question of how strong intervention strategies needs to be, or in other words, how strong the outflow from the monolingual population speaking the high-status language needs to be in order to maintain the bilingual population group at a certain level. Consequently, frameworks of the kinds described in this chapter provide a population-level view of the temporal and spatial dynamics of language shift and therefore allow for the inference of the average 'general' strength of the intervention strategies that is needed to obtain a certain outcome. However, language planners might additionally be interested in understanding which of two possible intervention strategies could prove most effective. In order to answer these kinds of questions a simulation framework which is able to mechanistically incorporate those different strategies might be more appropriate. Even though we focused here on differential equation based approaches we argue that different modeling approaches will add additional insights to the puzzle of language shift and there is no one 'right' model to describe the dynamics.

References

1. S.S. Mufwene, *The Ecology of Language Evolution* (Cambridge University Press, Cambridge, 2001)
2. M. Sereno, J. Theor. Biol. **151**, 467 (1991)
3. S. Greenhill, Q.D. Atkinson, A. Meade, R.D. Gray, Proc. R. Soc. B **277**, 2443 (2010)
4. C. Renfrew, *Archaeology and Language: The Puzzle of Indo-European Origin* (Johnatan Cape, London, 1987)
5. A. McMahon, *Understanding Language Change* (Cambridge University Press, Cambridge, 1994)
6. L.A. Grenoble, L.J. Whaley, *Saving Languages: An Introduction to Language Revitalisation* (Cambridge University Press, Cambridge, 2006)
7. M. Krauss, Language **68**, 4 (1992)
8. W.J. Sutherland, Nature **423**, 276 (2003)
9. T. Amano et al., Proc. R. Soc. B **281**, 20141574 (2014)
10. T. Tsunoda, *Language Endangerment and Language Revitalization* (Mouton De Gruyter, Berlin, 2005)
11. A. Kandler, R. Unger, J. Steele, Philos. Trans. R. Soc. B **365**, 3855 (2010)
12. M. Brenzinger, in *The Encyclopedia of Language and Linguistics,* ed. by K. Brown, Society and language, vol. 6 (Oxford, UK 2006)
13. N.C. Dorian, *Language Death: The Life Cycle of a Scottish Gaelic Dialect* (University of Pennsylvania Press, Philadelphia, 1981)
14. D. Nettle, S. Romaine, *Vanishing Voices: The Extinction of the World's Languages* (Oxford University Press, Oxford, 1999)
15. J.D. Murray, *Mathematical Biology I: An Introduction* (Springer, New York, 2002)
16. D.M. Abrams, S.H. Strogatz, Nature **424**, 900 (2003)
17. I. Baggs, H.I. Freedman, Math. Sociol. **16**, 51 (1990)

18. I. Baggs, H.I. Freedman, Math. Comput. Model. **18**, 9 (1993)
19. J. Fort, J. Pérez-Losada, Hum. Biol. **84**(6), 755 (2012)
20. A. Kandler, Hum. Biol. **81**(2–3), 181 (2009)
21. R.V. Solé, B. Corominas-Mutra, J. Fortuny, J. R. Soc. Interface **7**, 1647 (2010)
22. J. Steele, Hum. Biol. **81**(2–3), 121 (2009)
23. X. Castelló, L. Loureiro, V.M. Eguíluz, M. Miguel, in *Advancing Social Simulation: The First World Congress*, ed. by S. Takahashi, D. Sallach, J. Rouchier (Springer, New York, 2007)
24. K.J. Kosmidis, M. Halley, P. Argyrakis, Phys. A **353**, 595 (2005)
25. V. Schwämmle, Intl. J. Mod. Phys. C **17**, 103 (2005)
26. D. Stauffer, C. Schulze, Phys. Life Rev. **2**, 89 (2005)
27. D. Stauffer et al., Phys. A **374**, 835 (2007)
28. C. Schulze, D. Stauffer, S. Wichmann, Commun. Comput. Phys. **3**, 271 (2008)
29. R.V. Solé, J. Bascompte, *Self-organization in Complex Ecosystems* (Princeton University Press, Princeton NJ, 2006)
30. A. Turing, Philos. Trans. R. Soc. B **237**, 37 (1952)
31. M. Patriarca, T. Leppänen, Phys. A **338**, 296 (2004)
32. A. Kandler, J. Steele, Biol. Theor. **3**, 164 (2008)
33. C.E. Walters, Meccania **49**, 2189 (2014)
34. M. Patriarca, E. Heinsalu, Phys. A **388**, 174 (2009)
35. N. Isern, J. R. Soc. Interface **11**, 20140028 (2014)
36. M. Zhang, T. Gong, PNAS **110**(24), 9698 (2013)
37. J. Mira, Á. Paredes, Europhys. Lett. **69**, 1031 (2005)
38. J.W. Minett, W.S.-Y. Wang, Lingua **118**, 1945 (2008)
39. R.D. Parshad, S. Bhowmick, V. Chand, N. Kumari, N. Sinha, Phys. A **449**, 375 (2016)
40. I. Alves et al., Long distance dispersal shaped patterns of human genetic diversity in Eurasia. Mol. Biol. Evol. **33**(4), 946 (2015)
41. D. Brockman, L. Hufnagel, T. Geisel, Nature **439**, 462 (2005)
42. S. Fedotov, Phys. Rev. Lett. **85**(5), 926 (2001)
43. V. Mendez, T. Pujol, J. Fort, Phys. Rev. E **65**, 1 (2002)
44. J.A. Fishman, *Reversing Language Shift: Theoretical and Empirical Foundations of Assistance to Threatened Languages* (Multilingual Matters Ltd., Clevedon, 1991)
45. C.A. Ferguson, Word **15**, 325 (1959)
46. A. Hudson, Int. J. Sociol. Lang. **157**, 1 (2002)
47. S. Murdoch, *Language Politics in Scotland* (Aberdeen Universitie Scots Leid Quorum, Aberdeen, 1996)
48. C.W.J. Withers, *Gaelic in Scotland, 1698–1981: The Geographical History of a Language* (John Donald, Edinburgh, 1984)
49. C.W.J. Withers, *Gaelic Scotland: The Transformation of a Cultural Region* (Routledge, London, 1988)
50. General Register Office for Scotland, *Scotland's Census 2001, Gaelic Report* (General Register Office for Scotland, Edinburgh, 2005)
51. K. MacKinnon, in *The Celtic Languages*, ed. by M.J. Ball (Routledge, London, 1993)
52. R. Boyd, P.J. Richerson, J. Theor. Biol. **257**, 331 (2009)
53. J. Steele, A. Kandler, Theor. Biosci. **129**, 223 (2010)

Chapter 19
Human Mobility, Networks and Disease Dynamics on a Global Scale

Dirk Brockmann

19.1 Introduction

In early 2009, news accumulated in major media outlets about a novel strain of influenza circulating in major cities in Mexico [1]. This novel H1N1 strain was quickly termed "swine flu", in reference to its alleged origin in pig populations before jumping the species border to humans. Very quickly public health institutions were alerted and saw the risk of this local influenza epidemic becoming a major public health problem globally. The concerns were serious because this influenza strain was of the H1N1 subtype, the same virus family that caused one of the biggest pandemics in history, the Spanish flu that killed up to 40 million people in the beginning of the 20th century [2]. The swine flu epidemic did indeed develop into a pandemic, spreading across the globe in matters of months. Luckily, the strain turned out to be comparatively mild in terms of symptoms and as a health hazard. Nevertheless, the concept of emergent infectious diseases, novel diseases that may have dramatic public health, societal and economic consequences reached a new level of public awareness. Even Hollywood picked up the topic in a number of blockbuster movies in the following years [3]. Only a few years later, MERS hit the news, the Middle East Respiratory Syndrome, a new type of virus that infected people in the Middle East [4]. MERS was caused by a new species of corona virus of the same family of viruses that the 2003 SARS virus belonged to. And finally, the 2013 Ebola crisis in West African countries Liberia, Sierra Leone and Guinea that although it did not develop into a global crisis killed more than 10000 people in West Africa [5].

D. Brockmann (✉)
Institute for Theoretical Biology, Humboldt Universität zu Berlin, Philippstraße 13, 10115 Berlin, Germany
e-mail: dirk.brockmann@hu-berlin.de

D. Brockmann
Robert Koch-Institute, Nordufer 20, 13353 Berlin, Germany

© Springer International Publishing AG 2018
A. Bunde et al. (eds.), *Diffusive Spreading in Nature, Technology and Society*,
https://doi.org/10.1007/978-3-319-67798-9_19

Emergent infectious diseases have always been part of human societies, and also animal populations for that matter [6]. Humanity, however, underwent major changes along many dimensions during the last century. The world population has increased from approx. 1.6 billion in 1900 to 7.5 billion in 2016 [7]. The majority of people now live in so-called mega-cities, large scale urban conglomerations of more than 10 million inhabitants that live in high population densities [8] often in close contact with animals, pigs and fowl in particular, especially in Asia. These conditions amplify not only the transmission of novel pathogens from animal populations to human, high frequency human-to-human contacts yield a potential for rapid outbreaks of new pathogens.

Population density is only one side of the coin. In addition to increasing face-to-face contacts within populations we also witness a change of global connectivity [9]. Most large cities are connected by means of an intricate, multi-scale web of transportation links, see Fig. 19.1. On a global scale worldwide air-transportation dominates this connectivity. Approx. 4,000 airports and 50,000 direct connections span the globe. More than three billion passengers travel on this network each year. Every day the passengers that travel this network accumulate a total of more than 14 billion kilometers, which is three times the radius of our solar system [10, 11]. Clearly this amount of global traffic shapes the way emergent infectious diseases can spread across the globe. One of the key challenges in epidemiology is preparing for eventual outbreaks and designing effective control measures. Evidence based control measures, however, require a good understanding of the fundamental features and characteristics of spreading behavior that all emergent infectious diseases share. In this context this means addressing questions such as: If there is an outbreak at

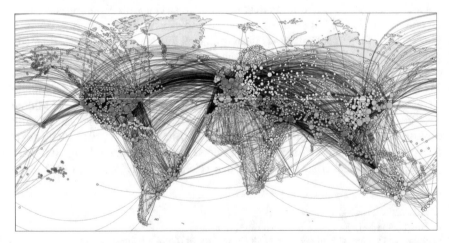

Fig. 19.1 The global air-transportation network. Each node represents one of approx. 4000 airports, each link one of approx. 50000 direct connections between airports. More than 3 billion passengers travel on this network each year. All in all every day more than 16 billion km are traversed on this network, three times the radius of our solar system

location X when should one expect the first case at a distant location Y? How many cases should one expect there? Given a local outbreak, what is the risk that a case will be imported in some distant country. How does this risk change over time? Also, emergent infectious diseases often spread in a covert fashion during the onset of an epidemic. Only after a certain number of cases are reported, public health scientists, epidemiologist and other professionals are confronted with cases that are scattered across a map and it is difficult to determine the actual outbreak origin. Therefore, a key question is also: Where is the geographic epicenter of an ongoing epidemic?

Disease dynamics is a complex phenomenon and in order to address these questions expertises from many disciplines need to be integrated, such as epidemiolgy, spatial statistics, mobility and medical research in this context. One method that has become particularly important during the past few years is the development of computational models and computer simulations that help address these questions. These are often derived and developed using techniques from theoretical physics and more recently complex network science.

19.2 Modeling Disease Dynamics

Modeling the dynamics of diseases using methods from mathematics and dynamical systems theory has a long history. In 1927 Kermack and McKenrick [12] introduced and analyzed the "Suceptible-Infected-Recovered" (SIR) model, a parsimoneous model for the description of a large class of infectious diseases that is also still in use today [13]. The SIR model considers a host population in which individuals can be susceptible (S), infectious (I) or recovered (R). Susceptible individuals can aquire a disease and become infectious themselves and transmit the disease to other susceptible individuals. After an infectious period individuals recover, acquire immunity, and no longer infect others. The SIR model is an abstract model that reduces a real world situation to the basic dynamic ingredients that are believed to shape the time course of a typical epidemic. Structurally, the SIR model treats individuals in a population in much the same way as chemicals that react in a well-mixed container. Chemical reactions between reactants occur at rates that depend on what chemicals are involved. It is assumed that all individuals can be represented only by their infectious state and are otherwise identical. Each pair of individuals has the same likelihood of interacting. Schematically, the SIR model is described by the following reactions

$$S + I \overset{\alpha}{\longrightarrow} 2I \qquad I \overset{\beta}{\longrightarrow} R \qquad (19.1)$$

where α and β are transmission and recovery rates per individual, respectively. The expected duration of being infected, the infectious period is given by $T = \beta^{-1}$ which can range from a few days to a few weeks for generic diseases. The ratio of rates $R_0 = \alpha/\beta$ is known as the basic reproduction ratio, i.e. the expected number of secondary infections caused by a single infected individual in a fully susceptible population. R_0 is the most important epidemiological parameter because the value of R_0 determines whether an infectious disease has the potential for causing an epidemic or not. When $R_0 > 1$ a small fraction of infected individuals in a susceptible population will cause an exponential growth of the number of infections. This epidemic rise will continue until the supply of susceptibles decreases to a level at which the epidemic can no longer be sustained. The increase in recovered and thus immune individuals dilutes the population and the epidemic dies out. Mathematically, one can translate the reaction scheme (19.1) into a set of ordinary differential equations. Say the population has $N \gg 1$ individuals. For a small time interval Δt and a chosen susceptible individual the probability of that individual interacting with an infected is proportional to the fraction I/N of infected individuals. Because we have S susceptibles the expected change of the number susceptibles due to infection is

$$\Delta S \approx -\Delta t \times \alpha \times S \times \frac{I}{N} \qquad (19.2)$$

where the rate α is the same as in (19.1) and the negative sign accounts for the fact that the number of susceptibles decreases. Likewise the number of infected individuals is increased by the same amount $\Delta I = +\Delta t \times \alpha \times S \times I/N$. The number of infecteds can also decrease due to the second reaction in (19.1). Because each infected can spontaneouly recover the expected change due to recovery is

$$\Delta I \approx -\Delta t \times \beta \times I. \qquad (19.3)$$

Based on these assumptions Eqs. (19.2) and (19.3) become a set of differential equations that describe the dynamics of the SIR model in the limit $\Delta t \to 0$:

$$ds/dt = -\alpha s j \qquad (19.4)$$
$$dj/dt = \alpha s j - \beta j$$
$$r = 1 - s - j$$

where $s(t) = S(t)/N$, $j(t) = I(t)/N$ and $r(t) = R(t)/N$ are the fractions of susceptibles, infecteds and recovereds in the population as a function of time. The last equation in (19.4) is a consequence of the conservation of individuals, $S(t) + I(t) + R(t) = N$. Solutions to this set of equations for a small initial fraction of infecteds $j(0) = j_0$, $r(0) = 0$, and $s(0) = 1 - j_0$ exhibit a typical epi-curve, i.e. an initial exponential increase of infecteds with a subsequent decline if the basic reproduction ratio $R_0 > 1$. Typical solutions of the SIR model are shown in Fig. 19.2. A more realistic approach accounts for fluctuations that are caused by the intrinsic randomness of the

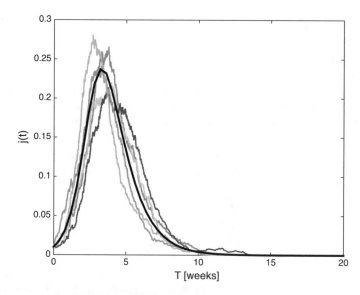

Fig. 19.2 The SIR model. The curves depict the generic time course of the fraction of infected individuals $j(t)$ generated by the SIR model defined by reactions (19.1, colored trajectories) and Eq. (19.4, black line). Initially only a small fraction of 1% of the population is infected. When $R_0 > 1$ (here $R_0 = 2.5$ and $\beta^{-1} = 1$ week) an exponential growth is followed by an exponential decay, leading to the generic epidemic curve. The fluctuations in the colored trajectories are generated by a stochastic generalization of the deterministic system defined by Eq. (19.4) in which in a finite population of $N = 1000$ individuals transmission and recovery events (reactions (19.1)) occur randomly

probabilistic reactions (19.1) and the finite number N of individuals in a population. Depending on the magnitude of N a model in which reactions occur randomly at rates α and β a stochastic system generally exhibits solutions that fluctuate around the solutions to the deterministic system of Eq. (19.4).

Both, the deterministic SIR model and the more general particle kinetic stochastic model are designed to model disease dynamics in a single population, spatial dynamics or movement patterns of the host population are not accounted for. These systems are thus known as well-mixed systems in which the analogy is one of chemical reactants that are well-stirred in a chemical reaction container as mentioned above.

19.2.1 Spatial Models

When a spatial component is expected to be important in natural scenario, several methodological approaches exist to account for space. Essentially the inclusion of a spatial component is required when the host is mobile and can transport the state of infection from one location to another. The combination of local proliferation of an infection and the disperal of infected host individuals then yields a spread along the spatial dimension [13, 14].

One of the most basic ways of incorporating a spatial dimension and host dispersal is by assuming that all quantities in the SIR model are also functions of a location $\mathbf{x}$, so the state of the system is defined by $s(\mathbf{x}, t)$, $j(\mathbf{x}, t)$ and $r(\mathbf{x}, t)$. Most frequently two spatial dimensions are considered. The simplest way of incorporating dispersal is by an ansatz following Eq. (2.19) in Chap. 2 which assumes that individuals move diffusively in space which yields the reaction-diffusion dynamical system

$$\partial s/\partial_t = -\alpha js + D\nabla^2 s \tag{19.5}$$

$$\partial j/\partial_t = \alpha js - \beta j + D\nabla^2 j \tag{19.6}$$

$$\partial r/\partial_t = \beta j + D\nabla^2 r \tag{19.7}$$

where e.g. in a two-dimensional system with $\mathbf{x} = (x, y)$ the Laplacian is $\nabla^2 = \partial^2/\partial_x^2 + \partial^2/\partial_y^2$ and the parameter D is the diffusion coefficient. The reasoning behind this approach is that the net flux of individuals of one type from one location to a neighboring location is proportional to the gradient or the difference in concentration of that type of individuals between neighboring locations. The key feature of diffusive dispersal is that it is local, in a discretized version the Laplacian permits movements only within a limited distance.

In reaction diffusion systems of this type the combination of initial exponential growth (if $R_0 = \alpha/\beta > 1$) and diffusion ($D > 0$) yields the emergence of an epidemic wavefront that progresses at a constant speed if initially the system is seeded with a small patch of infected individuals [15]. The advantage of parsimoneous models like the one defined by Eq. (19.7) is that properties of the emergent epidemic wavefront can be computed analytically, e.g. the speed of the wave in the above system is related to the basic reproduction number and diffusion coefficient by

$$v \sim \sqrt{(R_0 - 1) D} \tag{19.8}$$

in which we recognize the relation of Eq. (2.17). Another class of models considers the reaction of Eq. (19.1) to occur on two-dimensional (mostly square) lattices. In these models each lattice site is in one of the states S, I or R and reactions occur only with nearest neighbors on the lattice. These models account for stochasticity and spatial extent. Given a state of the system, defined by the state of each lattice site, and a small time interval Δt, infected sites can transmit the disease to neighboring sites that are susceptible with a probability rate α. Infected sites also recover to the

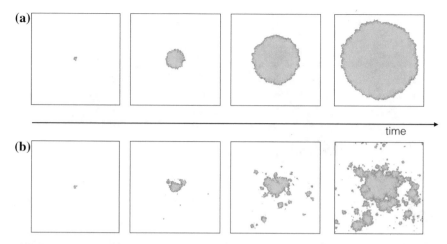

Fig. 19.3 Stochastic lattice SIR models. **a** From left to right the images are temporal snapshots of a stochastic SIR model in which an infected lattice site (red) can transmit an infection to a susceptible (white) neighboring site with probability rate α. At rate β infected sites recover and become immune (grey). Initially a single site in the center is infected. Asmyptotically a concentric pattern emerges. The infection front spreads at a constant speed. Stochastic effects at the wavefront caused the ragged structure of the interface. **b** The system is identical to the system depicted in (**a**). However, in addition to the generic next neighbor transmission, with a small but significant probability a transmission to a distant site can occur. This probability also decreases with distance as an inverse power-law, e.g. $p(d) \sim d^{-(1+\mu)}$ where the exponent is in the range $0 < \mu < 2$. Because the rare but significant occurance of long-range transmissions, a more complex pattern emerges, the concentric nature observed in system a is gone. Instead, a fractal, multiscale pattern emerges

R state and become immune with probability $\beta\Delta t$. Figure 19.3a illustrates the time course of the lattice-SIR model. Seeded with a localized patch of infected sites, the system exhibits an asymptotic concentric wave front that progresses at an overall constant speed if the ratio of transmission and recovery rate is sufficiently large. Without the stochastic effects that yield the irregular interface at the infection front, this system exhibits similar properties to the reaction diffusion system of Eq. (19.7). In both systems transmission of the disease in space is spatially restricted per unit time.

19.2.2 The Impact of Long-Distance Transmissions

The stochastic lattice model is particularly useful for investigating the impact of permitting long-distance transmissions. Figure 19.3b depicts temporal snapshots of a simulation that is identical to the system of Fig. 19.3a apart from a small but significant difference. In addition for infected sites to transmit the disease to neighboring susceptible lattice sites, every now and then (with a probability of 1%) they can also

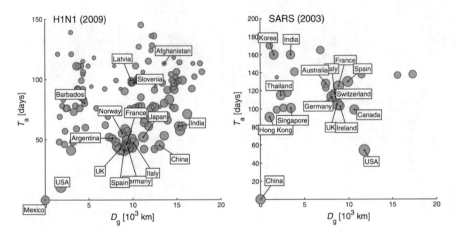

Fig. 19.4 Arrival time and geographic distance. Each panel depicts the relation of epidemic arrival time and geographic distance to the initial outbreak location (country of origin) for two different recent epidemics, the H1N1 pandemic 2009 (left) and the SARS epidemic 2003 (right). Because of the complexity of connectivity of the worldwide air-transportation network (see Fig. 19.1) geographic distance to the initial outbreak location is no longer a good predictor of arrival time, unlike in systems with local or spatially limited host mobility

infect randomly chosen lattice sites anywhere in the system. The propensity of infecting a lattice site at distance r decreases as an inverse power-law as explained in the caption to Fig. 19.3. The possibility of transmitting to distant locations yields new epidemic seeds far away that subsequently turn into new outbreak waves and that in turn seed second, third, etc. generation outbreaks, even if the overall rate at which long-distance transmission occur is very small. The consequence of this is that the spatially coherent, concerntric pattern observed in the reaction diffusion system is lost, and a complex spatially incoherent, fractal pattern emerges [16–18]. Practically, this implies that the distance from an initial outbreak location can no longer be used as a measure for estimating or computing the time that it takes for an epidemic to arrive at a certain location. Also, given a snapshot of a spreading pattern, it is much more difficult to reconstruct the outbreak location from the geometry of the pattern alone, unlike in the concentric system where the outbreak location is typically near the center of mass of the pattern.

A visual inspection of the air-transportation system depicted in Fig. 19.1 is sufficiently convincing that the significant fraction of long-range connections in global mobility will not only increase the speed at which infectious diseases spread but, more importantly, also cause the patterns of spread to exhibit high spatial incoherence and complexity caused by the intricate connectivity of the air-transportation network. As a consequence we can no longer use geographic distance to an emergent epidemic epicenter as an indicator or measure of "how far away" that epicenter is and how long it will take to travel to a given location on the globe. This type of decorrelation is shown in Fig. 19.4 for two examples: The 2003 SARS epidemic and the 2009 influenza H1N1 pandemic. On a spatial resolution of countries, the figure depicts scatter plots of the epidemic arrival time as a function of geodesic (shortest distance on the surface of the Earth) distance from the initial outbreak location. As expected, the correlation between distance and arrival time is weak.

19.3 Modeling Disease Dynamics on a Global Scale

Given that models based on local or spatially limited mobility are inadequate, improved models must be developed that account for both, the strong heterogeneity in population density, e.g. that human populations accumulate in cities that vary substantially in size, and the connectivity structure between them that is provided by data on air traffic. In a sense one needs to establish a model that captures that the entire population is a so-called meta-population, a system of $m = 1, \ldots, M$ subpopulation, each of size N_m and traffic between them, e.g. specifying a matrix F_{nm} that quantifies the amount of host individuals that travel from population m to population n in a given unit of time [19, 20]. For example N_n could correspond to the size of city n and F_{nm} the amount of passengers the travel by air from m to n. One of earliest and most employed models for disease dynamics using the meta-population approach is a generalization of Eq. (19.4) in which each population's dynamics is governed by the ordinary SIR model, e.g.

$$dS_n/dt = -\alpha S_n I_n / N_n \qquad (19.9)$$
$$dI_n/dt = \alpha S_n I_n / N_n - \beta I_n$$
$$dR_n/dt = \beta I_n$$

where the size $N_n = R_n + I_n + S_n$ of population n is a parameter. In addition to this, the exchange of individuals between populations is modeled in such a way that hosts of each class move from location m to location n with a probability rate ω_{nm} which yields

$$dU_n/dt = \sum_m \left(\omega_{nm} U_m - \omega_{mn} U_n \right) \qquad (19.10)$$

where U_m is a placeholder for S_m, I_m and R_m. The first term corresponds to the flux into location n from all other locations, the second term the flux in the opposite direction. Combining Eqs. (19.9) and (19.10) yields:

$$dS_n/dt = -\alpha S_n I_n/N_n + \sum_m \left(\omega_{nm}S_m - \omega_{mn}S_n\right) \qquad (19.11)$$

$$dI_n/dt = \alpha S_n I_n/N_n - \beta I_n + \sum_m \left(\omega_{nm}I_m - \omega_{mn}I_n\right)$$

$$dR_n/dt = \beta I_n + \sum_m \left(\omega_{nm}R_m - \omega_{mn}R_n\cdot\right)$$

which is a generic metapopulation SIR model. In principle one is required to fix the infection-related parameters α and β and the population sizes N_m as well as the mobility rates ω_{nm}, i.e. the number of transitions from m to n per unit time. However, based on very plausible assumptions [11], the system can be simplified in such a way that all parameters can be gauged against data that is readily available, e.g. the actual passenger flux F_{nm} (the amount of passengers that travel from m to n per day) that defines the air-transportation network, without having to specify the absolute population sizes N_n.

First the general rates ω_{nm} have to fulfill the condition

$$\omega_{nm}N_m = \omega_{mn}N_n$$

if we assume that the N_n remain constant. If we assume, additionally, that the total air traffic flowing out of a population n obeys

$$F_n = \sum_m F_{mn} \sim N_n,$$

i.e. it is proportional to the size of the population (e.g. the supply is proportional to the demand), the model defined by Eq. (19.11) can be recast into

$$ds_n/dt = -\alpha s_n j_n + \gamma \sum_m P_{mn} \left(s_m - s_n\right) \qquad (19.12)$$

$$dj_n/dt = \alpha s_n j_n - \beta j_n + \gamma \sum_m P_{mn} \left(j_m - j_n\right)$$

$$r_n = 1 - s_n - j_n.$$

where the dynamic variables are, again, fractions of the population in each class: $s_n = S_n/N_n, j_n = I_n/N_n$, and $r_n = R_n/N_n$. In this system the new matrix P_{mn} and the new rate parameter γ can be directly computed from the traffic matrix F_{nm} and the total population involved $\mathcal{N} = \sum_m N_m$ according to

$$P_{nm} = \frac{F_{nm}}{\sum_k F_{km}}$$

and

$$\gamma = \mathscr{F}/\mathscr{N}$$

where $\mathscr{F} = \sum_{n,m} F_{mn}$ is the total traffic in the network. The matrix P_{nm} is therefore the fraction of passengers that are leaving node m with destination n. Because passengers must arrive somewhere we have $\sum_n P_{nm} = 1$.

An important first question is concerning the different time scales, i.e. the parameters α, β and γ that appear in system (19.12). The inverse $\beta^{-1} = T$ is the infectious period, that is the time individuals remain infectious. If we assume $T \approx 4$–6 days and $R_0 = \alpha/\beta \approx 2$ both rates are of the same order of magnitude. How about γ? The total number of passengers $\mathscr{F}$ is approximately 8×10^6 per day. If we assume that $\mathscr{N} \approx 7 \times 10^9$ people we find that

$$\gamma \approx 0.0015\,\mathrm{d}^{-1}.$$

It is instructive to consider the inverse $T_{\text{travel}} = \gamma^{-1} \approx 800$ days. On average a typical person boards a plane every 2–3 years or so. Keep in mind though that this is an average that accounts for both a small fraction of the population with a high frequency of flying and a large fraction that almost never boards a plane. The overall mobility rate γ is thus a few orders of magnitude smaller than those rates related to transmissions and recoveries. This has important consequences for being able to replace the full dynamic model by a simpler model discussed below.

Figure 19.5 depicts a numerical solution to the model defined by Eq. (19.12) for a set of initial outbreak locations. At each location a small seed of infected individuals initializes the epidemic. Global aspects of an epidemic can be assessed by the total fraction of infected individuals $j_G(t) = \sum_n c_n j_n(t)$ where c_n is the relative size population n with respect to the entire population size $\mathscr{N}$. As expected the time course of a global epidemic in terms of the epicurve and duration depends substantially on the initial outbreak location.

A more important aspect is the spatiotemporal pattern generated by the model. Figure 19.6 depicts temporal snapshots of simulations initialized in London and Chicago, respectively. Analogous to the qualitative patterns observed in Fig. 19.3b, we see that the presence of long-range connections in the worldwide air-transportation network yields incoherent spatial patterns much unlike the regular, concentric wavefronts observed in systems without long-range mobility. Figure 19.7 shows that also the model epidemic depicts only a weak correlation between geographic distance to the outbreak location and arrival time. For a fixed geographic distance arrival times at different airports can vary substantially and thus the traditional geographic distance is useless as a predictor.

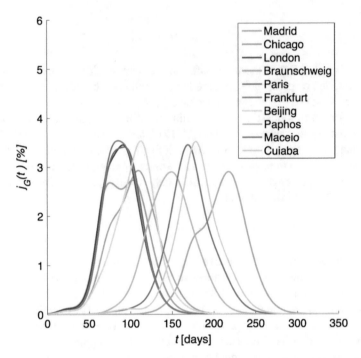

Fig. 19.5 Global epi-curves. Each curve depicts the global fraction of infected individuals as a function of time for different outbreak locations as predicted by the metapopulation model defined by Eq. (19.12). Depending on the initial outbreak location curves differ in epidemic maximum, curve shape and epidemic duration

Fig. 19.6 Properties of spatiotemporal patterns of global disease dynamics. Each panel from left to right depicts temporal snapshots of the spread of a computer-simulated hypothetical pandemic. Red nodes denote locations with a high fraction of infecteds. Each row corresponds to a different initial outbreak location (London (LHR), top and Chicago (ORD), bottom). The patterns are spatially incoherent, especially for larger times. It is thus difficult to assess which locations are affected next in the sequence of locations

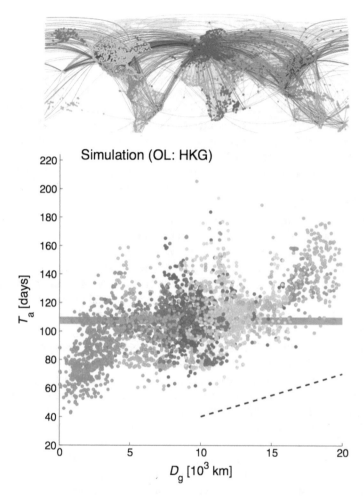

Fig. 19.7   Arrival time and geographic distance. For a simulated pandemic based on the dynamical system of Eq. (19.12) and the worldwide air-transportation network (top) the bottom panel depicts the arrival time at each location as a function of the geographic distance to the initial outbreak location Hong Kong. Airports are colored according to geographic location. Only a weak correlation between arrival time and geographic distance exists (dashed line). For a fixed small range of geographic distances a wide range of arrival times exists, geographic distance is thus not a good predictor

19.4   Issues with Computational Models

The system defined by Eq. (19.12) is one of the most parsimoneous models that accounts for strongly heterogeneous population distributions that are coupled by traffic flux between them and that can be gauged against actual population size distributions and traffic data. Surprisingly, despite its structural simplicity this type of model has been quite successful in accounting for actual spatial spreads of past epi- and pandemics [19]. Based on early models of this type and aided by the exponential increase of computational power, very sophisticated models have been developed that account for factors that are ignored by the deterministic metapopulation SIR model. In the most sophisticated approaches, e.g. GLEAM [21], the global epidemic and mobility computational tool, not only traffic by air but other means of transportation are considered, more complex infectious dynamics is considered and in hybrid dynamical systems stochastic effects caused by random reactions and mobility events are taken into account. Household structure, available hospital beds, seasonality have been incorporated as well as disease specific features, all in order to make predictions more and more precise. The philosophy of this type of research line heavily relies on the increasing advancement of both computational power as well as more accurate and pervasive data often collected in natural experiments and webbased techniques [21–25].

Despite the success of these quantitative approaches, this strategy bears a number of problems some of which are fundamental. First, with increasing computational methods it has become possible to implement extremely complex dynamical systems with decreasing effort and also without substantial knowledge of the dynamical properties that often nonlinear dynamical systems can possess. Implementing a lot of dynamical detail, it is difficult to identify which factors are essential for an observed phenomenon and which factors are marginal. Because of the complexity that is often incorporated even at the beginning of the design of a sophisticated model in combination with the lack of data modelers often have to make assumptions about the numerical values of parameters that are required for running a computer simulation [26]. Generically many dozens of unknown parameters exist for which plausible and often not evidence-based values have to be assumed. Because complex computational models, especially those that account for stochasticity, have to be run multiple times in order to make statistical assessments, systematic parameter scans are impossible even with the most sophisticated supercomputers.

Finally, all dynamical models, irrespective of their complexity, require two ingredients to be numerically integrated: (1) fixed values for parameters and (2) initial conditions. Although some computational models have been quite successful in describing and reproducing the spreading behavior of past epidemics and in situations where disease specific parameters and outbreak locations have been assessed, they are difficult to apply in situations when novel pathogens emerge. In these situations, when computational models from a practical point of view are needed most, little is known about these parameters and running even the most sophisticated models "in the dark" is problematic. The same is true for fixing the right initial conditions. In many cases, an emergent infectious disease initially spreads unnoticed and the public becomes aware of a new event after numerous cases occur in clusters at different locations. Reconstructing the correct initial condition often takes time, more time than is usually available for making accurate and valueable predictions that can be used by public health workers and policy makers to devise containment strategies.

19.5 Effective Distance

Given the issues discussed above one can ask if alternative approaches exist that can inform about the spread without having to rely on the most sophisticated highly detailed computer models. In this context one may ask whether the complexity of the observed patterns that are solutions to models like the SIR metapopulation model of Eq. (19.12) are genuinely complex because of the underlying complexity of the mobility network that intricately spans the globe, or whether a simple pattern is really underlying the dynamics that is masked by this complexity and our traditional ways of using conventional maps for displaying dynamical features and our traditional ways of thinking in terms of geographic distances.

In a recent approach Brockmann and Helbing [11] developed the idea of replacing the traditional geographic distance by the notion of an *effective distance* derived from the topological structure of the global air-transportation network. In essence the idea is very simple: If two locations in the air-transportation network exchange a large number of passengers they should be effectively close because a larger number of passengers implies that the probability of an infectious disease to be transmitted from A to B is comparatively larger than if these two locations were coupled only by a small number of traveling passengers. Effective distance should therefore decrease with traffic flux. What is the appropriate mathematical relation and a plausible ansatz to relate traffic flux to effective distance? To answer this question one can go back to the metapopulation SIR model, i.e. Eq. (19.12). Dispersal in this equation is governed by the flux fraction P_{nm}. Recall that this quantity is the fraction of all passengers that leave node m and arrive at node n. Therefore P_{nm} can be operationally defined as the probability of a randomly chosen passenger departing node m arriving at node n. If, in a thought experiment, we assume that the randomly selected person is infectious, P_{nm} is proportional to the probability of transmitting a disease from airport m to airport n. We can now make the following ansatz for the effective distance:

$$d_{nm} = d_0 - \log P_{nm} \qquad (19.13)$$

where $d_0 \geq 0$ is a non-negative constant to be specified later. This definition of effective distance implies that if all traffic from m arrives at n and thus $P_{nm} = 1$ the effective distance is $d_{nm} = d_0$ which is the smallest possible value. If, on the other hand P_{nm} becomes very small, d_{nm} becomes larger as required. The definition (19.13) applies to nodes m and n that are connected by a link in the network. What about pairs of nodes that are not directly connected but only by paths that require intermediate steps? Given two arbitrary nodes, an origin m and a destination n, an infinite amount of paths (sequence of steps) exist that connect the two nodes. We can define the shortest effective route as the one for which the accumulation of effective distances along the legs is minimal. So for any path we sum the effective distance along the legs according to Eq. (19.13) adding up to an effective distance D_{nm}. This approach also explains the use of the logarithm in the definition of effective distance. Adding effective distances along a route implies the multiplication of the probabilities P_{nm} along the involved steps. Therefore the shortest effective distance D_{nm} is equivalent to the most probable path that connect origin and destination. The parameter d_0 is a free parameter in the definition and quantifies the influence of the number of steps involved in a path. Typically it is chosen to be either 0 or 1 depending on the application.

One important property of effective distance is its asymetry. Generally we have

$$d_{nm} \neq d_{mn}.$$

This may seem surprising at first sight, yet it is plausible. Consider for example two airports A and B. Let's assume A is a large hub that is strongly connected to many other airports in the network, including B. Airport B, however, is only a small airport with only as a single connection leading to A. The effective distance $B \rightarrow A$ is much smaller (equal to d_0) than the effective distance from the hub A to the small airport B. This accounts for the fact that if, again in a thought experiment, a randomly chosen passenger at airport B is most definitely going to A whereas a randomly chosen passenger at the hub A is arriving at B only with a small probability.

Given the definition of effective distance one can compute the shortest effective paths to every other node from a chosen and fixed reference location. Each airport m thus has a set of shortest paths $\mathscr{P}_m$ that connect m to all other airports. This set forms the shortest path tree T_m of airport m. Together with the effective distance matrix D_{nm} the tree defines the perspective of node m. This is illustrated qualitatively in the Fig. 19.8 that depicts a planar random triangular weighted network.

One can now employ these principles and compute the shortest path trees and effective distances from the perspective of actual airports in the worldwide air-transportation network based on actual traffic data, i.e. the flux matrix F_{nm}. Figure 19.9 depicts the shortest path tree of one of the Berlin airports (Tegel, TXL). The radial distance of all the other airports in the network is proportional to their effective distance from TXL. One can see that large European hubs are effectively close to TXL as expected. However, also large Asian and American airports are effectively close to TXL. For example the airports of Chicago (ORD), Beijing (PEK), Miami (MIA) and New York (JFK) are comparatively close to TXL. We can also see that from the perspective of TXL, Germany's largest airport FRA serves as a gateway to a considerable fraction of the rest of the world. Because the shortest path tree also represents the most probable spreading routes one can use this method to identify airports that are particularly important in terms of distributing an infectious disease throughout the network.

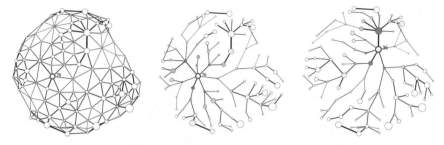

Fig. 19.8 Shortest paths and shortest path trees in complex networks. *Left*: A random planar weighted network consisting of 100 nodes and 283 links. Links vary in strength. The size of the nodes quantifies the total link weight per node. *Center*: For a chosen node (no. 76) the shortest path tree is shown. Color depicts effective distance. *Right*: The shortest path tree of node no. 36. The shortest path trees are also those paths that correspond to the most probable paths of a random walker that starts at the reference location and terminates at the respective target node

19.6 Recovery of Concentric Patterns

The use of effective distance and representing the air-transportation network from the perspective of chosen reference nodes and making use of the more plausible notion of distance that better reflects how strongly different locations are coupled in a networked system is helpful for "looking at" the world. Yet, this representation is more than a mere intuitive and plausible spatial representation. What are the dynamic consequences of effective distance? The true advantage of effective distance is illustrated in Fig. 19.10. This figure depicts the identical computer-simulated hypothetical pandemic diseases as Fig. 19.6. Unlike the latter, that is based on the traditional geographic representation, Fig. 19.10 employs the effective distance and shortest path tree representation from the perspective of the outbreak location as discussed above. Using this method, the spatially incoherent patterns in the traditional representation are transformed into concentric spreading patterns, similar to those expected for simple reaction diffusion systems.

This shows that the complexity of observed spreading patterns is actually equivalent to simple spreading patterns that are just convoluted and masked by the underlying network's complexity. This has important consequences. Because only the topological features of the network are used for computing the effective distance and no dynamic features are required, the concentricy of the emergent patterns are a generic feature and independent of dynamical properties of the underlying model. It also means that in effective distance, contagion processes spread at a constant speed, and just like in the simple reaction diffusion model one can much better predict the arrival time of an epidemic wavefront, knowing the speed and effective distance. For example if shortly after an epidemic outbreak the spreading commences and the initial spreading speed is assessed, one can forecast arrival times without having to run computationally expensive simulations. Even if the spreading speed is unknown, the

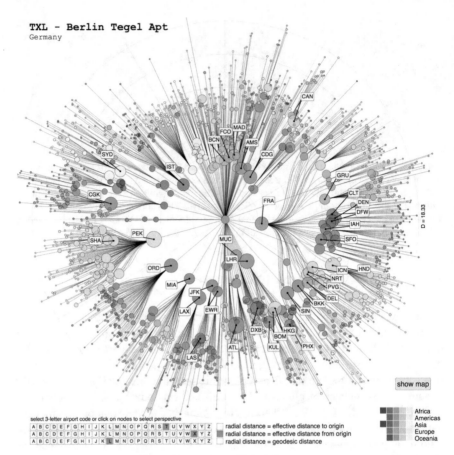

Fig. 19.9 Shortest path trees and effective distance from the perspective of airport Tegel (TXL) in Berlin. TXL is the central node. Radial distance in the tree quantifies the effective distance to the reference node TXL. As expected large European hubs like Frankfurt (FRA), Munich (MUC) and London Heathrow (LHR) are effective close to TXL. However, also hubs that are geographically distant such as Chicago (ORD) and Beijing (PEK) are effectively closer than smaller European airports. Note also that the tree structure indicates that FRA is a gateway to a large fraction of other airports as reflected by the size of the tree branch at FRA. The illustration is a screenshot of an interactive effective distance tool available online [27]

effective distance which is independent of dynamics can inform about the sequence of arrival times, or relative arrival times.

The benefit of the effective distance approach can also be seen in Fig. 19.11 in which arrival times of the 2003 SARS epidemic and the 2009 H1N1 pandemic in affected countries are shown as a function of effective distance to the outbreak origin. Comparing this figure to Fig. 19.7 we see that effective distance is a much better predictor of arrival time, a clear linear relationship exists between effective distance

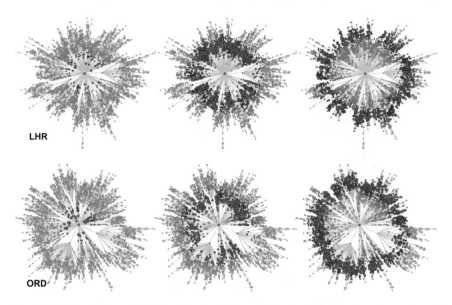

Fig. 19.10 Simulations and effective distance. The panels depict the same temporal snapshots of computer simulated hypothetical pandemic scenarios as in Fig. 19.6. The top row corresponds to a pandemic initially seeded at LHR (London) the bottom row at ORD (Chicago). The networks depict the shortest path tree effective distance representation of the corresponding seed airports as in Fig. 19.9. The simulated pandemics that exhibit spatially incoherent complex patterns in the traditional representation (Fig. 19.6) are equivalent to concentric wave fronts that progress at constant speeds in effective distance space. This method thus substantially simplifies the complexity seen in conventional approaches and improves quantitative predictions

and epidemic arrival. Thus, effective distance is a promising tool and concept for application in realistic scenarios, being able to provide a first quantitative assessment of an epidemic outbreak and its potential consequences on a global scale.

19.7 Reconstruction of Outbreaks

In a number of situation epidemiologists are confronted with the task of reconstructing the outbreak origin of an epidemic. When a novel pathogen emerges in some cases the infection spreads covertly until a substantial case count attracts attention and public health officials and experts become aware of the situation. Quite often cases occur much like the patterns depicted in Fig. 19.3b in a spatially incoherent way because of the complexity of underlying human mobility networks. When cases emerge at apparently randomly distributed locations it is a difficult task to assess where the event initially started. The computational method based on effective distance can also be employed in these situations provided that one knows the underlying mobility network. This is because the concentric pattern depicted in Fig. 19.10 is

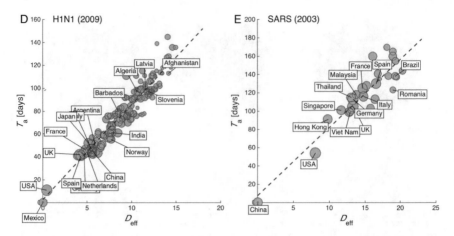

Fig. 19.11 Correlation of arrival time with effective distance. *Left*: the relationship of epidemic arrival time and effective distance for the H1N1 pandemic 2009. Compared to the conventional use of geographic distance effective distance is a much better predictor of epidemic arrival time as is reflected by the linear relationship between arrival time and effective distance, e.g. compare to Fig. 19.7. *Right*: The same analysis for the 2003 SARS epidemic. Also in this case effective distance is much more strongly correlated with arrival time than geographic distance

only observed if and only if the actual outbreak location is chosen as the center perspective node. In other words, if the temporal snapshots are depicted using a different reference node the concentric pattern is scrambled and irregular. Therefore, one can use the effective distance method to identify the outbreak location of a spreading process based on a single temporal snapshot. This method is illustrated in a proof-of-concept example depicted in Fig. 19.12. Assume that we are given a temporal snapshot of a spreading process as depicted in Fig. 19.12a and the goal is to reconstruct the outbreak origin from the data. Conventional geometric considerations are not sucessful because the network-driven processes generically do not yields simple geometric patterns. Using effective distance, we can now investigate the pattern from the perspective of every single potential outbreak location. We could for example pick a set of candidate outbreak locations (panel (b) in the figure). If this is done we will find that only for one candidate outbreak location the temporal snapshot has the shape of a concentric circle. This must be the original outbreak location. This process, qualitatively depicted in the figure, can be applied in a quantitative way and has been applied to actual epidemic data such as the 2011 EHEC outbreak in Germany [28].

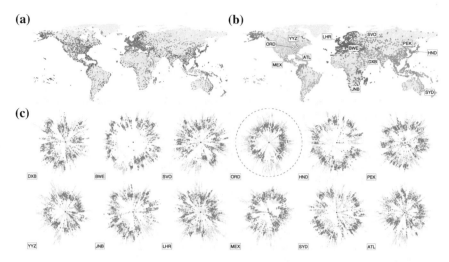

Fig. 19.12 Outbreak reconstruction using effective distance. **a** The panel depicts a temporal snapshot of a computer simulated hypothetical pandemic, red dots denote airports with a high prevalence of cases. From the snapshot alone it is difficult to assess the outbreak origin which in this case is ORD (Chicago). **b** A choice of 12 potential outbreak locations as candidates. **c** For these candidate locations the pattern is depicted in the effective distance perspective. Only for the correct outbreak location the pattern is concentric. This method can be used quantitatively to identify outbreaks of epidemics that initially spread in a covert way

19.8   Conclusions

Emergent infectious diseases that bear the potential of spreading across the globe are an illustrative example of how connectivity in a globalized world has changed the way human mediated processes evolve in the 21st century. We are connected by complex networks of interaction, mobility being only one of them. With the onset of social media, the internet and mobile devices we share information that proliferates and spreads on information networks in much the same way (see also Chap. 20). In all of these systems the scientific challenge is understanding what topological and statistical features of the underlying network shape particular dynamic features observed in natural systems. The examples addressed above focus on a particular scale, defined by a single mobility network, the air-transportation network that is relevant for this scale. As more and more data accumulates, computational models developed in the future will be able to integrate mobility patterns at an individual resolution, potentially making use of pervasive data collected on mobile devices and paving the way towards predictive models that can account very accurately for observed contagion patterns. The examples above also illustrate that just feeding better and faster computers with more and more data may not necessarily help understanding the fundamental processes and properties of the processes that underlie a specific dynamic phenomenon. Sometimes we only need to change the conventional and traditional ways of looking at patterns and adapt our viewpoint appropriately.

References

1. M.P. Girard, J.S. Tam, O.M. Assossou, M.P. Kieny, Vaccine **28**, 4895 (2010)
2. P.R. Saunders-Hastings, D. Krewski, Pathogens (Basel, Switzerland) **5**, 66 (2016)
3. Note 1, examples are *Contagion*, a surprisingly accurate depiction of the consequences of a severe pandemic, and *Rise of the planet of the apes* that concludes with ficticious explanation for the extinction of mankind due to a man made virus in the future
4. M. Cotten, S.J. Watson, A.I. Zumla, H.Q. Makhdoom, A.L. Palser, S.H. Ong, A.A. Al Rabeeah, R.F. Alhakeem, A. Assiri, J.A. Al-Tawfiq, et al., mBio **5** (2014)
5. J.A. Backer, J. Wallinga, PLoS Comput. Biol. **12**, e1005210 (2016)
6. J. Diamond, *Guns, Germs, and Steel: The Fates of Human Societies* (W. W. Norton & Company, 1999). ISBN 0-393-03891-2
7. Current world population, http://www.worldometers.info/world-population/. Accessed 15 Jan 2017
8. United Nations (2014)
9. J.J. Lee, H. Seebens, B. Blasius, D. Brockmann, Eur. Phys. J. B **84**, 589 (2011)
10. A. Barrat, M. Barthélemy, R. Pastor-Satorras, A. Vespignani, Proc. Natl. Acad. Sci. USA **101**, 3747 (2004)
11. D. Brockmann, D. Helbing, Science (New York, N.Y.) **342**, 1337 (2013)
12. W.O. Kermack, A.G. McKendrick, Proc. R. Soc. Lond. A: Math. Phys. Eng. Sci. **115**, 700 (1927), http://rspa.royalsocietypublishing.org/content/115/772/700.full.pdf, http://rspa.royalsocietypublishing.org/content/115/772/700. ISSN 0950-1207
13. R.M. Anderson, R.M. May, *Infectious Diseases of Humans: Dynamics and Control* (Oxford University Press, Oxford, 1992)
14. M.J. Keeling, P. Rohani, *Modeling Infectious Diseases in Humans and Animals* (Princeton University Press, 2008)
15. J.D. Murray, *Mathematical Biology. II Spatial Models and Biomedical Applications Interdisciplinary Applied Mathematics*, vol. 18 (Springer New York Incorporated, 2001)
16. D. Brockmann, *Eur* (Phys. J. Spec, Top, 2008)
17. D. Brockmann, L. Hufnagel, *Phys* (Rev, Lett, 2007)
18. D. Brockmann, *Human Mobility and Spatial Disease Dynamics* (Wiley-VCH, 2009), pp. 1–24
19. L. Hufnagel, D. Brockmann, T. Geisel, Proc. Natl. Acad. Sci. USA **101**, 15124 (2004)
20. O. Ovaskainen, I. Hanski, Am. Nat. **164**, 364 (2004)
21. W. Van den Broeck, C. Gioannini, B. Gonçalves, M. Quaggiotto, V. Colizza, A. Vespignani, B.M.C. Infect, Dis. **11**, 37 (2011)
22. V. Colizza, A. Barrat, M. Barthelemy, A.-J. Valleron, A. Vespignani, PLoS Med. **4**, e13 (2007)
23. D. Balcan, V. Colizza, B. Gonçalves, H. Hu, J.J. Ramasco, A. Vespignani, Proc. Natl. Acad. Sci. USA **106**, 21484 (2009)
24. M. Ajelli, B. Gonçalves, D. Balcan, V. Colizza, H. Hu, J.J. Ramasco, S. Merler, A. Vespignani, B.M.C. Infect, Dis. **10**, 190 (2010)
25. P. Bajardi, C. Poletto, J.J. Ramasco, M. Tizzoni, V. Colizza, A. Vespignani, PLoS One **6**, e16591 (2011)
26. R.M. May, Science (New York, N.Y.) **303**, 790 (2004)
27. *Effective Distance and Shortest Paths in Global Mobility*, http://rocs.hu-berlin.de/D3/waneff/. Accessed 15 Jan 2017
28. J. Manitz, T. Kneib, M. Schlather, D. Helbing, D. Brockmann, PLoS Curr. **6** (2014)

Chapter 20
Spreading of Failures in Interdependent Networks

Louis M. Shekhtman, Michael M. Danziger and Shlomo Havlin

20.1 Introduction

Most studies of spreading focus on how some physical objects move in space, yet spreading can also involve other phenomenon. Recent research has explored the spreading of *failures* in various complex systems like power grids, communications networks, financial networks and others. In these systems, when one failure occurs it can trigger a cascade wherein that failure spreads to other parts of the system. Failure spreading can have dramatic results leading to blackouts, economic collapses, and other catastrophic events.

In order to combat this problem, it is often useful to model and understand the physical mechanisms of failure spreading. While it would be ideal if failures could be prevented entirely, this is unlikely since every system will experience failure at one time or another. Rather, the approach and models reviewed in this chapter focus on the mechanisms of how initial failures spread and how this information can be used to mitigate the spreading. It is also noteworthy that many of the same models used here in the context of failures in infrastructure networks can also be used to identify influential individuals who are capable of spreading a message to a large audience in social networks.

A specific focus here will be on the mechanism involved in the realistic situation where two systems are interdependent such that components of one system cannot function without components of the other. This could be for example, the case in the context of a communication tower that needs to receive power from a

L. M. Shekhtman (✉) · M. M. Danziger · S. Havlin
Department of Physics, Bar Ilan University, Ramat Gan, Israel
e-mail: lsheks@gmail.com

M. M. Danziger
e-mail: mmdanziger@gmail.com

S. Havlin
e-mail: havlins@gmail.com

© Springer International Publishing AG 2018   397
A. Bunde et al. (eds.), *Diffusive Spreading in Nature, Technology and Society*,
https://doi.org/10.1007/978-3-319-67798-9_20

nearby power station. If the power station fails, that failure immediately spreads to the communication tower. The failure of the communication tower can then lead to additional failures that also impact the power grid (either directly or indirectly) and so on. Understanding how these failures spread in both time and space is critical in order to ensure that large-scale complex systems remain functional.

A further challenge examined here involves the question of how to optimally repair and recover a system after it has experienced some failures. While it may seem simple to repair failed components of a system, it will be ineffective if the spreading of the failures has not first been contained. Making repairs is useless if the failures quickly spread to the repaired components once again. Instead, repairs must be made in a clear and purposeful way in order to restore the system to a functional state.

Many complex systems such as power grids, communication systems, the internet, and biological systems have recently been modeled as *complex networks*. This representation of the system involves defining sets of nodes that are connected to one another through links. Precisely what constitutes a node and/or link will depend on the exact system being analyzed. For example, in power grids the nodes are typically defined as the power stations and the links are powerlines that connect the power stations. In communication networks, nodes could be antenna towers and towers that are in range of one another are linked. Many different systems can be modeled in such a manner and researchers have discovered that while different systems have unique properties, many global network properties remain true across multiple systems.

Most of the properties discovered in complex networks relate to the structure of the connections between the nodes. The number of connections of a particular node is known as its degree, k. Networks where connections are assigned purely randomly (Erdős-Rényi networks) have a degree distribution that is Poisson. The most important feature of Poisson distributions, at least in the context of complex networks, is that they have a typical mean degree, $< k >$ and it is highly unlikely for any node to have a degree that is substantially larger or smaller than $< k >$. Explicitly the likelihood of a node to have degree k, is given by $P(k) = < k >^k e^{-k}/k!$. (Note also, that in the text we will simply refer to the mean degree of an Erdős-Rényi network as k rather than $< k >$ and that it is common to do so in the literature). Early research found that the distribution of the degree in real networks often takes the form of a power law. This means that the likelihood of a node to have degree k is proportional to $k^{-\gamma}$, i.e. $P(k) \sim k^{-\gamma}$. Notably, if $\gamma < 3$, then some nodes end up with far more connections than others and the variance tends to infinity. Networks with this property are known as scale-free networks [1]. Another unique feature present in many networks is the existence of tightly connected communities that have many links to other nodes in the same community (module), but few links to nodes outside of the community [2]. This property is often referred to as modularity and it is highly ubiquitous in many networks. Lastly, many networks, like power grids, are embedded in physical space (spatial networks) and the expense of creating long-range links forces most links to be of short length [3]. There are many other significant structures that exist in networks which are more fully reviewed in one of the recent books on the subject [4, 5]. More information on diffusion in complex networks can be found in [6] or in [7, 8].

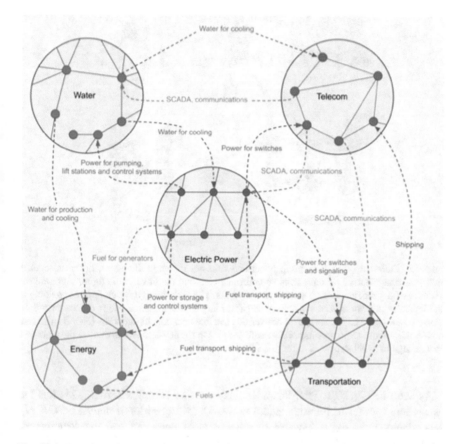

Fig. 20.1 Interdependence in modern infrastructure causes failures to spread between systems. This is the result of multiple systems needing power for switches, supervisory control and data acquisition (SCADA), fuel transport, and other resources. After [20]

Further research has led to the recognition that many networks do not exist in isolation, but rather a network is often only one of several interdependent networks [9–14]. This situation refers to the case where a node in one network, say a communication antenna tower, depends on a node in another network, say a power station. This relationship can be described through the existence of a new type of link known as a *dependency* link [15–17]. Whereas connectivity links represent the idea that some sort of flow occurs between the two connected nodes (e.g. flow of electricity in power grids, flow of information in communication networks), dependency links mean that if the node being depended upon fails then the dependent node also fails. Such situations are especially common in infrastructure, but they can also arise in biological systems [18] and financial networks [19]. An example detailing the interdependence between different infrastructure networks can be found in Fig. 20.1.

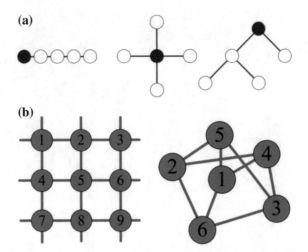

Fig. 20.2 Different types of interdependent networks are shown in the figure. In this case, each node shown above actually represents an entire network and the links in the figure represent the existence of dependency links between two networks. The path through which failure spreading in interdependent networks occurs is determined according to which networks have dependency links between them. (**a**) The top structures are various treelike networks of networks (NON) structures and (**b**) the bottom structures are NONs with loops. On the left is a lattice and on the right is a random-regular NON structure. After [21, 22]

As seen in Fig. 20.1, interdependencies can take complex forms. This has led researchers to refer to interdependent networks as 'networks of networks' (NON). Dependency links exist between specific pairs of networks and the structure of the network of networks is defined according to which pairs of networks have dependency links. A few examples of networks of networks are shown in Fig. 20.2. Simple examples include cases where the NON dependencies form a tree, a single loop, and a random-regular configuration where all networks depend on the same number of networks.

One of the most important properties of many networks and systems in general is their robustness to failures. As discussed briefly above, power stations can become overloaded or fail for other reasons and communication antenna towers can have problems due to bad weather or other issues. While efforts are always made to minimize the frequency of these failures, they are bound to occur. When analyzing the system as a whole, it is desirable to optimize the network such that it can still continue functioning even if some of the nodes fail. In many cases the functioning of the system can be quantified by asking how many nodes remain connected after some nodes fail. For example, for communications networks it is often most relevant to ask, "How many nodes can communicate after some others fail?" or in the context of power grids, "How many power stations are still linked to the grid after some stations fail?" The failure of one node can cause other nodes to become disconnected from the network as a whole and fail as well, thus the initial failures are magnified and can spread throughout the network.

The question of what fraction of a system remains connected after some set of failures can be answered through percolation theory from physics. Percolation theory essentially determines clusters of nodes that are connected to one another such that flow can occur between them. The largest cluster, which contains the largest number of nodes, is referred to as the *giant connected component* and is described by P_∞ [4, 5, 23, 24]. Explicitly, P_∞ is defined as the fraction of nodes remaining in the largest connected component (or equivalently, the likelihood of a node to be in the largest component) at some point in a percolation process. For our purposes, only nodes that are part of this largest cluster are considered functional whereas all other nodes are considered to have failed. The goal in designing resilient systems is to maximize the size of the giant component for any case of failures.

Percolation theory was able to discover that scale-free networks (i.e. those whose degree distribution follows a power-law) are far more resilient to random failures than random networks. In other words, if the same number of failures occur in both random and scale-free networks, a larger fraction of a scale-free network will remain connected. More precisely, in contrast to random networks where only a finite fraction of nodes must be removed, for scale-free networks only if nearly all of the nodes are randomly removed, will the network become totally fragmented [25]. In any case, for both isolated random and isolated scale-free networks, slightly increasing the number of initial failures only slightly increases the number of total failures. In other words, the transition from a functioning to non-functioning state is continuous.

Failures in interdependent networks occur and spread through two different mechanisms. The first is the same as in single networks, i.e. failures of nodes lead further nodes to become disconnected from the giant component. The second mechanism is through failures spreading due to the dependency links. As mentioned previously, a node at one end of a dependency link relies on the node at the other end of the link to function. If a node on one end of a dependency link fails, the node on the other end of the link also fails. As we will see in the next section, such mechanisms can lead to abrupt collapse.

20.2 Robustness of Interdependent Networks

Percolation methods were widely applied to solve problems in single networks [4, 5, 25–27] and recent research has expanded these methods to interdependent networks [15, 16, 21, 28, 29]. In interdependent networks, when some nodes fail, they cause other dependent nodes to fail [15]. The failure of these dependent nodes then disconnects other nodes from the giant component and leads to the failure of more dependent nodes. In this manner, failures spread through the system until a steady state is reached. It is noteworthy that because of the cascade, removing a single additional node can cause the system to collapse entirely, i.e. the transition is abrupt and first-order [15, 16, 28, 30]. This is significantly different from isolated networks where the transition is continuous.

In the initial work on interdependent networks, Buldyrev et al. [15] calculated the final fraction of functional nodes after the cascade analytically. They also carried out numerical simulations to verify their results. In explaining the results from [15], it is important to note the result for percolation of a single Erdős-Rényi network, namely $P_\infty = p(1 - e^{-kP_\infty})$, where p is the fraction of the network that survives the initial failures and k is the average degree of the network [31–33]. It is also noteworthy that since P_∞ appears on both sides of the equation and no additional simplification is possible, the equation is transcendental and can only be solved numerically. If p nodes survive the initial failures in a system of two interdependent Erdős-Rényi networks, the size of the giant component is described by [15, 28],

$$P_\infty = p(1 - e^{-kP_\infty})^2. \tag{20.1}$$

While the difference between the formulas for P_∞ for single and interdependent networks may seem small, namely a power of 2 instead of 1 on the right side of the equation, this small change has dramatic consequences leading to the long cascades and abrupt failures described throughout this review. Essentially, the power of 2 can be understood by recognizing that nodes now must be both in the giant component of their own network and have their dependent node be in the giant component of the second network. Gao et al. [21, 22, 28, 34] later solved several cases that involved more than two networks. In the case of n interdependent Erdős-Rényinetworks with full dependency such that they form a tree (Fig. 20.2a), the size of the giant component is given by [28]

$$P_\infty = p(1 - e^{-kP_\infty})^n. \tag{20.2}$$

Gao et al. [21, 28] also solved several other simple structures of networks of networks analytically.

Other papers by Bianconi et al. [35, 36], Baxter et al. [30, 37], Hu et al. [38], Kim et al. [39], Lee et al. [40] and Cellai et al. [41] have also obtained further analytic results for interdependent networks.

20.3   Interdependent Networks with Realistic Features

In this section we will provide a brief review on more realistic models of failure spreading in interdependent networks. The internet and power-grids, as well as other networks, are not purely random and instead contain non-random structure. This structure influences the spreading of failures.

One common feature is a degree distribution that is scale-free. It was found for interdependent scale-free networks, that a broader degree distribution makes the networks *more* vulnerable to the spreading of failures [15, 42].

Another realistic feature that has recently been found to influence failure spreading in single and interdependent networks is modularity [43, 44]. Both of those studies [43, 44] examined the case of attacks on interconnected nodes, i.e. nodes that

(a)

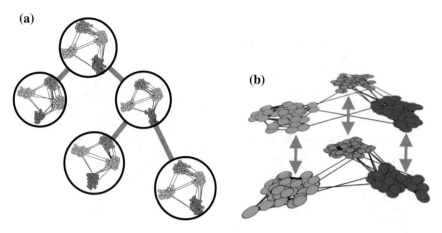

(b)

Fig. 20.3 Here we show examples of modular structure in interdependent networks. In this case, each of the interdependent networks has a modular structure, i.e. they are segregated into distinct tightly connected communities. This is seen by the fact that inside each black circle in (**a**) there is a network with modular structure. Further, each of the communities is highlighted with a different color. The specific structure shown in (**a**) is a treelike network of modular networks. The dependency links are restricted such that a node in a particular module in one network will depend on a node that is in the same module in the second network (i.e. dependency links are between nodes of the same color). This is illustrated clearly in (**b**). After [43]

connect between two communities. Another study using a similar model, showed that attacks on interconnected nodes lead to very fast spreading of failures, especially in the Western U.S. power grid [45]. Shekhtman et al. [43] solved analytically the case where there are several networks each of which has the same number of modules of the same size. Dependency links were also restricted to be between corresponding modules in different layers. An example of where this model is realistic is the case of infrastructure within and between cities. Each city has its own infrastructure and the interdependence occurs within the city. At the same time, different infrastructure networks will connect both within and across several cities. Consider the example of a coupled system of a power grid and a communications system. Most likely, a power station and a communication tower that depend on one another will be in the same city. This is true even though both the communication tower and the power station have connections to other cities. The model is visualized in Fig. 20.3. Failures can lead to collapse that occurs in one or two stages, i.e. there can be two transitions. When two transitions occur, the first is the result of modules separating but continuing to function independently. After additional failures, the modules themselves also collapse.

Failure spreading is also influenced by spatial features. It is well accepted that many infrastructures are embedded in space, including power grids and many communication networks [3, 46]. This embeddedness has significant influence on how failures spread. To simplify studies of spatial networks, 2D lattices are often used as models and it is noted that any other embedded network is in the same universality class [24, 47]. An early study on spatially embedded networks found that they

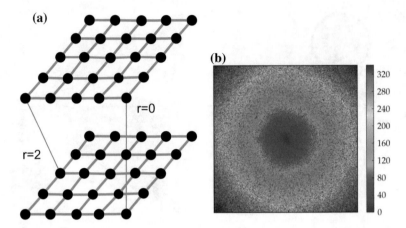

Fig. 20.4 **a** Dependency links can be restricted such that only pairs of nodes within some distance r are allowed to be interdependent. In the figure the cases of $r = 0$ and $r = 2$, where the pairs are zero and two lattice spaces apart are shown. After [51]. **b** The radial spreading at criticality is shown. The redder regions end up failing at later times (as shown in the colorbar on the right) in comparison to the central regions. After [50]

are extremely vulnerable in the sense that two interdependent networks can collapse abruptly even if only a few of the nodes in each network are interdependent [48].

Another study examined a case where the dependency links are restricted to a maximal length, r (demonstrated in Fig. 20.4a). This accounts for the fact that dependencies are most likely to be short range and that it is highly unlikely for example that a communication tower in the Eastern United States is dependent on a power grid in the Western United States. For short range dependency lengths, i.e. low r, the percolation transition is continuous, but for larger r, the transition is abrupt. The shift between the behaviors occurs when r reaches a critical value, $r_c \approx 8$ [49]. Above this critical dependency length, the percolation transition occurs in such a way that failures spread radially outward from an initial damage site until they end up finally consuming the entire network [50], see Fig. 20.4b.

Later works incorporated additional spatial features in order to move towards even more realistic models of interdependent spatial infrastructure including considering the case of NON formed of more than just two networks [50–53].

The cascade in interdependent networks can be mapped to other cascades like blackouts in power grids [54, 55]. Most blackouts and other failures spread in a predictable manner. Understanding the spatio-temporal spreading is crucial in order to understand and contain such failure spreading. Specifically, it has been found that the spatio-temporal dynamics of the cascade can be used to identify a specific dependency correlation distance that determines how failures spread [56]. This dependency correlation distance defines how far the failures are likely to spread. In that work, Zhao et al. [56] studied the case of overload failures in spatially embedded networks and examined how failures propagate in space and time. The authors defined load according to the well-known *betweenness* centrality, which measures how many

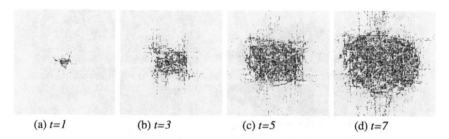

| (a) $t=1$ | (b) $t=3$ | (c) $t=5$ | (d) $t=7$ |

Fig. 20.5 The propagation of failures in a synthetic overloaded system are shown. The red nodes in the center represent the initial failures. At each time step, additional nodes that fail due to overloads are shown in blue. Nodes that have already failed are shown in black. As seen, the spread occurs almost radially outward from the location of the initial failures. After [56]

shortest paths go through a particular link [54]. The initial load depends on the structure of the network and nodes that have more shortest paths going through them have a higher load. After an initial set of localized failures, the paths between nodes change and load becomes redistributed, especially around the failed nodes. However, due to this redistribution, some other nodes will also become overloaded and fail. This process will continue either until the load manages to rebalance or until the entire network collapses. The dependency correlation distance describes how far the direct effects of the initial redistribution are felt. Zhao et al. [56] studied how the failures spread as a function of the tolerance, α. The quantity α is defined such that $1+\alpha$ times the original load is the maximal load above which the node becomes overloaded and fails. They found that for all values of α the spreading of failures occurs radially from the initial failures and spreads at approximately constant velocity. As α increases, the velocity of the spreading of failures decreases. This is intuitive as it means that the system is able to accept a higher increased load without failing. An example of the spreading in a synthetic power grid can be seen in Fig. 20.5. These results support the model of interdependent spatial networks studied in previous works [49, 57, 58] where now the velocities can be mapped to the length of the dependency links [56]. As explained earlier, the length of the dependency links represents the distance between two nodes that rely on one another. The velocity of the failure spreading in the model from [56] has the similar meaning of how quickly failures from a node in one location reach a node in another location. The specific procedure for mapping between these two quantities is described in [56].

Other aspects relating to more realistic models of interdependent networks have also been analyzed in many further works which consider many types of network structures and conditions on dependency links [38, 41, 59–72].

20.4 Localized Attacks on Interdependent Networks

Another realistic feature that has recently been incorporated into understanding the resilience of both single and interdependent networks is localized attack [57, 73, 74]. For localized attack on a pair of spatially embedded networks it was found that

(a)   **(b)**

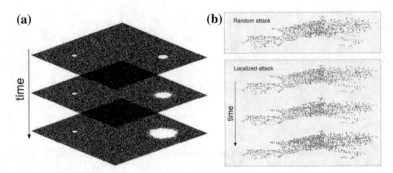

Fig. 20.6 **a** Depending on the initial size of the hole, it may either spread through a system of interdependent networks (the hole on the right) or not (the hole on the left). Whether the hole spreads, depends only on the degree of the networks and not on the number of total nodes in the system. **b** Here a localized attack is shown to spread on an interdependent system with a layout according to the European Power grid, whereas a random attack does not spread. After [57]

a 'hole' above a certain critical size must be made in one of the networks in order for the failures to spread throughout the entire system (see Fig. 20.6). The researchers in [57] found that even though a network may be robust to random attacks, it can be vulnerable to localized attacks. In addition, the critical size of the 'hole,' denoted r_h^c, that must be made to collapse the system is independent of the size of the system and instead depends only on the degree. This behavior is vastly different from the case of a single spatially embedded network where the size of the hole necessary for total system failure scales with the size of the system. Localized attacks are particularly relevant in the realistic case of an Electromagnetic Pulse (EMP) detonation, i.e. a short burst of electromagnetic energy that damages all electronic devices within some radius.

20.5 Recovery in Single Networks and Interdependent Networks

In order to understand the spreading of failures, it is also important to consider how to repair failures as they spread throughout a system. This question is of course highly relevant since while the goal is always to prevent failures from occurring, all systems will experience failure at some point. To address this question, researchers have begun studying how to optimally repair and recover a system like a power grid or the internet. It was found in [75] that when node recoveries are introduced in a simple dynamic cascade model [76], the system can spontaneously recover. The model contains three key parameters: one describes the fraction of internally failed nodes (p^*), a second governs the time for recovery to occur (τ), and the third describes the probability of failure due to lack of support from external nodes (r). For the case of small networks, r and p^* in the system, due to stochasticity will not be fixed, but will instead wander in phase space near their average values. This exploration of phase

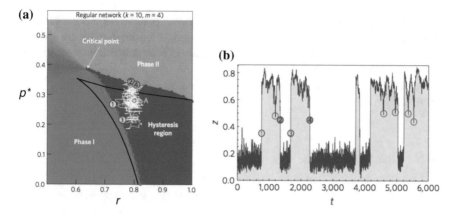

Fig. 20.7 **a** Here we show the phase space describing the state of the system according to the number of internally failed nodes (p^*) and the likelihood of external failure (r). The white line shows the trajectory of the system as it diffuses in phase space between failed (Phase II) and functional states (Phase I). When the system crosses the red line (points 1 and 3) it transits to a recovered state and when it crosses the blue line (points 2 and 4) it moves to a failed state. **b** Here we show the time evolution of the system in accordance with the diffusion in phase space according to the trajectory shown in (**a**) where the y-axis, z represents the fraction of active nodes at a given time t. As seen in the figure there are two clear states, either a failed or recovered state. The system changes between these two states, but never exists in an intermediate state. After [75]

space causes the system to dynamically recover or fail over time. In Fig. 20.7a, when the system crosses the blue line it reaches a failed state and when it crosses the red line it recovers. The crossing of these points can also be observed in Fig. 20.7b according to the corresponding numbered transitions. When the goal is to repair the system, a global planner will make repairs such that they reduce the likelihood of external failure, r, and help the system to pass the red line which represents the transition to a repaired state.

One example of a real system where this model was applied is stock-market networks. In such networks, each stock represents a node and it is connected to other stocks based on correlations between their stock prices. Stocks that are going up can be considered to be in a functional state and stocks that are falling can be considered in a failed state. Each stock (company) has an internal probability of failure (p^*) which could occur due to internal problems that are inherent to the company. Next there is a time (τ) it takes the company to fix the problems that caused the stock price drop, i.e. to recover. Lastly, because stocks are connected to one another they require support from one another and thus there is a probability (r) for a stock to drop if other stocks in the same or related sectors are falling. Naturally, the probability of a stock to fail, p^*, and the probability for failure due to the collapse of other stocks, r, will change based on overall market conditions, recent shocks to the markets, and for other reasons as well. The explicit application of this model to stock-market networks and a comparison to real data for the S&P500 can be found in Majdandzic et al. [75].

It was recently shown that a similar but much richer phenomenon occurs in interdependent networks [77]. The study in [77] found an optimal repairing strategy, which describes how many repairs should be made in each network in order to move the system towards a functional state.

In addition there have been several other studies on restoration of interdependent networks [78, 79].

20.6   Conclusions

Modern systems are becoming more and more interdependent especially through the use of SMART technologies, which require information from both their own system and from other systems. This information is then used to optimize the performance of each system based on the functioning of the other systems. Examples are SMART grids, SMART cities, and the internet of things (IOT). Understanding how failures spread both within and between the different systems that form SMART cities is crucial in order to ensure the stability of these highly interdependent systems. Methods from diffusion, percolation and physics in general can serve as useful tools to contain and predict the spreading of failures in these systems. Furthermore, models of interdependent networks have also explained the spreading of failures in other areas like finance [19, 80]. Continuing to study how failures spread in real-world systems is a crucial area of research and will likely provide many additional interesting results.

Acknowledgements We acknowledge the Israel-Italian collaborative project NECST, Israel Science Foundation, ONR, Japan Science Foundation, BSF-NSF, and DTRA (Grant no. HDTRA-1-10-1- 0014) for financial support.

References

1. A.L. Barabási, R. Albert, Science **286**, 5439 (1999)
2. M. Girvan, M. Newman, PNAS **99**, 12 (2002)
3. M. Barthélemy, Phys. Rep. **499**, 1 (2011)
4. M. Newman, *Networks: An Introduction* (OUP Oxford, 2010)
5. R. Cohen, S. Havlin, *Complex Networks: Structure, Robustness and Function* (Cambridge University Press, 2010)
6. M. Newman, A.L. Barabási, D.J. Watts, *The Structure and Dynamics of Networks* (Princeton University Press, 2011)
7. S. Havlin, E. López, S.V. Buldyrev, H.E. Stanley, Diffus. Fundam. **2**, 4 (2005)
8. S. Havlin, E. López, S.V. Buldyrev, H.E. Stanley, in *Leipzig, Einstein, Diffusion*, ed. by J. Kärger (Leipziger Universitätsverlag, 2007), p. 144
9. M. Kivelä, A. Arenas, M. Barthélémy, J.P. Gleeson, Y. Moreno, M.A. Porter, J. Complex Netw. **2**, 3 (2014)
10. G. Bianconi, Phys. Rev. E **87**, 062806 (2013)
11. S. Boccaletti, G. Bianconi, R. Criado, C.I. Del Genio, J. Gómez-Gardeñes, M. Romance, I. Sendina-Nadal, Z. Wang, M. Zanin, Phys. Rep. **424**, 4 (2014)

12. A. Vespignani, Nature **464**, 7291 (2010)
13. S.M. Rinaldi, J.P. Peerenboom, T.K. Kelly, Control systems. IEEE **21**, 11 (2001)
14. E.A. Leicht, R.M. D'Souza, arXiv e-prints (2009)
15. S.V. Buldyrev, R. Parshani, G. Paul, H.E. Stanley, S. Havlin, Nature **464**, 08932 (2010)
16. R. Parshani, S.V. Buldyrev, S. Havlin, Phys. Rev. Lett. **105**(4), 048701 (2010)
17. S.W. Son, G. Bizhani, C. Christensen, P. Grassberger, M. Paczuski, EPL (Europhys. Lett.) **97**, 1 (2012)
18. A. Bashan, R.P. Bartsch, J.W. Kantelhardt, S. Havlin, P.C. Ivanov, Nat. Commun. **3**, 702 (2012)
19. D.Y. Kenett, S. Havlin, Mind Soc. **1** (2015)
20. J. Gao, D. Li, S. Havlin, Natl. Sci. Rev. **1**, 346 (2014)
21. J. Gao, S.V. Buldyrev, H.E. Stanley, S. Havlin, Nat. Phys. **8**, 40–48 (2012)
22. J. Gao, S.V. Buldyrev, H.E. Stanley, X. Xu, S. Havlin, Phys. Rev. E **88**, 062816 (2013)
23. D. Stauffer, A. Aharony, *Introduction to Percolation Theory* (Taylor & Francis, 1994)
24. A. Bunde, S. Havlin, *Fractals and Disordered Systems* (Springer, New York, 1991)
25. R. Cohen, K. Erez, D.B. Avraham, S. Havlin, Phys. Rev. Lett. **85**, 4626 (2000)
26. D.S. Callaway, M. Newman, S.H. Strogatz, D.J. Watts, Phys. Rev. Lett. **85**, 5468 (2000)
27. M. Newman, I. Jensen, R.M. Ziff, Phys. Rev. E **65**, 021904 (2002)
28. J. Gao, S.V. Buldyrev, S. Havlin, H.E. Stanley, Phys. Rev. Lett. **107**, 195701 (2011)
29. F. Radicchi, Nat. Phys. **11**, 7 (2015)
30. G.J. Baxter, S.N. Dorogovtsev, A.V. Goltsev, J.F.F. Mendes, Phys. Rev. Lett. **109**, 248701 (2012)
31. B. Bollobás, *Modern Graph Theory*, vol. 184 (Springer, 1998)
32. P. Erdős, A. Rényi, Publ. Math. Debrecen **6**, 290 (1959)
33. P. Erdős, A. Rényi, Acta Math. Hung. **12**, 1 (1961)
34. J. Gao, S.V. Buldyrev, S. Havlin, H.E. Stanley, Phys. Rev. E **85**, 066134 (2012)
35. G. Bianconi, S.N. Dorogovtsev, J. Mendes, Phys. Rev. E **91**, 012804 (2015)
36. G. Bianconi, S.N. Dorogovtsev, Phys. Rev. E **89**, 062814 (2014)
37. G.J. Baxter, S.N. Dorogovtsev, A.V. Goltsev, J.F.F. Mendes, in *Networks of Networks: The Last Frontier of Complexity, Understanding Complex Systems*, ed. by G. D'Agostino, A. Scala (Springer, 2014)
38. Y. Hu, D. Zhou, R. Zhang, Z. Han, C. Rozenblat, S. Havlin, Phys. Rev. E **88**, 052805 (2013)
39. J.Y. Kim, K.-I. Goh, Phys. Rev. Lett. **111**, 058702 (2013)
40. K.M. Lee, J.Y. Kim, W.K. Cho, K.-I. Goh, I.-M. Kim, New J. Phys. **14**, 033027 (2012)
41. D. Cellai, E. López, J. Zhou, J.P. Gleeson, G. Bianconi, Phys. Rev. E **88**, 052811 (2013)
42. D. Zhou, J. Gao, H.E. Stanley, S. Havlin, Phys. Rev. E **87**, 052812 (2013)
43. L.M. Shekhtman, S. Shai, S. Havlin, New J. Phys. **17**, 123007 (2015)
44. S. Shai, D.Y. Kenett, Y.N. Kenett, M. Faust, S. Dobson, S. Havlin, Phys. Rev. E **92**, 062805 (2015)
45. B.R. da Cunha, J.C. González-Avella, S. Gonçalves, PloS One **10**, 11 (2015)
46. P. Hines, S. Blumsack, E. Cotilla Sanchez, C. Barrows, in *System Sciences (HICSS)* (2010)
47. L. Daqing, K. Kosmidis, A. Bunde, S. Havlin, Nat. Phys. **7**, 481–484 (2011)
48. A. Bashan, Y. Berezin, S.V. Buldyrev, S. Havlin, Nat. Phys. **9**, 667 (2013)
49. W. Li, A. Bashan, S.V. Buldyrev, H.E. Stanley, S. Havlin, Phys. Rev. Lett. **108**, 228702 (2012)
50. M.M. Danziger, A. Bashan, Y. Berezin, S. Havlin, J. Complex Netw. **2**, 460 (2014)
51. L.M. Shekhtman, Y. Berezin, M.M. Danziger, S. Havlin, Phys. Rev. E **90**, 012809 (2014)
52. M.M. Danziger, A. Bashan, Y. Berezin, S. Havlin, in *Signal-Image Technology Internet-Based Systems (SITIS)* (2013)
53. M.M. Danziger, L.M. Shekhtman, Y. Berezin, S. Havlin, EPL (Europhys. Lett.) **115**, 36002 (2016)
54. A.E. Motter, Y.C. Lai, Phys. Rev. E **66**, 065102 (2002)
55. I. Dobson, B.A. Carreras, V.E. Lynch, D.E. Newman, Chaos: an interdisciplinary. J. Nonlinear Sci. **17**, 026103 (2007)
56. J. Zhao, D. Li, H. Sanhedrai, R. Cohen, S. Havlin, Nat. Commun. **7**, 10094 (2016)
57. Y. Berezin, A. Bashan, M.M. Danziger, D. Li, S. Havlin, Sci. Rep. **5**, 8934 (2015)

58. M.M. Danziger, A. Bashan, S. Havlin, New J. Phys. **17**, 043046 (2015)
59. S. Shao, X. Huang, H.E. Stanley, S. Havlin, Phys. Rev. E **89**, 032812 (2014)
60. X. Huang, S. Shao, H. Wang, S.V. Buldyrev, H.E. Stanley, S. Havlin, EPL (Europhys. Lett.) **101**, 18002 (2013)
61. D. Zhou, H.E. Stanley, G. D'Agostino, A. Scala, Phys. Rev. E **86**, 066103 (2012)
62. C.M. Schneider, N. Yazdani, N.A.M. Araújo, S. Havlin, H.J. Herrmann, Sci. Rep. **3**, 1969 (2013)
63. C. Buono, L.G. Alvarez-Zuzek, P.A. Macri, L.A. Braunstein, PloS One **9**, 3 (2014)
64. S. Watanabe, Y. Kabashima, Phys. Rev. E **89**, 012808 (2014)
65. R. Parshani, C. Rozenblat, D. Ietri, C. Ducruet, S. Havlin, EPL (Europhys. Lett.) **92**, 68002 (2010)
66. M. Li, R.R. Liu, C.X. Jia, B.-H. Wang, New J. Phys. **15**, 093013 (2013)
67. S.V. Buldyrev, N.W. Shere, G.A. Cwilich, Phys. Rev. E **83**, 016112 (2011)
68. Y. Kornbluth, S. Lowinger, G. Cwilich, S.V. Buldyrev, Phys. Rev. E **89**, 032808 (2014)
69. L.D. Valdez, P.A. Macri, H.E. Stanley, L.A. Braunstein, Phys. Rev. E **88**, 050803 (2013)
70. L.D. Valdez, P.A. Macri, L.A. Braunstein, J. Phys. A: Math. Theoret. **47**, 055002 (2014)
71. X. Huang, J. Gao, S.V. Buldyrev, S. Havlin, H.E. Stanley, Phys. Rev. E **83**, 065101 (2011)
72. G. Dong, J. Gao, R. Du, L. Tian, H.E. Stanley, S. Havlin, Phys. Rev. E **87**, 052804 (2013)
73. S. Shao, X. Huang, H.E. Stanley, S. Havlin, New J. Phys. **17**, 023049 (2015)
74. X. Yuan, S. Shao, H.E. Stanley, S. Havlin, Phys. Rev. E **92**, 032122 (2015)
75. A. Majdandzic, B. Podobnik, S.V. Buldyrev, D.Y. Kenett, S. Havlin, H.E. Stanley, Nat. Phys. **10**, 34 (2013)
76. D.J. Watts, PNAS **99**, 9 (2002)
77. A. Majdandzic, L.A. Braunstein, C. Curme, I. Vodenska, S. Levy-Carciente, H.E. Stanley, S. Havlin, Nat. Commun. **7**, 10850 (2015)
78. M.A. Di Muro, C.E. La Rocca, H.E. Stanley, S. Havlin, L.A. Braunstein, Sci. Rep. **6**, 22834 (2016)
79. M. Stippinger, J. Kertész, Phys. A: Stat. Mech. Appl. **416**, 481–487 (2014)
80. L. Bargigli, G. di Iasio, L. Infante, F. Lillo, F. Pierobon, Quant. Finan. **15**, 4 (2015)

Index

A
Abrams & Strogatz model, 353
Active sites, 205
Activity, 218
Adaptive dynamics, 335, 337–340
Adsorbents, 210
Adsorption, 180
Adsorption isotherm, 182
Advection, 17, 83
AFM, 76, 79, 80
Agent-based model, 22, 335
Agriculture, 320, 329
Albatross, 51
Allee effect, 37
Ambrosia artemisiifolia (common ragweed), 42
Amorphization, 272
Amorphous, 268
Anisotropic Geometries, 249
Anomalous diffusion, 106
Anthropocene, 333
Apparent Diffusion Coefficient (ADC), 239
Arrhenius law, 268
Asymmetric simple exclusion process, 148
Atmosphere, 115
Atmospheric stratification, 119
Attenuation, 174
Autothermal reforming of methane, 223
Average value, 117

B
Basic reproduction ratio, 378
Basic research, 299
Bethe ansatz, 154
Bidisperse, 220
Bilingual, 357–359, 365, 371
Bilingualism, 361, 363
Bimodal, 220
Binding, 234, 246
Biological organisms, 212

Biology, 229
Bipolar plates, 225
Blackouts, 404
Blood-brain Barrier (BBB), 98, 112
Boltzmann distribution, 161
Bone, 213
Boundary conditions, 212
Bovine Serum Albumin (BSA), 107
Brain, 93
Brogniard, Adolphe, 134
Broken detailed balance, 141
Broken symmetries, 143
Brownian motion, 31, 53, 128, 233
Brownian motor, 149
Brownian pump, 149
Brownian thermospectrometry, 133
Brown, Robert, 134
Bulk current, 152
Bulk flow, 98
Bumblebee, 60
Buoyancy, 119
Burger's turbulence, 158

C
Ca^{2+} ions, 109
Carrying capacities, 355
Cascading failures, 397, 404
Catalysis, 171, 191
Catalyst pellets, 206
Cellular automaton, 22
Cellular grids, 195
Central limit theorem, 50, 51, 58
    generalized, 50, 59
Centre-of-mass diffusivity, 199
Ceramics, 261
Chains, 270
Characteristic traits, 337–340, 342, 343
Chemical conversions, 20
Chemical engineering, 229

© Springer International Publishing AG 2018
A. Bunde et al. (eds.), *Diffusive Spreading in Nature, Technology and Society*,
https://doi.org/10.1007/978-3-319-67798-9

Chemical potential, 18, 181
Chondroitin sulfate, 108
City infrastructure, 403
Climate change, 34
CO_2, 72, 74
Coarse-graining, 131
Coefficient of self-diffusion, 13
Co-evolution, 304
Coexistence, 355, 358, 364, 365
Cold Brownian motion, 143
Community structure, 403
Complexity, 212
Complex networks, 398
Complex systems, 398
Computer simulations, 133
Concentration distribution, 116
Concentration gradient, 156
Concentrations of pollutants, 119
Concerntric pattern, 382
Conditional probability, 152
Configurational diffusion, 205
Conformation, 270
Consequentialist approach, 303
Conspecifics, 190
Constrained exponential growth, 280
Continuity, 32
Continuity equation, 16, 83
Continuous chimney emission, 118
Continuum models, 207
Control parameter, 153
Convection, 17
Convection-enhanced delivery, 112
Convective flow, 205
Converging technologies, 300
Conversion, 191
Cooperative, 269
Correlation, 63
Co-transport, 79, 90
Coulomb scattering, 137
Covariance, 63
Creep, 265
Critical micelle concentration, 243
Crystallization, 268, 269
Cultural diffusion, 321, 323, 329, 343–345,
    347
Cultural transmission, 325
Current-density relation, 152
Cuticle, 72
Cutin, 72
Cyborgs, 298

D
Darcy's law, 18, 209
Deactivation of the catalyst by fouling, 223

Dead-space microdomains, 105
Demic diffusion, 320, 323, 326, 343–345, 347
Density fluctuation, 149
Density functional theory, 166
Density oscillation, 162
Density profile, 155
Dependency Links, 399, 401
Deposition of emitted substance, 119
Desiccation, 72
Desorption, 180
Detailed balance, 129, 140
Detailed balance, violated, 143
Determinism, 32
Dextran, 107
Diffusion, 31, 172, 205
    anomalous, 60
    chamber, 77–79
Diffusion coefficient, 318, 355, 359, 380
Diffusion constant, 158
Diffusion current, 156
Diffusion ellipsoid, 254
Diffusion equation, 130
Diffusion in networks, 398
Diffusion matrix, 189
Diffusion path, 13
Diffusion propagator, 249
Diffusion-reaction models, 359
Diffusion-reaction system, 354
Diffusion tensor, 253
Diffusion Tensor Imaging (DTI), 252
Diffusion Weighted Imaging (DWI), 252
Diffusive diffraction, 250
Diffusivity, 14, 130, 207, 266
Diglossia, 364, 368, 371
Discrete models, 207
Discretization, 32
Disease dynamic, 377
Dislocations, 271
Dispersal, 29, 371
    mode of, 30
    pattern, 30
Dispersal kernel, 356, 363
Dissipation, 141
Distributed plumes, 120
Distribution
    posterior, 44
    prior, 44
Distribution function, 197
Domain wall, 148
DOSY, 248
Drift current, 156
Driven diffusion, 149
Driven lattice gas, 148
Driving force, 18

Driving force of diffusion, 181
Drug delivery, 112
Dynamic equilibrium, 179
Dynamic imaging, 175
Dystopia, 302

E
Ebola, 375
Ecological niche, 30, 34
Ecosystem, 45
Effectiveness factor, 193, 218
Effective one-dimensional model, 220
Effective temperature, 132
EHEC outbreak, 394
Einstein, Albert, 127, 128
Einstein's diffusion equation, 17
Ejection rate, 165
Electro-catalysis, 227
Electrodes, 225
Electron micrographs, 107
Electrophoretic mobilities, 247
Electrophoretic NMR, 247
Emergence, 56, 66
Enabling technologies, 297
Energy, 148
English-Gaelic shift, 354, 371
Entropy, 45, 148
Entropy production, 140, 141, 143
Environmental conditions, 23
Epi-curve, 386
Epicuticular, 71, 75, 76, 79, 81, 82, 85–88, 90
Epidemic arrival time, 382
Epidemic wavefront, 380
Epidermis, 72
Equal hydraulic path lengths, 226
Equilibrium, 33
Equilibrium dissociation constant, 110
Ergodicity, 200
Evaporation, 80
Evolution, 71
Exchange, 246
Exchange probabilities, 195
Exponential growth, 319
Extinct, 354
Extinction, 355, 361, 363, 365, 370
Extracellular matrix, 94, 108
Extracellular space, 94

F
Failure spreading, 397, 406
Farmers, 320, 325, 329
Farming, 313, 320

Fast-equilibrium reaction, 110
Fast exchange, 195
Fast transport, 219
Fermi's golden rule, 165
Ferrierite, 178
Feynman, Richard, 127
Fiber Tracking Mapping, 252
Fibonacci sequence, 149
Fibonacci's model, 36
Fick, Adolf, 136
Fick's laws, 116, 129
Fick's 1st law, 14, 179
Fick's 2nd law, 16, 99, 179
Financial networks, 407
Finger-printing, 177
First-order phase transition, 157
Fitness gradient, 338
Fitness landscape, 339
Fixed-bed reactor, 206
Flow channels, 226
Fluctuating force, 22
Fluctuation-dissipation relation, 129, 140
Fluctuation theorem, 140–142
Fluorescence correlation spectroscopy, 138
Focal plane array detector, 176
Foraging, 49, 53, 61
Fourier, Jean-Babtiste Joseph, 136
Fourier's law, 179
Fractal, 130, 211
Fractal dimension, 216
Fractal distributor, 216, 228
Fractal geometry, 216
Fractal pattern, 382
Fractional loading, 184
Free energy, 148
Frequentist interpretation of probability, 43
Frictional force, 182
Friction coefficient, 22
Front, 319, 322, 324, 325

G
Gaelic-English language shift, 365
Gas diffusion layers, 225
Gaussian, 197
Gaussian white noise, 22
Generalized Einstein relation, 132
Generalized Linear Model, 34
Glauber rates, 160
GLEAM, 388
Glia, 93
Globalization, 352
Glymphatic, 98

Golden mean, 149
Gold nanoparticle, 136
Grand stories, 292
Grape berry morphology, 256
Great American Interchange, 29
Grid cells, 24
Growth
　bounded, 36
　Malthusian (unbounded), 36
Growth rate, 19, 36

H
Habitat, 21, 180
Hahn spin-echo, 239
Heat of adsorption, 195
Heat reservoir, 160
Heparan sulfate, 108
Heparin, 109
Hermeneutic approach, 303
Hermeneutic circle, 304
Hetero-diffusion, 263
Hierarchical channel system, 219
Hierarchical design, 214
Hierarchically structuring, 213
Hierarchical network, 214
High-density phase, 154
High-frequency pulses, 174
Homogeneity, 32
Homogeneous, 262
Horizontal homogeneity, 121
Host-guest systems, 173
Hot Brownian motion, 128, 131
Hot Brownian particle, 130
Hot Brownian swimmer, 128, 142
Hot Brownian thermometry, 133
Hot microswimmer, 134
Human enhancement, 301
Human vascular network, 217
Humidity, 71, 72, 84, 89
Hunter-gatherers, 325
Hunting-gathering, 313
Hyaluronan, 108
Hydraulic conductivity, 84
Hydrodynamic memory, 139
Hydrogenation, 191
Hydrogenation of benzene to cyclohexane, 224
Hydrogen bonded environment, 242
Hydrophobic, 72

I
Ideal Fermi gas, 165
Immunoglobulin G (IgG), 107
Incompressible fluid, 117
Infection, 381

Infestation probability, 23
Infestation spreading, 23
Influenza H1N1 pandemic, 383
Infrastructure, 397
Injection rate, 165
Innovation process, 287, 298
Innovation stories, 296
Instantaneous source, 117
Integrative Optical Imaging (IOI) method, 107
Intentional design, 291
Interdependence, 399
Interdependent networks, 397, 401, 402, 408
Interfaces, 271
Interference Microscopy (IFM), 176, 251
Intergenerational transmission, 370
Intermittency, 57
Interparticle interactions, 234
Interstitial, 267
Interstitial space, 94
Intervention, 364, 369, 371
Invasive species, 31
Inversion, 120
Ion pump, 167
Ion-selective Microelectrode (ISM), 101
Iontophoretic release, 100
IR Microscopy (IRM), 176, 251
Irreversible thermodynamics, 183
Ising model, 161
Isomers, 242
Isotope, 262
Isotropy, 32

J
Jamming, 147
Janus particle, 136
Joint probability, 150
Jump length, 195

K
Kärger equations, 247
Kepler ratio, 149
Kernel, 321, 324
Kinetic Monte Carlo simulation, 157
Knudsen diffusion, 205

L
Lactoferrin, 109
Langevin equation, 22, 65, 135
Language diffusion principle, 357
Language inheritance principle, 357
Language shift, 351
Laplacian, 380
Leaf, 219
Leptokurtic, 40

Lèvy, 49
    Environmental Hypothesis, 56, 66
    flight, 49, 58
        hypothesis, 40, 49, 52, 66
        paradigm, 54, 56, 66
    motion, 49, 53, 58
    noise, 59
    Search Hypothesis, 56, 66
    stable distribution, 49, 58
    walk, 52, 59
        truncated, 52
Light scattering, 137
Likelihood, 43
Linearity, 32
Link function, 34
Li PFG NMR, 186
Liquid crystals, 243
Lithium cations, 186
Localized failures, 406
Logarithmic wind profile, 123
Logistic growth, 20, 319, 355
Long-distance economic migration, 352
Long-distance transmissions, 381
Loschmidt's number, 129
Lotka-Volterra model, 37
Lotus effect, 74
Low-density phase, 154
LSX zeolites, 186
Lungs, 215

M
Macropore, 206
Macroporosity, 220
Magnetic field gradient, 237
Magnetization, 174
Manmade systems, 212
Marginal benefit, 338, 339
Markov limit, 130, 131
Markov process, 59
Mass conservation, 31
Mass transfer limitations, 225
Mass transport, 229
Master equation, 151
Maximum current phase, 154
Maximum current principle, 157
Maximum likelihood estimation, 43
MCell, 104
Mean-field approach, 199
Mean-field approximation, 152
Mean free path, 122
Mean square displacement, 12, 130, 136, 235,
    264, 318
Mechanistic models, 33
Membrane, 71, 187

Memory, 63, 197
MERS, 375
Mesolithic, 313
Mesopore, 195, 206
Mesoporous $deNO_x$ catalyst, 222
Mesoporous Ni/Al_2O_3 catalyst, 223
Mesopotamia, 180
Mesoscopic, 131
Metals, 261, 264
Metapopulation, 384
Metapopulation models, 41
Metastable, 268
Micelles, 243
Microimaging, 176
Micropore, 195, 206
Microstate, 161
Microswimmer, 134
Minimization of entropy production, 217
Minimum current principle, 157
Mixing path length, 124
Mixtures, 186
Mobility, 183
Modularity, 402
Molecular assembler, 306
Molecular diffusion, 205
Molecular machine, 149
Molecular modelling, 200
Molecular uptake and release, 179
Monodisperse, 220
Monte Carlo, 104
Monte Carlo simulations, 22
Motor protein, 147
Movement ecology, 49, 66, 67
MRI, 251
Multiphase catalyst, 224
Multiphase reactions, 224
Multiple length scales, 203
Multiscale architectures, 204
Multiscale structure, 229
Multiscale transport, 206
Murray's law, 227
Muskrats, 39
Mutual diffusion, 263

N
NaCaA, 176, 195
Nanocrystallization, 272
Nanoethics, 299
Nanoparticle, 130, 298
Nanoporous, 171
Nanoporous particle, 281
Nanostructured catalyst, 220
Nanotechnology, 297
Narratives, 306

Nature, 204
Nature-inspired Chemical Engineering (NICE),
    205, 213
Nature-inspired design, 216
Nature-inspired Optimization, 203
Nearest-neighbor, 266
Nearest-neighbor interaction, 149
Neolithic, 313, 323
Neolithic transition, 21
Net flux, 264
Network extraction algorithms, 209
Network recovery, 407, 408
Network resilience, 400, 405
Network robustness, 400
Networks of networks, 400, 402
Neurons, 93
NMR, 237
NMR Diffusography, 242
NMR diffusometry, 237
Nonequilibrium processes, 140
Non-equilibrium steady state, 147
Normal distribution, 281
Nuclear Magnetic Resonance (NMR), 173
Nuclear spin, 174

O

Obstruction, 234
Obstruction factor, 235
Occupation number, 150
Occupation probability, 199
Open innovation, 289
Open laboratory, 291
Operations research, 49
Optical pathways, 177
Optimal design, 214
Optimality, 229
Order parameter, 157
Origin of an epidemic, 393
Ostwald, Wilhelm, 127
Overshooting, 188

P

Pandemics, 375
Particle flux, 14
Particle-hole symmetry, 152
Particle reservoir, 149
Pd/Al_2O_3 catalyst pellet, 224
Péclet number, 216
Percolation theory, 401
Permeability, 207
Perrin, Jean, 127
Persistence length, 135
Persistent random walk, 136
Phase transition, 147, 272

Phenomenological description, 33
Phospholipid double layer, 71
Photon-nudging, 139
Photothermal Correlation Spectroscopy
    (PhoCS), 138
Photothermal microscopy, 137, 138
Photothermal scanning microscopy, 138
Photothermal spectroscopy, 137
Physiological principle of minimum work, 217
Plasmon resonance, 136
Platinum catalyst, 228
Point-source paradigm, 101, 111
Pollution, 116
Polymers, 261
Polymer surface, 78
Population dynamics, 30
Pore Hierarchies, 194
Pore network models, 208
Poren-Nutzungsgrad, 194
Pore size distribution, 207
Porous catalysts, 210, 218
Porous glass, 191
Porous media, 203
Porous system, 233
Post-human future, 297
Power grids, 398, 404
Power law, 49, 54, 68
    truncated, 54, 60
Prandtl mixing length, 122
Precipitates, 273
Probability distribution, 17
Process-based methods, 208
Propagator, 175
Propagule, 32
Proton Exchange Membrane (PEM) fuel cells,
    225
Public perception, 296
Pulsed Field Gradient (PFG), 173
Pulsed Field Gradient (PFG) NMR, 237
Pulsed Gradient Spin-Echo (PGSE), 237

Q

Quantum ratchet, 167
Quasi-elastic Neutron Scattering (QENS), 200

R

Ragweed, 23
Randomness of pore networks, 211
Random walk, 11, 31, 58, 130, 158, 197, 264
    continuous time, 60
Random sphere model, 224
Range shift, 30, 34
Rate equation, 21
Reactant, 191

Reaction, 20, 31
Reaction-diffusion, 318
Reaction diffusion equations, 21
Reaction-diffusion model, 38
Real-time Iontophoresis (RTI) method, 101
Real-time Pressure (RTP), 109
Real-time Pressure (RTP) method, 104
Refractive index, 177
Repair process, 80
Reproduction, 317, 321
Restricted diffusion in pores, 107
Röntgen, Wilhelm, 127
Rhynie Chert, 73
Robustness, 224
Root mean square, 236
Rutherford scattering, 138

S
SARS epidemic, 383
Scale-free networks, 402
Scenarios, 302
Science fiction, 301
Scottish Gaelic, 365
Search, 49, 57, 66
    efficiency, 52
    optimality, 52
    strategy, 49, 52
        intermittent, 57
        optimal, 56
Second law of thermodynamics, 141
Second-order (bimolecular) reaction, 110
Second-order phase transition, 157
Selectivity, 187, 218
Self-diffusion, 262
Self-diffusion constant, 13
Self-diffusivity, 13
Self-organized structure, 166
Self-propulsion, 136
Self-repair, 76
Semiconductors, 261
Sensitivity, 223
Separation, 171
Severe plastic deformation, 271
Shock front, 158
Shortest path, 391
Shortest path tree, 391
Signal attenuation, 238
Silicon, 269
Simulation, 195
Simulation-based approaches, 354
Single-file diffusion, 197
Single-particle tracking, 200
Sintering, 265
Skellam, J.G., 38

Slow exchange, 195
Spaceflight, 300
Spanish flu, 375
Spanish-quechua shift, 354
Spatial dispersal, 361
Spatial networks, 403
Spatial resolution, 177
Spatiotemporal gradient, 334, 335
Species distribution model, 34
Spectral turbulent diffusivities, 125
Spread kernel, 40
Spread of diseases, 22
Spread of neobiota, 22
Stability, 218
State space, 68
Stationarity, 121
Stationary point source, 118
Stationary state, 150
Statistical methods, 208
Status, 353, 360, 362
Steady-state, 273
Steady-state conditions, 187
Steering, force-free, 139
Stochasticity, 32, 41
Stochastic lattice SIR models, 381
Stokes-Einstein-Sutherland, 235
Stokes' law, 129
Stock market, 407
Stomata, 74
Structure-function relation, 213
Structure-function relationships, 204
Structure of porous media, 209
Subdiffusion, 60
Substitutional, 265, 266
Superdiffusion, 60
Superposition of turbulent eddies, 125
Supersymmetry, 144
Surface diffusion, 205
Surface permeabilities, 179
Surface roughness, 211
Survival probabilities, 23
Sutherland, William, 129
Swarms, 143
Swine flu, 375
Symmetry breaking, 143
Synapses, 93
System Recovery, 398, 406
System-reservoir coupling, 162

T
Techno-visionary futures, 304
Temperature spectrum, 133
Tetramethylammonium (TMA), 101
Thermodynamic factor, 182

Thermodynamic limit, 152
The SIR model, 379
Thiele modulus method, 222
Time-reversal symmetry, 140
Tortuosity, 96, 210
Tracer diffusivity, 15
Tracheae, 74
Tractography, 252
Transfer matrix technique, 161
Transition rate, 151
Transpiration, 80
Transport, 14, 203
Transport equation, 86
Transport mechanism, 204
Transport networks, 207
Transport number, 100
Transport properties, 209
Trapping, 269
Travelling wave, 319, 355, 361, 366
Tree crowns, 214
Trees, 216
Turbulent diffusion, 115, 121
Turbulent momentum fluxes, 121
Twin-PhoCS, 138, 139
Two-component adsorption, 188
Two-dimensional model, 220

U
Uncertainty, 296
Universality class, 149
"Uphill" diffusion, 190
Urbanization, 352
Utopia, 302

V
Vacancy, 265, 271
Value judgments, 302

Velocity autocorrelation function, 63, 64
Verhulst model, 36
Viscous flow, 205
Visions, 295
V. Karman constant, 122
Volume fraction, 95, 96
Volume transmission, 95

W
Washcoat, 223
Water
    flux, 84
    potential, 84
    vapor, 71, 72, 74, 84
Water molecules, 186
Water-repellency, 74, 76
Wave of advance, 20, 315, 319, 334
Wax, 72
    concentration, 82, 83, 86, 87
    crystal, 74–76, 80, 82, 84, 89
    layer, 76, 77, 80–82, 85, 88–90
    production rate, 83
    regeneration, 77, 79, 81
    restoration, 88
    transport, 76
Wind velocity, 123
Worldwide air-transportation, 376

X
X-ray microtomography, 208

Z
Zeolite DDR, 189
Zeolites, 171
ZIF-8, 188

Printed in the United States
By Bookmasters